Martingale Methods in Statistics

Recently Published Titles

Multistate Models for the Analysis of Life History Data
Richard J. Cook and Jerald F. Lawless 158

Nonparametric Models for Longitudinal Data
with Implementation in R
Colin O. Wu and Xin Tian 159

Multivariate Kernel Smoothing and Its Applications
José E. Chacón and Tarn Duong 160

Sufficient Dimension Reduction
Methods and Applications with R
Bing Li 161

Large Covariance and Autocovariance Matrices
Arup Bose and Monika Bhattacharjee 162

The Statistical Analysis of Multivariate Failure Time Data: A Marginal Modeling Approach
Ross L. Prentice and Shanshan Zhao 163

Dynamic Treatment Regimes
Statistical Methods for Precision Medicine
Anastasios A. Tsiatis, Marie Davidian, Shannon T. Holloway, and Eric B. Laber 164

Sequential Change Detection and Hypothesis Testing
General Non-i.i.d. Stochastic Models and Asymptotically Optimal Rules
Alexander Tartakovsky 165

Introduction to Time Series Modeling
Genshiro Kitigawa 166

Replication and Evidence Factors in Observational Studies
Paul R. Rosenbaum 167

Introduction to High-Dimensional Statistics, Second Edition
Christophe Giraud 168

Object Oriented Data Analysis
J.S. Marron and Ian L. Dryden 169

Martingale Methods in Statistics
Yoichi Nishiyama 170

For more information about this series please visit: https://www.crcpress.com/Chapman--HallCRC-Monographs-on-Statistics--Applied-Probability/book-series/CHMONSTAAPP

Martingale Methods in Statistics

Yoichi Nishiyama

CRC Press
Taylor & Francis Group
Boca Raton London New York

CRC Press is an imprint of the
Taylor & Francis Group, an **informa** business
A CHAPMAN & HALL BOOK

First edition published 2022
by CRC Press
6000 Broken Sound Parkway NW, Suite 300, Boca Raton, FL 33487-2742

and by CRC Press
2 Park Square, Milton Park, Abingdon, Oxon, OX14 4RN

CRC Press is an imprint of Taylor & Francis Group, LLC

ISBN: 9781466582811 (hbk)
ISBN: 9781032146041(pbk)
ISBN: 9781315117768 (ebk)

DOI: 10.1201/9781315117768

Publisher's note: This book has been prepared from camera-ready copy provided by the authors.

Contents

Preface

The martingale theory is known as a powerful tool for researchers in statistics to analyze both financial data based on the theory of stochastic differential equations which was created by Kiyosi Itô in 1944, and life-time data based on the counting process approach which was initiated by the pioneering work of Odd O. Aalen in 1975. I try to write a monograph which should be helpful for readers including

those who hope to build up mathematical bases
to deal with high-frequency data in mathematical finance

and

those who hope to learn the theoretical background
for Cox's regression model in survival analysis.

A highlight of the monograph may be Chapters 8–10 dealing with Z-estimators and related topics. Section A1.1 in Appendices contains some new inequalities for maxima of finitely many martingales.

Besides these topics, I also try to explain an opinion of mine to readers—mastering martingale methods is useful not only for constructing and analyzing statistical models of stochastic processes but also for getting better understanding a common mechanism of randomness appearing in various procedures in statistics, including the analysis of i.i.d. models as the most important case.

This monograph consists of three parts—"Introduction", "A User's Guide to Martingale Methods" and "Asymptotic Statistics with Martingale Methods". I try to write what is not available in other textbooks in each of the three parts, with different spirits. In the first part, I try to write a novel, terse introduction to statistics of stochastic processes. The second part gives a systematic exposition of martingale methods in details in a rather standard format, although the proofs of some of the theorems are skipped; I dare to avoid "copying" the well-known proofs from the other authoritative textbooks to the current small monograph, by making clear citations instead. On the other hand, I try to present some intuitive explanations or concrete usages instead of the formal proofs of many theorems. The formal proofs will be given only to theorems whose proofs are thought to be helpful for readers at first reading to learn some important methods, concepts or techniques in the martingale theory. In contrast, the third part concerning the asymptotic statistics is written in a completely self-contained way in principle. Each section or chapter in this part is a step by step exposition starting from elementary examples towards some general results that can be applied to complex situations.

Some sections and chapters of this monograph are revised and extended versions of my preceding book, written in Japanese, and published in 2011 from Kindaikagakusha (KKS), which is originally based on my lecture notes at Kitasato University, The University of Osaka Prefecture, The University of Tokyo, The Graduate University for Advanced Studies (SOKENDAI), and Waseda University, as well as open lectures at The Institute of Statistical Mathematics for audiences working in the society. I especially thank Tohru Koyama of KKS for his helpful comments and advices that improved both the Japanese and the current versions, and enthusiastic, sharp-eyed and patient students of SOKENDAI and Waseda University for stimulating discussions with them.

Along the way of this work, I have learned a lot of things from my teachers including Nobuyuki Ikeda, Nobuo Inagaki and Richard D. Gill, as well as Nakahiro Yoshida. I also really thank Jean Jacod for reading some parts of earlier versions of this monograph and giving me very helpful comments and remarks; all remaining errors are, of course, due to me. My special thanks also go to Sigeo Aki, Takayuki Fujii, Kou Fujimori, Yu Hayakawa, Toshiharu Hayashi, Satoshi Inaba, Yoshihide Kakizawa, Hiroyuki Kanazu, Satoshi Kuriki, Yury A. Kutoyants, Sangyeol Lee, Shuhei Mano, Hiroki Masuda, Junko Murakami, Yoshifumi Muroi, Ilia Negri, Yosihiko Ogata, Yasutaka Shimizu, Takaaki Shimura, Peter Spreij, Takeru Suzuki, Yoshiharu Takagi, Masanobu Taniguchi, Koji Tsukuda, Masayuki Uchida, Masao Urata, Sara A. van de Geer, Aad W. van der Vaart, Harry van Zanten and Jiancang Zhuang for comments, discussions, instructions, lectures, preprints and encouragement. The advices for the Japanese version of this monograph from Genshiro Kitagawa, Tomoyuki Higuchi and Junji Nakano have been useful also for the preparation of the current monograph.

I am very grateful to John Kimmel for advices, support, patience and encouragement for many years. I also greatly acknowledge the anonymous reviewers for their careful reading and insightful comments and suggestions.

Last but not least, I thank my wife Rei for her support and patience during the period of this work; it may have been long or short for us, but thanks to her dedication, the last nine years have been a truly happy and wonderful time for our family.

Tokyo, May 2021 Yoichi Nishiyama

Notations and Conventions

Basic Notations

$\mathbb{R}$	Totality of real numbers.
$\mathbb{Q}$	Totality of rational numbers.
$\mathbb{N}$	Totality of positive integers: $\mathbb{N} = \{1,2,...\}$.
$\mathbb{N}_0$	Totality of non-negative integers: $\mathbb{N}_0 = \{0,1,2,...\}$.
i	$\sqrt{-1}$.
$\xrightarrow{a.s.}$	The almost sure convergence.
$\xrightarrow{\mathsf{P}}$ or $\xrightarrow{\mathsf{P}^n}$	The convergence in probability.
$\xRightarrow{\mathsf{P}^n}$ or $\xRightarrow{\mathsf{P}}$	The convergence in distribution (or the *weak convergence*).
$o_{\mathsf{P}}(1)$ or $o_{\mathsf{P}^n}(1)$	The convergence in probability to zero.
$O_{\mathsf{P}^n}(1)$ or $O_{\mathsf{P}}(1)$	Bounded in probability.
$\xrightarrow{\mathsf{P}^*}$ or $\xrightarrow{\mathsf{P}^{n*}}$	The convergence in outer-probability.
$o_{\mathsf{P}^*}(1)$ or $o_{\mathsf{P}^{n*}}(1)$	The convergence in outer-probability to zero.
$O_{\mathsf{P}^{n*}}(1)$ or $O_{\mathsf{P}^*}(1)$	Bounded in outer-probability.
$\mathcal{N}_p(\mu,\Sigma)$	p-dimensional Gaussian distribution.
$\mathcal{N}(\mu,\sigma^2)$	1-dimensional Gaussian distribution.
$1\{A\}$ or 1_A	The indicator function which is 1 if A is true and 0 otherwise.
∂_i	The abbreviation of $\frac{\partial}{\partial\theta_i}$.
$\partial_{i,j}$	The abbreviation of $\frac{\partial^2}{\partial\theta_i\partial\theta_j}$.
D_i	The abbreviation of $\frac{\partial}{\partial x_i}$.
$D_{i,j}$	The abbreviation of $\frac{\partial^2}{\partial x_i\partial x_j}$.
A^{tr}	The transpose of a matrix or vector A.
$\vert\vert\cdot\vert\vert$	The Euclidean norm: $\vert\vert x\vert\vert = \sqrt{\sum_{i=1}^p (x_i)^2}$.
$\wedge$, $\vee$	$x\wedge y = \min\{x,y\}$, $x\vee y = \max\{x,y\}$.
$:=$	Defining the left-hand side by the right-hand side.
$=:$	Defining the right-hand side by the left-hand side.
$\stackrel{d}{=}$	The distributions of the both sides are the same.
a.s.	The abbreviation of "almost surely".
$\emptyset$	The empty set.
A^c	The complement of the set A.
$t \rightsquigarrow X_t$	The stochastic process $(X_t)_{t\in[0,\infty)}$ or $(X_t)_{t\in[0,T]}$.
$t \mapsto x(t)$	The (non-random) function $(x(t))_{t\in[0,\infty)}$ or $(x(t))_{t\in[0,T]}$ of t.

Some Conventions

In this monograph, a mapping from a probability space $(\Omega, \mathcal{F}, \mathsf{P})$ to a measurable space $(\mathcal{X}, \mathcal{A})$ is said to be an *$\mathcal{X}$-valued random variable* if it is $\mathcal{F}/\mathcal{A}$-measurable. The terminology "$\mathcal{X}$-valued random *element*" is used when we do not assume any measurability of a mapping from a probability space to a set $\mathcal{X}$ (with no σ-field). We will treat only the cases where $\mathcal{X}$ is a metric space and $\mathcal{A}$ is the corresponding Borel σ-field in the cases where we assume the measurability of $\mathcal{X}$-valued random elements.

An *$\mathcal{X}$-valued random field* indexed by $\mathbb{T}$ is a family $\{X(t); t \in \mathbb{T}\}$ of $\mathcal{X}$-valued random variables, defined on a common probability space, indexed by a non-empty set $\mathbb{T}$ (with no ordering). In particular, instead of the terminology "random field" for $\{X(t); t \in \mathbb{T}\}$ indexed by $\mathbb{T}$, we use other terminologies in some special cases for $\mathbb{T}$ (with or without ordering) as follows.

- We use "discrete-time stochastic process" instead of "random field" if $\mathbb{T}$ is a subset of integers such as $\mathbb{N} = \{1, 2, ...\}$ or $\mathbb{N}_0 = \{0, 1, 2, ...\}$ with the natural ordering. In such cases, the notations like $(X_n)_{n \in \mathbb{N}_0}$ will be used instead of $\{X(n); n \in \mathbb{N}_0\}$.

- We use "stochastic process" instead of "random field" if $\mathbb{T} = [0, \infty)$, $[0, \infty]$, or $[0, T]$ for a constant $T > 0$, with the (usual) total ordering. In such cases, the notations like $(X_t)_{t \in [0,\infty)}$ will be used instead of $\{X(t); t \in [0, \infty)\}$.

- A random field $X = \{X(t); t \in \mathbb{T}\}$ and a stochastic process $X = (X_t)_{t \in [0,\infty)}$, etc., are sometimes denoted respectively by $t \rightsquigarrow X(t)$ and $t \rightsquigarrow X_t$.

- For a given stochastic process like $X = (X_t)_{t \in [0,\infty)}$, if we fix a $\omega \in \Omega$, then $t \mapsto X_t(\omega)$ can be regarded as a function from $[0, \infty)$ to $\mathcal{X}$, and it is called a *path* of X. For a deterministic function like $(x(t))_{t \in [0,\infty)}$ or a path like $(X_t(\omega))_{t \in [0,\infty)}$ for a fixed ω, we use the notation like $x \mapsto x(t)$ or $t \mapsto X_t(\omega)$, respectively.

- Readers should not be confused the notation "$t \rightsquigarrow X_t$", where $\omega \in \Omega$ is suppressed, with "$t \mapsto X_t(\omega)$ for a fixed ω".

Finally, note that a special attention should be paid to the usage of the phrase "increasing process", a term with a special meaning in the martingale theory. Readers should not confuse an *increasing process* $t \rightsquigarrow A_t$, which is an adapted process starting from zero, defined on a filtered space, such that all paths $t \mapsto A_t(\omega)$ are non-decreasing, right-continuous and have left-hand limits at every point $t \in (0, \infty)$, with a "non-decreasing process" which is just a process whose all paths are non-decreasing. See Definition 5.4.4 and the subsequent Remark for the details.

List of Figures

Part I

Introduction

1

Prologue

The most important keyword for this monograph is "martingale". One of the principal roles of the martingale in statistics is to serve as *an important tool for building semimartingale models in a variety of application areas*. This chapter aims to introduce readers to the "faces" of two useful statistical models based on semimartingales, so that the real image of our research subject is nurtured more clearly in our minds; this will be done in Section 1.2.

Before proceeding to the above objectives, some of the reasons why the martingale is so useful will be explained from two different perspectives in Section 1.1, using minimal mathematical formulas.

The semimartingale may be interpreted as a "stochastic process version" of a statistical regression model. To illustrate the implications of this interpretation, Section 1.2, which is the main part of this chapter, provides an overview of statistical modelling with semimartingales by building two types of practical models. The first model is the diffusion process model, which is constructed in a fairly intuitive way in Subsection 1.2.1. On the other hand, it will be explained in Subsection 1.2.2 that Cox's regression model, which is widely used in survival analysis, is indeed one of the semimartingale models based on the Doob-Meyer decomposition.

It has already been recognized that the diffusion process model and Cox's regression model are actually very useful. So the most primary purpose of this monograph is to provide a self-contained, detailed explanation of the statistical analysis (to be more specific, the derivation of the consistency and the asymptotic normality of some statistical estimators) in these two important models.

1.1 Why is the Martingale so Useful?

1.1.1 Martingale as a tool to analyze time series data in real time

One of the important characteristics of economic data is that such a data is a realization of random phenomena that varies from moment to moment depending on the events up to the present. In order to describe such a mechanism of randomness, a natural and promising method may be to consider the usual time series like

$$X_k = f(X_{k-1}) + \varepsilon_k, \tag{1.1}$$

or, more generally,

$$X_k = f(X_{k-1}, X_{k-2}, ..., X_{k-p}) + \varepsilon_k. \tag{1.2}$$

DOI: 10.1201/9781315117768-1

This kind of statistical models are especially useful when the data $X_0, X_1, X_2, \ldots$ is such as (converted) stock prices observed at equidistant time points (such as hourly or daily). In fact, time series analysis over the last several decades has a huge amount of successful research results. But if the stock prices are observed at irregular time points $t_0 < t_1 < t_2 < \cdots$ (that is, if the data is such as $Y_{t_0}, Y_{t_1}, Y_{t_2}, \ldots$), how should we analyze the data? Deeming $Y_{t_k} = X_k$ and pushing the data into a model like (1.1) or (1.2) may no longer be the most promising method for the right statistical analysis. As a matter of fact, it is often assumed that ε_k's are independently, identically distributed random variables, but the actual variances of Y_{t_k}'s are likely to vary depending on the length of the time intervals $t_k - t_{k-1}$, $k = 1, 2, \ldots$.

From this point of view, when explaining the mechanism of a random phenomena that varies from moment to moment like economic data, we sometimes had better prepare a *continuous-time stochastic process model*, like

$$Y_t = Y_0 + \int_0^t f(Y_s)ds + \sigma W_t, \tag{1.3}$$

so that we could naturally regard the data Y_{t_k}'s as the values of the stochastic process $(Y_t)_{t\in[0,\infty)}$ observed at discrete time points $t_0, t_1, t_2, \ldots$. The stochastic process $(W_t)_{t\in[0,\infty)}$ appearing above is called a *standard Wiener process*, and it has the property that $(W_{t_k} - W_{t_{k-1}})$'s are independently distributed to Gaussian distributions $\mathcal{N}(0, t_k - t_{k-1})$'s, respectively; see Definition 5.1.4 for the rigorous definition. This approach solves the sometimes controversial assumption that all variances of ε_k's are assumed to be the same, described above, in a natural way.

One of the special cases of the above general model is a *Vasicek process* given by

$$Y_t = Y_0 - \int_0^t \beta_1(Y_s - \beta_2)ds + \sigma W_t.$$

This model is widely applied as a statistical model that describes the random mech-

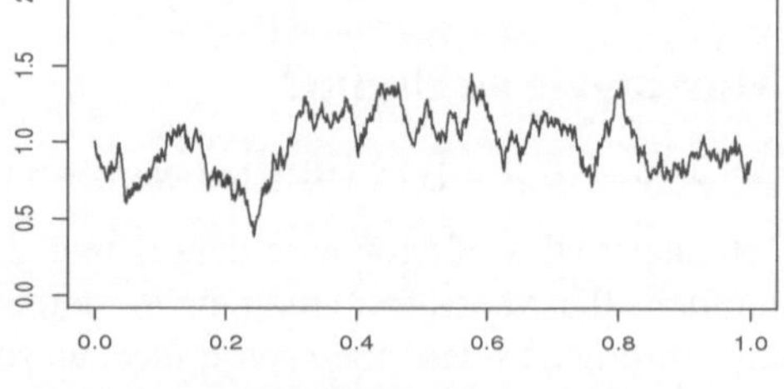

FIGURE 1.1
A path of a Vasicek process, where $\beta_1 = \beta_2 = 1$, $\sigma = 1$, $X_0 = 1$. Since we have set $\beta_2 = 1$, the stochastic process is taking values around 1.

anism of the short-term interest rate Y_t. One of the goals of this monograph is to *estimate* β_1, β_2, σ based on the data $Y_{t_0}, Y_{t_1}, Y_{t_2}, ...$ sampled at discrete-time points.

Having viewed the usefulness of continuous-time stochastic processes, the next question of readers' may be "Why martingales?" One of the answers to this question is that the property that the conditional expectations of increments given past events are zero is involved as the main condition in the definition of martingale. The property is a natural generalization of *orthogonality of noises,* which is an important component in the statistics theory. Let us thus have a preview on the orthogonality of noises briefly. For example, suppose that we intend to evaluate the fitness of the function f in the model (1.1) (or (1.2)) by the expected square risk

$$\mathsf{E}\left[\left(\sum_{k=1}^{n}(X_k - f(X_{k-1}))\right)^2\right].$$

This value is computed further as

$$= \mathsf{E}\left[\left(\sum_{k=1}^{n}\varepsilon_k\right)^2\right] = \sum_{k=1}^{n}\mathsf{E}[\varepsilon_k^2] + 2\sum_{k<l}\mathsf{E}[\varepsilon_k\varepsilon_l].$$

Due to the *independence* of ε_k's, the second terms of the last expression is computed as

$$\mathsf{E}[\varepsilon_k\varepsilon_l] = \mathsf{E}[\varepsilon_k]\mathsf{E}[\varepsilon_l] = 0\cdot 0 = 0,$$

and the calculation that this kind of terms vanish is called the *orthogonality of noises.* All readers those who have learned statistics in any sense must have understood how essential in statistics this computation is. Fortunately, it holds also in martingale cases that for every $k < l$,

$$\mathsf{E}[\varepsilon_k\varepsilon_l] = \mathsf{E}[\varepsilon_k\mathsf{E}[\varepsilon_l|\mathcal{F}_k]] = \mathsf{E}[\varepsilon_k \cdot 0] = 0. \tag{1.4}$$

We can readily expect that the martingale is an important concept for operations taking expectations of "squares", and therefore, it hits the essential points of statistics.

A standard Wiener process $(W_t)_{t\in[0,\infty)}$ appearing e.g., in (1.3), or more generally, a continuous-time martingale $(M_t)_{t\in[0,\infty)}$ possesses a property to make it possible to perform such computations in continuous-time, which correspond to (1.4) in discrete-time—this is one of the reasons why the martingale is so important.

1.1.2 Martingale as a tool to deal with censored data correctly

Another area where the martingale theory has been practically applied is the survival analysis. As an illustration, let us consider the problem estimating the approximate (or expected) value of the time length for a patient to have a cancer relapse after surgery. If here are 10 patients, and if all of the duration times $T_1, ..., T_{10}$ are observed, then it may be reasonable to use $\frac{1}{10}\sum_{k=1}^{10} T_k$ as an estimated value.

On the other hand, how should we construct an estimated value if the given data is the following:

Patient 1:	A relapse of cancer occurred 105 days after surgery;
Patient 2:	A relapse of cancer occurred 45 days after surgery;
Patient 3:	A relapse of cancer occurred 159 days after surgery;
Patient 4:	Died 253 days after surgery due to a disease not related to cancer;
Patient 5:	Died 652 days after surgery due to a disease not related to cancer;
Patient 6:	Left hospital 120 days after surgery;
Patient 7:	Changed hospital 120 days after surgery;
Patient 8:	Still under medical treatment 252 days after surgery;
Patient 9:	Still under medical treatment 145 days after surgery;
Patient 10:	Still under medical treatment 185 days after surgery.

Like the above example, a data which contains individuals for which the observation was dropped due to some reason even though they were still on the way of medical treatment for a disease of interest (i.e., cancer, in the current case) is called a *censored data*. In this case, in order to estimate the duration time until a relapse of cancer, it is *wrong* to use $\frac{1}{3}(T_1+T_2+T_3)$ based on the fact that only the patients 1,2, and 3 had a relapse of cancer. The reasons include that the patients 8, 9 and 10 are successful cases in view of medical treatment for cancer, and thus doing data analysis after removing these values leads to a wrong conclusion. Similarly, the patients 4 and 5 also had longer life as far as cancer is concerned, and thus removing these values results in a too small evaluation for the duration time until a relapse of cancer.

Then, how should we analyze this kind of data? An answer given by Odd O. Aalen in 1975 is the following: if we define

$N_t :=$ "the number of individuals who had a relapse of cancer up to time t"

and

$Y_t :=$ "the number of individuals who are under investigation just before time t",

then the stochastic process $(M_t)_{t\in[0,\infty)}$ given by

$$M_t = N_t - \int_0^t \alpha(s)Y_s ds$$

becomes a martingale, where α is the *hazard function* for the distribution of duration time of a relapse of cancer defined by

$$\alpha(t) = \frac{f(t)}{1-F(t)}$$

with F and f being the distribution and density functions, respectively, for the distribution of the duration time. Let us apply the fact from the martingale theory that

"if M is a martingale, then for any "predictable" process $H = (H_t)_{t\in[0,\infty)}$ the new stochastic process $\int_0^t H_s dM_s$ is a martingale starting from zero",

by setting "$H_s = \frac{1}{Y_s}$"[1] and "$dM_s = dN_s - \alpha(s)Y_s ds$", to have that

$$\int_0^t \frac{1}{Y_s}(dN_s - \alpha(s)Y_s ds)$$

[1]Precisely speaking, some careful discussion for the case of $Y_s = 0$ is necessary. This problem will be studied more rigorously in Example 3.5.5.

is a martingale starting from zero. Since the expectation of a martingale starting from zero is zero in general, we may conclude that

$$\int_0^t \frac{1}{Y_s} dN_s \quad \text{is an unbiased estimator for} \quad \int_0^t \alpha(s)ds.$$

It is proved that the estimator is good also from the viewpoint of statistical asymptotic theory.

This approach to survival analysis based on martingales was initiated by Odd O. Aalen's Ph.D. thesis in 1975, and the estimator $\int_0^t \frac{1}{Y_s} dN_s$ is now called the *Nelson-Aalen estimator*. The approach has made dramatic progress in a lot of directions since early 1980's, and it is now known that the martingale theory is essential for a deeper understanding of survival analysis.

1.2 Invitation to Statistical Modelling with Semimartingales

1.2.1 From non-linear regression to diffusion process model

In the non-linear regression model

$$X_k = f(X_{k-1}) + \varepsilon_k, \quad k = 1, 2, ...,$$

where ε_k's are usually assumed to be independently, identically distributed (i.i.d.) random variables with mean zero and a finite variance. In this kind of linear and non-linear regression models, an important feature is that the noise ε_k is added in the form of *summation*. Although it might not be impossible to consider the model of the product form

$$X_k = f(X_{k-1})\varepsilon_k, \quad k = 1, 2, ...,$$

by introducing some i.i.d. random variables ε_k with mean 1, various computations would not actually be performed smoothly in such models and we could not carry out a good statistical analysis. It is important that *the trend and noise terms are given in the form of summation.*

A *discrete-time stochastic process* or *time series* is a family of random variables indexed by a discrete set, that is, $X = (X_k)_{k \in \mathbb{N}}$, $(X_k)_{k \in \mathbb{N}_0}$, $(X_k)_{k \in \mathbb{Z}}$, and so on, where $\mathbb{N} = \{1, 2, ...\}$, $\mathbb{N}_0 = \{0, 1, 2, ...\}$ and $\mathbb{Z} = \{..., -1, 0, 1, 2, ...\}$. The 1st order linear auto-regressive (AR (1)) process model given by

$$X_k = \alpha X_{k-1} + \varepsilon_k, \quad k = 1, 2, ...,$$

with an appropriate initial value X_0, is one of the most elementary process models among time series. More generally, one may sometimes consider the 1st order non-linear auto-regressive process model

$$X_k = \widetilde{f}(X_{k-1}) + g(X_{k-1})\varepsilon_k, \quad k = 1, 2, ...$$

By setting $f(x) = x + \widetilde{f}(x)$ we have

$$X_k - X_{k-1} = f(X_{k-1}) + g(X_{k-1})\varepsilon_k, \quad k = 1, 2, ...,$$

and this can be represented by

$$X_k = X_0 + \sum_{j=1}^{k} f(X_{j-1}) + \sum_{j=1}^{k} g(X_{j-1})\varepsilon_j, \quad k = 1, 2, ... \tag{1.5}$$

In any case, the structure that the noise is added in the form of summation is essential, and we cannot drop this assumption in our discussion by some reasons in mathematical analysis.

Having approached to the target of our study, *continuous-time stochastic processes*, it seems that now is good time to explain what is *diffusion processes* in order to develop an image of the research object in our brains.

First, let us (briefly and informally) introduce a standard Wiener process as a stochastic process $(W_t)_{t\in[0,\infty)}$ such that for any finite number of time points $0 = t_0 < t_1 < \cdots < t_n$, the difference $W_{t_k} - W_{t_{k-1}}$, $k = 1, 2, ..., n$, are independently distributed to $\mathcal{N}(0, t_k - t_{k-1})$, $k = 1, 2, ..., n$, respectively[2]. While the time grids t_k's in the case of time series model (1.5) are equidistant, i.e., $t_k - t_{k-1} \equiv 1$, by considering a more general sampling scheme, which may not be equidistant, and assuming the noises ε_k's are independent Gaussian random variables $W_{t_k} - W_{t_{k-1}}$'s, let us introduce the model

$$X_{t_n} = X_0 + \sum_{k=1}^{n} f(X_{t_{k-1}})|t_k - t_{k-1}| + \sum_{k=1}^{n} g(X_{t_{k-1}})(W_{t_k} - W_{t_{k-1}}).$$

Here, let us make a leap of our idea!; a leap of ideas is absolutely necessary for scientific progress! By making the time span $|t_k - t_{k-1}|$'s converge to zero, we take a limit to replace $\sum_{k=1}^{n}$ with $\int_0^{t_n}$, which leads to the continuous-time stochastic process model

$$X_t = X_0 + \int_0^t f(X_s)ds + \int_0^t g(X_s)dW_s \tag{1.6}$$

of the autoregression type. This is the so-called *diffusion process* or *Itô process*. Although beginners might have an impression that "integration" is more difficult than "summation", readers those who have already learned some advanced mathematics must already know that the integration is a natural generalization of the summation and that discussing with integration often make our understanding clearer than doing with summation. Moreover, the last term $\int_0^t g(X_s)dW_s$ of the above model is called the *Itô integral*, with which we have a big merit that a lot of powerful tools of stochastic calculus are available.

Now, let us generalize the above discussion further to make our idea clearer. A *semimartingale* is a stochastic process $(X_t)_{t\in[0,\infty)}$ such that when we rewrite it into the additive form

$$X_t = X_0 + A_t + M_t$$

[2]Some more conditions are necessary; see Definition 5.1.4. A Wiener process may be regarded as an infinite-dimensional Gaussian random "vector", in the sense that it has infinitely many Gaussian random variables as its components.

of the regression (trend) term $(A_t)_{t\in[0,\infty)}$ and the noise term $(M_t)_{t\in[0,\infty)}$, the latter term $(M_t)_{t\in[0,\infty)}$ is a good stochastic process called *martingale*. Here, roughly speaking, a martingale is a stochastic process whose "conditional trend" is zero[3]. The Itô integral $\int_0^t g(X_s)dW_s$ is the limit of the summation of infinitely many, tiny, zero mean Gaussian random variables, and it is a martingale; this will be explained later in this monograph. The diffusion process model (1.6) is a *stochastic process of autoregression type with the noise terms of weighted summation of tiny, mean-zero Gaussian random variables*. The reason why we assume that the noise term is a martingale is that a lot of mathematical tools are available with martingales, and a mission of this monograph is to give an exposition about them.

The equations like (1.6) are called *stochastic differential equations*. The construction of diffusion processes based on stochastic differential equations was established by Kiyosi Itô in 1944. Since the path-break work of Black and Scholes in 1973, the Itô calculus has now become a system of fundamental tools for the study of mathematical finances.

1.2.2 Cox's regression model as a semimartingale

Let some events in which a statistician is interested occur at random time points $0 < \tau_1 < \tau_2 < \cdots$. For example, let τ_j be the time when the j-th fit of a patient occurs after a medication.

A counting process $(N_t)_{t\in[0,\infty)}$ is a stochastic process such that N_t denotes the number of events that occurred up to time t:

$$N_t := \#\{j : \tau_j \leq t\}, \quad \forall t \in [0,\infty).$$

The realization of a counting process $t \rightsquigarrow N_t$ is a non-decreasing step function, starting from zero, whose jumps are $+1$; see Figure 1.2.

As we will learn later, a counting process is a *submartingale*. Thus, by using the most fundamental theorem called the *Doob-Meyer decomposition theorem*, it is proved that the counting process $(N_t)_{t\in[0,\infty)}$ can be decomposed into the sum of a predictable, increasing process $(A_t)_{t\in[0,\infty)}$ and a martingale $(M_t)_{t\in[0,\infty)}$:

$$N_t = A_t + M_t, \quad \forall t \in [0,\infty).$$

Here, a sufficient condition for being a predictable process is that it is an "adapted process" whose paths are left-continuous". At this stage of the first reading, the "predictable" process may be interpreted as a "so smooth" stochastic process that the value A_t can be predicted just before the time t, namely, $A_t = A_{t-}$, where $A_{t-} := \lim_{s\nearrow t} A_s$. Roughly speaking, we may interpret the decomposition as follows: *since $t \rightsquigarrow N_t$ is non-decreasing, it is written as the sum of a smooth trend $t \rightsquigarrow A_t$ and a noise $t \rightsquigarrow M_t$*. The trend $t \rightsquigarrow A_t$ is called the *compensator* of the original counting

[3]This explanation is not exact. A more rigorous explanation using a formula is that the martingale $(M_t)_{t\in[0,\infty)}$ is a stochastic process satisfying that $\mathsf{E}[M_t|\mathcal{F}_s] = M_s$, a.s., for every $0 \leq s \leq t < \infty$. The phrase "a.s." is the abbreviation to "almost surely, and we refer to Section 2.1.2 for the meaning.

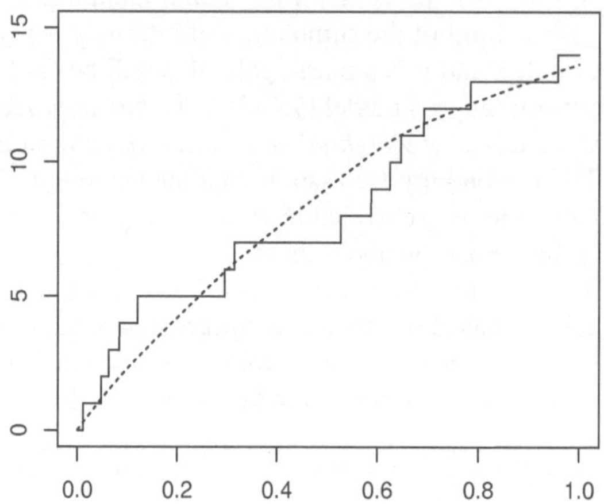

FIGURE 1.2
A path of a counting process $t \rightsquigarrow N_t$ (solid), and its compensator $t \rightsquigarrow A_t$ (dashed). The former is of the form of a step function, starting from zero whose jumps are $+1$.

process $t \rightsquigarrow N_t$, and it may be considered as representing an approximate history of the increase of the original process; see Figure 1.2 again.

Thus, by investigating the value of the derivative

$$\lambda_t := \frac{dA_t}{dt}$$

we can grasp the degree of the increase of the original counting process. This stochastic process $t \rightsquigarrow \lambda_t$ is called the *intensity process* of $t \rightsquigarrow N_t$. If the value of the intensity λ_t is large at time t then many events occur around time t, while if small then the time points are sparse around there. Note that by substituting $A_t = \int_0^t \lambda_s ds$ into the above decomposition, we get

$$N_t = \int_0^t \lambda_s ds + M_t, \quad \forall t \in [0, \infty).$$

Now, let the index $k = 1, 2, \ldots, n$ denotes the individual (e.g., the reference numbers of patients). For every k, let us consider a sequence $0 < \tau_1^k < \tau_2^k < \cdots$ of occurring times of some events and the corresponding counting process $(N_t^k)_{t \in [0,\infty)}$ defined by $N_t^k = \#\{j : \tau_j^k \leq t\}$. If the patient k dies or leaves hospital, then the intensity λ_t^k has to become zero; thus, for later use let us introduce the indicator process

$$Y_t^k = \begin{cases} 1, & \text{if the individual } k \text{ is under investigation just before time } t, \\ 0, & \text{otherwise.} \end{cases}$$

Next, we assume that besides the characteristics of each individual k there is a "risk" of death which is common for all individuals and denote it by $\alpha(t)$. This is assumed to be deterministic and called a *hazard function*. Based on these preparations, we shall assume that the intensity of individual k is given by

$$\lambda_t^k = e^{\theta^{\mathrm{tr}} Z_t^k} \alpha(t) Y_t^k,$$

where $t \rightsquigarrow Z_t^k = (Z_t^{k,1}, ..., Z_t^{k,p})^{\mathrm{tr}}$ is an $\mathbb{R}^p$-valued stochastic process called *covariate*, which represents some own characteristics of individual k such as age, gender, weight, blood pressure, dose of medicine, and so on. In this way, the statistical model, where the exponential of the linear combination of covariates is assumed to reflect the intensity, which has been introduced in order to investigate the effects of covariates by estimating the parameter θ of interest, with the baseline hazard function α being an infinite-dimensional nuisance parameter, is called *Cox's regression model*. In view of the expression with components of the form

$$\begin{aligned} \lambda_t^k &= \prod_{i=1}^p e^{\theta_i Z_t^{k,i}} \alpha(t) Y_t^k \\ &= e^{\theta_1 Z_t^{k,1}} \cdots e^{\theta_p Z_t^{k,p}} \alpha(t) Y_t^k, \end{aligned}$$

the model may be interpreted as the one given of the product form of effects of all components of covariates and the baseline hazard function, by representing on/off of observing the individual k at time t by Y_t^k.

This model was proposed by David R. Cox in 1972, and via the establishment of the asymptotic theory based on martingales by Per K. Andersen and Richard D. Gill in 1982, it has been playing an important role in survival analysis.

2

Preliminaries

In this chapter, we prepare three things. The probability theory, including the martingale theory, is developed based on the measure theory. We thus first recall the usages of some important limit operations in the measure theory.

Secondly, an intuitive explanation for the conditional expectation is presented. Since the martingale is defined by using the conditional expectation, it would be very helpful for readers of this monograph to understand what is the conditional expectation at this stage. We also summarise some important properties of the conditional expectation.

The main theme of this monograph is the asymptotic theory in statistics. Thirdly, we introduce the definitions, relationships, and related theorems concerning the almost sure convergence, the convergence in probability and the convergence in law for random sequences that will be the bases for our study on asymptotic statistics.

2.1 Remarks on Limit Operations in Measure Theory

2.1.1 Limit operations for monotone sequence of measurable sets

Observe the following elementary facts in the measure theory.

Theorem 2.1.1 (i) Let $(\mathcal{X},\mathcal{A},\mu)$ be a measure space. If a sequence $(A_n)_{n=1,2,\ldots}$ of elements in $\mathcal{A}$ satisfies either

$$A_1 \subset A_2 \subset \cdots$$

or

$$A_1 \supset A_2 \supset \cdots \quad \text{and} \quad \mu(A_{n_*}) < \infty \text{ for some } n_* \in \mathbb{N},$$

then it holds that

$$\lim_{n\to\infty} \mu(A_n) = \mu\left(\lim_{n\to\infty} A_n\right).$$

(ii) Let $(\Omega,\mathcal{F},\mathsf{P})$ be a probability space. For any monotone (non-decreasing or non-increasing) sequence $(A_n)_{n=1,2,\ldots}$ of events in $\mathcal{F}$, it holds that

$$\lim_{n\to\infty} \mathsf{P}(A_n) = \mathsf{P}\left(\lim_{n\to\infty} A_n\right).$$

We remark that the assumption "$\mu(A_{n_*}) < \infty$ for some $n_* \in \mathbb{N}$" in the second condition of (i) is indeed necessary, as we see in the following counter example.

DOI: 10.1201/9781315117768-2

Example 2.1.2 Let μ be the Lebesgue measure on $(\mathbb{R}, \mathfrak{B}(\mathbb{R}))$, and put $A_n = [n, \infty)$ for every $n \in \mathbb{N}$. Although $A_1 \supset A_2 \supset \cdots$, since $\mu(A_n) = \infty$ for all $n \in \mathbb{N}$ the sequence $(\mu(A_n))_{n=1,2,...}$ is not convergent, while $\mu(\lim_{n\to\infty} A_n) = \mu(\emptyset) = 0$.

Related to (ii) of the above theorem, here is another important remark.

Discussion 2.1.3 Let $(X_n)_{n=1,2,...}$ be an *non-decreasing* sequence of real-valued random variables on a probability space $(\Omega, \mathcal{F}, \mathsf{P})$, and let c be a real number.

(a) The following claim is **false**: $\lim_{n\to\infty} \mathsf{P}(X_n \geq c) = \mathsf{P}(\lim_{n\to\infty} X_n \geq c)$.

(b) The following claim is **true**: $\lim_{n\to\infty} \mathsf{P}(X_n > c) = \mathsf{P}(\lim_{n\to\infty} X_n > c)$.

For example, put $X_n = c - \frac{1}{n}$ for every $n \in \mathbb{N}$ (a deterministic sequence). Then, it holds that (a) $\mathsf{P}(X_n \geq c) = 0$ for all $n \in \mathbb{N}$ and $\mathsf{P}(\lim_{n\to\infty} X_n \geq c) = 1$, and that (b) $\mathsf{P}(X_n > c) = 0$ for all $n \in \mathbb{N}$ and $\mathsf{P}(\lim_{n\to\infty} X_n > c) = 0$.

The mistake (a) is due to the confusion between the limit operations for monotone sequences of events and those of random variables. To understand this point more clearly, observe that

$$\lim_{n\to\infty} \{\omega : X_n(\omega) \geq c\} = \bigcup_{n=1}^{\infty} \{\omega : X_n(\omega) \geq c\} \subset \left\{\omega : \lim_{n\to\infty} X_n(\omega) \geq c\right\},$$

and that

$$\lim_{n\to\infty} \{\omega : X_n(\omega) > c\} = \bigcup_{n=1}^{\infty} \{\omega : X_n(\omega) > c\} = \left\{\omega : \lim_{n\to\infty} X_n(\omega) > c\right\}.$$

Thus, the mistake (a) can be corrected as follows.

(a') The following claim is **true**: $\lim_{n\to\infty} \mathsf{P}(X_n \geq c) \leq \mathsf{P}(\lim_{n\to\infty} X_n \geq c)$.

2.1.2 Limit theorems for Lebesgue integrals

Let $(\mathcal{X}, \mathcal{A}, \mu)$ be a measure space. The sales point of the Lebesgue integral theory is that various *limit operations* concerning the integrals work smoothly. The most important tool among them is called *Lebesgue's convergence theorem* or the *dominated convergence theorem*, and it may give an answer to the following question: *When a sequence $(f_n)_{n=1,2,...}$ of measurable functions converges to a measurable function f in the sense that*

$$\lim_{n\to\infty} f_n(x) = f(x), \quad \forall x \in \mathcal{X}, \tag{2.1}$$

may we conclude that

$$\lim_{n\to\infty} \int_{\mathcal{X}} f_n(x)\mu(dx) = \int_{\mathcal{X}} f(x)\mu(dx)? \tag{2.2}$$

The answer to this question is **"No"** in general. However, when we can check an "additional condition", the convergence (2.2) appearing in the conclusion becomes true; thus the answer to the above question should be **"Conditionally Yes!"**

Theorem 2.1.4 (Lebesgue's convergence theorem) For given sequence $(f_n)_{n=1,2,\ldots}$ of measurable functions and a measurable function f, suppose that there exists an $A \in \mathcal{A}$ such that $\mu(A^c) = 0$ and that

$$\lim_{n\to\infty} f_n(x) = f(x), \quad \forall x \in A. \tag{2.3}$$

If, moreover, it is possible to find an integrable function φ not depending on $n \in \mathbb{N}$ such that

$$|f_n(x)| \le \varphi(x), \quad \forall x \in A, \quad \forall n \in \mathbb{N}, \tag{2.4}$$

then f is integrable and it holds that

$$\lim_{n\to\infty} \int_{\mathcal{X}} f_n(x)\mu(dx) = \int_{\mathcal{X}} f(x)\mu(dx).$$

Compared with the condition (2.1) that was announced just for explanation, the condition (2.3) that we should actually check is a little weaker, because the domain $\mathcal{X}$ of the convergence assumption has been replaced by a smaller set $A \subset \mathcal{X}$. Furthermore, the condition (2.4) is also sufficient to hold not for all $x \in \mathcal{X}$ but only for $x \in A$. Hereafter, this kind of expression, say,

"there exists $A \in \mathcal{A}$ such that $\mu(A^c) = 0$ and that CONDITION holds for all $x \in A$",

is shortly written as

"CONDITION, μ-a.e. x",

where "a.e." stands for "almost everywhere", or as

"CONDITION, a.s.",

where "a.s." does for "almost surely", when μ is a probability measure.

An immediate corollary to Lebesgue's convergence theorem is the following.

Corollary 2.1.5 (Bounded convergence theorem) Let $(\mathcal{X}, \mathcal{A}, \mu)$ be a finite measure space; that is, it is a measure space such that $\mu(\mathcal{X}) < \infty$[1]. In this case, the condition (2.4) in Theorem 2.1.4 may be replaced with

there is a constant $\quad K > 0 \quad$ s.t. $\quad |f_n(x)| \le K, \quad \mu$-a.e. $x, \quad \forall n \in \mathbb{N}$.

The crucial point of Lebesgue's convergence theorem is that the dominating function φ appearing in (2.4) has to be found not depending on n. Even when it is difficult to find such a dominating function, if $(f_n)_{n=1,2,\ldots}$ is *non-negative* and *non-decreasing* in n, then another tool called the *monotone convergence theorem* may work well.

[1] An example of a finite measure space is a probability space $(\Omega, \mathcal{F}, \mathsf{P})$, where the condition $\mathsf{P}(\Omega) = 1$ is presumed.

Theorem 2.1.6 (Monotone convergence theorem) Suppose that given sequence $(f_n)_{n=1,2,...}$ of measurable function and a measurable function f satisfy that

$$\lim_{n\to\infty} f_n(x) = f(x), \quad \mu\text{-a.e. } x.$$

If, moreover, it is satisfied also that

$$0 \le f_1(x) \le f_2(x) \le \ldots \le f_n(x) \le \ldots, \quad \mu\text{-a.e. } x,$$

then it holds that

$$\lim_{n\to\infty} \int_{\mathcal{X}} f_n(x)\mu(dx) = \int_{\mathcal{X}} f(x)\mu(dx),$$

allowing the possibility that the both sides are ∞.

Needless to say, no theorem is omnipotent. Let us observe two examples for which both of the above two theorems do not work; these may be regarded as counter examples to the opening question stated at the beginning of this subsection.

Example 2.1.7 Put

$$f_n(x) = \sum_{k=0}^{n} \frac{(-x^2/2)^k}{k!}, \quad \forall x \in \mathbb{R}, \quad \forall n \in \mathbb{N}.$$

Then, the condition (2.1) is satisfied; indeed, it holds that

$$\lim_{n\to\infty} f_n(x) = e^{-x^2/2} (= f(x)), \quad \forall x \in \mathbb{R}.$$

The values of the integrals $\int_{\mathbb{R}} f_n(x)dx$ are ∞ for even n and $-\infty$ for odd n, while the limit f is integrable: $\int_{\mathbb{R}} f(x)dx = \sqrt{2\pi}$; thus the conclusion (2.2) does not hold.

Example 2.1.8 Put

$$f_n(x) = 1_{(n-1,n]}(x), \quad \forall x \in \mathbb{R}, \quad \forall n \in \mathbb{N}.$$

Then, the condition (2.1) is satisfied; indeed, it holds that

$$\lim_{n\to\infty} f_n(x) = 0 (= f(x)), \quad \forall x \in \mathbb{R}.$$

However, it holds that $\int_{\mathbb{R}} f_n(x)dx = 1$ for all $n \in \mathbb{N}$, while $\int_{\mathbb{R}} f(x)dx = 0$; thus the conclusion (2.2) does not hold.

Note that it is impossible to find any dominating function φ that is integrable for these sequences $(f_n)_{n=1,2,...}$, and that the sequences do not meet the condition "non-negative" and "non-decreasing in n". This is why both of the above theorems do not work for these two sequences.

Here, let us notice also that either of the above two theorems gives only a set of sufficient conditions for the convergence (2.2), and that either of the sets of conditions is not necessary.

Example 2.1.9 Put

$$f_n(x) = \frac{1}{n} \cdot 1_{(n-1,n]}(x), \quad \forall x \in \mathbb{R}, \quad \forall n \in \mathbb{N}.$$

Then, the condition (2.1) is satisfied; indeed, it holds that

$$\lim_{n\to\infty} f_n(x) = 0 (= f(x)), \quad \forall x \in \mathbb{R}.$$

It also holds that $\int_{\mathbb{R}} f_n(x)dx = 1/n$, which converges as $n \to \infty$ to $\int_{\mathbb{R}} f(x)dx = 0$; thus the convergence (2.2) holds true. However, the sequence $(f_n)_{n=1,2,\dots}$ is not dominated by any integrable function[2], and the sequence is not non-decreasing. Therefore, the convergence (2.2) holds true, but it is not a consequence of either of the two theorems.

Another important device concerning limit operations for Lebesgue integrals is *Fatou's lemma*. However, we omit the exposition of the lemma because it will not be used anywhere in this monograph.

2.2 Conditional Expectation

The common set-up for the materials treated in this section is the following. Let $(\Omega, \mathcal{F}, \mathsf{P})$ be a probability space. Let a sub-σ-field $\mathcal{G}$ of $\mathcal{F}$ be given; that is, it holds that

$$\{\emptyset, \Omega\} \subset \mathcal{G} \subset \mathcal{F}$$

and that $\mathcal{G}$ itself is a σ-field on Ω.

2.2.1 Understanding the definition of conditional expectation

Let us start with an intuitive explanation of the operation called "conditional expectation" given a sub-σ-field $\mathcal{G}$ of $\mathcal{F}$. It is an operation for a given $\mathcal{F}$-measurable real-valued random variable X to reduce it by a new random variable, namely $\mathsf{E}[X|\mathcal{G}]$, which is measurable with respect to the "poorer" σ-field $\mathcal{G}$ than the original σ-field $\mathcal{F}$. In particular, when $\mathcal{G} = \{\emptyset, \Omega\}$, the operation is nothing else than reducing X by a constant, which is the usual expectation $\mathsf{E}[X]$. For a general $\mathcal{G}$ which has some "information", the operation is to make a random reduction $\mathsf{E}[X|\mathcal{G}]$ of X by using the given information contained in $\mathcal{G}$.

We may also have the following interpretation.

$\mathsf{E}[X]$	$\leftarrow$--	$\mathsf{E}[X\|\mathcal{G}]$	$\leftarrow$--	X
$\{\emptyset, \Omega\}$-measurable		$\mathcal{G}$-measurable		$\mathcal{F}$-measurable

[2]The dominating function φ should satisfy that $\varphi(x) \geq \sup_{n\in\mathbb{N}} |f_n(x)| = \sum_{n=1}^{\infty} f_n(x)$, and thus $\int_{\mathbb{R}} \varphi(x)dx \geq \sum_{n=1}^{\infty}(1/n) = \infty$.

Here, the notation "$\widetilde{A} \leftarrow\text{--}\ A$" may be read as "$\widetilde{A}$ is an object obtained by reducing the information that A has"; readers who like mathematical expressions may read it as "$\widetilde{A}$ is a projection of A"; readers who like rough, informal explanations may dare to read it as "$\widetilde{A}$ is an approximate value of A based on more restricted information".

Now, let us describe the rigorous definition of the conditional expectation.

Theorem 2.2.1 (Conditional expectation) Let $(\Omega, \mathcal{F}, \mathsf{P})$ be a probability space, and $\mathcal{G}$ a sub-σ-field of $\mathcal{F}$. For any given real-valued, $\mathcal{F}$-measurable, integrable random variable X, there exists a real-valued, $\mathcal{G}$-measurable, integrable random variable $\widetilde{X}$ such that

$$\int_G \widetilde{X}(\omega)\mathsf{P}(d\omega) = \int_G X(\omega)\mathsf{P}(d\omega), \quad \forall G \in \mathcal{G}.$$

This $\widetilde{X}$ is unique up to the P-almost sure sense; that is, if $\widetilde{X}'$ is a random variable satisfying the same properties as $\widetilde{X}$, then $\widetilde{X} = \widetilde{X}'$, P-almost surely.

This equivalent class $\widetilde{X}$ is called the *conditional expectation* of X given $\mathcal{G}$, and it is denoted by $\mathsf{E}[X|\mathcal{G}]$.

We dare to skip the proof of this theorem, because just following the usual proof based on the Radon-Nikodym theorem that can be seen in the standard textbooks on probability theory may not be so helpful for us to get a good understanding of the meaning of the theorem. Alternatively, let us try to get an intuitive interpretation of the theorem with an illustrative example.

Discussion 2.2.2 Some readers might have an experience to be explained something like that a sub-σ-field $\mathcal{G}$ is an "information". Let us interpret the meaning of this kind of explanations as follows: for any $G \in \mathcal{G}$, it is known to observer whether $\omega \in G$ or not for any ω.

Now, when $\mathcal{G} = \{\emptyset, A, A^c, \Omega\}$ for example, let us call A and A^c the *atoms* of $\mathcal{G}$. More generally, when a finite disjoint partition $\Omega = A_1 \cup \cdots \cup A_p$ is given and $\mathcal{G}$ is the σ-field consists of sets in the form of finite unions of some of $A_1, ..., A_p$, let us call sets that are not able to be divided any more except for the empty set (that is, sets $A_1, ..., A_p$) the *atoms* of $\mathcal{G}$. In this case, if we recall the definition of the concept of "measurability", it is easily seen that any $\mathcal{G}$-measurable function is constant on each of the atoms[3]. Since the conditional expectation $\mathsf{E}[X|\mathcal{G}]$ has to be $\mathcal{G}$-measurable, it has to be constant on each atom. Moreover, it should "approximate" X in the sense of "expectation". In summary, the conditional expectation is a random variable $\mathsf{E}[X|\mathcal{G}]$ of the form

$$\mathsf{E}[X|\mathcal{G}](\omega) = x_i, \quad \forall \omega \in A_i, \quad i = 1, ..., p,$$

where x_i's are some constants, which satisfies that

$$\sum_{i=1}^{p} x_i \mathsf{P}(G \cap A_i) = \sum_{i=1}^{p} \int_{G \cap A_i} X(\widetilde{\omega})\mathsf{P}(d\widetilde{\omega}), \quad \forall G \in \mathcal{G}.$$

[3]The reason is the following. In order for a random variable Y to be $\mathcal{G}$-measurable, it is necessary to satisfy that $\{\omega; Y(\omega) \geq a\} \in \mathcal{G}$ holds for any $a \in \mathbb{R}$. If Y takes different values y_1 and y_2 on an atom A_i, then the above condition is not satisfied for $a = (y_1 + y_2)/2$.

This equation has to hold in particular for $G = A_i$, thus it holds that

$$x_i \mathsf{P}(A_i) = \int_{A_i} X(\tilde{\omega}) \mathsf{P}(d\tilde{\omega}), \quad i = 1, \ldots, p. \tag{2.5}$$

This implies that if $\mathsf{P}(A_i) > 0$, then

$$x_i = \frac{\int_{A_i} X(\tilde{\omega}) \mathsf{P}(d\tilde{\omega})}{\mathsf{P}(A_i)};$$

in case of $\mathsf{P}(A_i) = 0$, any value of x_i satisfies the equation appearing in (2.5), and this is why we need the statement concerning the uniqueness in the P-almost sure sense in the theorem.

Consequently, the true identity of the conditional expectation is that

$$\mathsf{E}[X|\mathcal{G}](\omega) = \begin{cases} \frac{\int_{A_i} X(\tilde{\omega}) \mathsf{P}(d\tilde{\omega})}{\mathsf{P}(A_i)}, & \forall \omega \in A_i \text{ such that } \mathsf{P}(A_i) > 0, \\ \text{any constant } x_i, & \forall \omega \in A_i \text{ such that } \mathsf{P}(A_i) = 0, \end{cases} \quad i = 1, \ldots, p.$$

This is a step function of ω which is constant on each A_i. When $\mathcal{G}$ is a fine-grained σ-field, the function $\omega \mapsto \mathsf{E}[X|\mathcal{G}](\omega)$ becomes a "fine" approximation of $\omega \mapsto X(\omega)$.

When we pick up $\omega \in \Omega$ at random, if the given information is merely $\mathcal{G}$, we do not know the exact value of $X(\omega)$; however, since we know at least which A_i the chosen ω belongs to, we can compute the approximate value x_i of $X(\omega)$ based on the information $\mathcal{G}$. We call this value the realization of the conditional expectation, and it is denoted by $\mathsf{E}[X|\mathcal{G}](\omega)$.

So far we have discussed the case of finite events because the intuitive explanation is possible for such a case. The general case should be regarded as a natural extension of this illustrative example.

2.2.2 Properties of conditional expectation

Now, let us list up some important properties of the conditional expectation; the proofs can be seen in the standard textbooks for the probability theory. All random variables appearing from now on are assumed to be real-valued, integrable ones defined on a probability space $(\Omega, \mathcal{F}, \mathsf{P})$. Let $\mathcal{G}, \mathcal{H}$ be some sub-σ-fields of $\mathcal{F}$.

- If X is $\mathcal{G}$-measurable, then $\mathsf{E}[X|\mathcal{G}] = X$, a.s.
- $\mathsf{E}[aX + bY|\mathcal{G}] = a\mathsf{E}[X|\mathcal{G}] + b\mathsf{E}[Y|\mathcal{G}]$, a.s., where a and b are any constants.
- If $X \geq Y$, then $\mathsf{E}[X|\mathcal{G}] \geq \mathsf{E}[Y|\mathcal{G}]$, a.s.
- **(Tower property)** If $\mathcal{H} \subset \mathcal{G}$, then $\mathsf{E}[\mathsf{E}[X|\mathcal{G}]|\mathcal{H}] = \mathsf{E}[X|\mathcal{H}]$, a.s.; in particular, $\mathsf{E}[\mathsf{E}[X|\mathcal{G}]] = \mathsf{E}[X]$.
- If Y is $\mathcal{G}$-measurable and YX is integrable, then $\mathsf{E}[YX|\mathcal{G}] = Y\mathsf{E}[X|\mathcal{G}]$, a.s.

- **(Jensen's inequality)** If the function $\varphi : \mathbb{R} \to \mathbb{R}$ is convex, then

$$\varphi(\mathsf{E}[X|\mathcal{G}]) \leq \mathsf{E}[\varphi(X)|\mathcal{G}], \quad \text{a.s.}$$

- **(Hölder's inequality)** For $p,q > 1$ such that $\frac{1}{p} + \frac{1}{q} = 1$, if $\mathsf{E}[|X|^p] < \infty$ and $\mathsf{E}[|Y|^q] < \infty$, then

$$\mathsf{E}[|XY||\mathcal{G}] \leq (\mathsf{E}[|X|^p|\mathcal{G}])^{1/p}(\mathsf{E}[|Y|^q|\mathcal{G}])^{1/q}, \quad \text{a.s.;}$$

in particular, when $p = q = 2$, this is called **the Cauchy-Schwarz inequality**.

- **(Minkowski's inequality)** For $p \geq 1$, if $\mathsf{E}[|X|^p] < \infty$ and $\mathsf{E}[|Y|^p] < \infty$, then

$$(\mathsf{E}[|X+Y|^p|\mathcal{G}])^{1/p} \leq (\mathsf{E}[|X|^p|\mathcal{G}])^{1/p} + (\mathsf{E}[|Y|^p|\mathcal{G}])^{1/p}, \quad \text{a.s.}$$

Exercise 2.2.1 Let $X_1, X_2, ..., X_n$ be independent random variables, defined on a probability space $(\Omega, \mathcal{F}, \mathsf{P})$, such that $\mathsf{E}[X_k] = 0$ and $\mathsf{E}[X_k^2] = \sigma^2 < \infty$ for all $k = 1, ..., n$. Put $\mathcal{G}_k = \sigma(X_1, ..., X_k)$ for every $k = 1, ..., n$.

(i) Find $\mathsf{E}[(X_1 + X_2)^2]$ and $\mathsf{E}[(X_1 + X_2)^2|\mathcal{G}_1]$.

(ii) Find $\mathsf{E}[(X_1 + \cdots + X_n)^2]$ and $\mathsf{E}[(X_1 + \cdots + X_n)^2|\mathcal{G}_k]$ for every $k = 1, ..., n$.

Exercise 2.2.2 Let $X_1, X_2, ...$ be a sequence of (not necessarily independent) random variables, defined on a probability space $(\Omega, \mathcal{F}, \mathsf{P})$, such that $\mathsf{E}[X_k|\mathcal{G}_{k-1}] = 1$ a.s. all $k \in \mathbb{N}$, where $\mathcal{G}_k = \sigma(X_1, ..., X_k)$ and $\mathcal{G}_0 = \{\emptyset, \Omega\}$. Put $L_n = \prod_{k=1}^n X_k$ for every $n \in \mathbb{N}$. Find $\mathsf{E}[L_n|\mathcal{G}_k]$ for every $k = 0, 1, 2, ..., n$.

2.3 Stochastic Convergence

In this section, we summarise the definitions of "almost sure convergence", "convergence in probability" and "convergence in law"[4] for sequences of random variables taking values in a metric space and their relationships, which are needed for our purpose in this monograph. See, e.g., Chapters 2 and 18 of van der Vaart (1998) for more complete summaries.

- $(\mathbb{D}, d)$ is called a *metric space* if the function $d : \mathbb{D} \times \mathbb{D} \to [0, \infty)$ satisfies the following:

  (i) $d(x,y) = d(y,x)$;

  (ii) $d(x,z) \leq d(x,y) + d(y,z)$;

  (iii) $d(x,y) = 0 \Leftrightarrow x = y$.

[4] Another important mode of stochastic convergence is "L^p-convergence".

- Examples of metric spaces:

  (1) $\mathbb{D} = \mathbb{R}$, $d(x,y) = |x-y|$;

  (2) $\mathbb{D} = \mathbb{R}^p$, $d(x,y) = ||x-y|| = \sqrt{\sum_{i=1}^p (x^i - y^i)^2}$;

  (3) $\mathbb{D} = L^q(\mathcal{X},\mu)$, $q \geq 1$ (the space of equivalent classes of functions whose q-th power are μ-integrable), $d(f,g) = \left(\int_{\mathcal{X}} |f(z)-g(z)|^q \mu(dz)\right)^{1/q}$;

  (4) $\mathbb{D} = \ell^\infty(\mathbb{T})$ (the space of bounded functions on $\mathbb{T}$), $d(x,y) = \sup_{t\in\mathbb{T}} |x(t) - y(t)|$.

Definition 2.3.1 (Almost sure convergence) Let $(X_n)_{n=1,2,...}$ and X be $\mathbb{D}$-valued random variables defined on a probability space $(\Omega, \mathcal{F}, \mathsf{P})$. In this setting, $X_n \xrightarrow{a.s.} X$ means $\mathsf{P}(\lim_{n\to\infty} d(X_n, X) = 0) = 1$.

Definition 2.3.2 (Convergence in probability) Let $(X_n)_{n=1,2,...}$ and X be $\mathbb{D}$-valued random variables defined on a probability space $(\Omega, \mathcal{F}, \mathsf{P})$. In this setting, $X_n \xrightarrow{\mathsf{P}} X$ means that for any $\varepsilon > 0$ it holds that $\lim_{n\to\infty} \mathsf{P}(d(X_n,X) > \varepsilon) = 0$.

Definition 2.3.3 (Convergence in outer-probability) Let $(X_n)_{n=1,2,...}$ and X be $\mathbb{D}$-valued random elements, with no measurability, defined on a probability space $(\Omega, \mathcal{F}, \mathsf{P})$. In this setting, $X_n \xrightarrow{\mathsf{P}^*} X$ means that for any $\varepsilon > 0$ it holds that $\lim_{n\to\infty} \mathsf{P}^*(d(X_n,X) > \varepsilon) = 0$, where P^* denotes the *outer-probability meaure* of P defined by $\mathsf{P}^*(A) = \inf\{\mathsf{P}(B) : A \subset B,\ B \in \mathcal{F}\}$ for any (possibly, non-measurable) set $A \subset \Omega$.

Definition 2.3.4 (Convergence in law) For every $n \in \mathbb{N}$, let X_n be $\mathbb{D}$-valued random variables defined on a probability space $(\Omega^n, \mathcal{F}^n, \mathsf{P}^n)$. Let $\mathcal{L}$ be a Borel probability measure on $(\mathbb{D}, d)$. In this setting, $X_n \overset{\mathsf{P}^n}{\Longrightarrow} \mathcal{L}$ in $\mathbb{D}$ means that for any bounded d-continuous function $f : \mathbb{D} \to \mathbb{R}$, it holds that

$$\lim_{n\to\infty} \mathsf{E}^n[f(X_n)] = \int_{\mathbb{D}} f(x)\mathcal{L}(dx). \tag{2.6}$$

When the limit $\mathcal{L}$ is given as the law $\mathcal{L}_X$ of a random variable X, we use also the notation $X_n \overset{\mathsf{P}^n}{\Longrightarrow} X$ in $\mathbb{D}$. The convergence in law is called also the *convergence in distribution* or the *weak convergence*[5].

[5] This footnote remark may be skipped at first reading. Hoffmann-Jørgensen and Dudley's weak convergence theory (HJ-D theory), where no measurability of X_n's is assumed, defines the weak convergence "$X_n \overset{\mathsf{P}^n}{\Longrightarrow} \mathcal{L}$ in $\mathbb{D}$" by replacing "$\mathsf{E}^n[f(X_n)]$" in (2.6) of the standard definition with "$\mathsf{E}^{n*}[f(X_n)]$", where the outer-integral $\mathsf{E}^*[X]$ of a non-measurable random element X is defined by $\mathsf{E}^*[X] := \inf\{\mathsf{E}[Y] : X \leq Y$, and Y is measurable and $\mathsf{E}[Y]$ exists$\}$. An important point is that the limit $\mathcal{L}$ has to be a *Borel* probability measure also in the HJ-D theory; that is, the measurability of the limit is required. See van der Vaart and Wellner (1996) for the details.

Remark. As it is clear from the definitions, when we deal with the almost sure convergence or the convergence in probability, all random variables X_n, $n = 1, 2, ...$ and X have to be defined on the same probability space. In contrast, in the cases of the convergence in law and the convergence in probability to a constant, the probability spaces on which X_n's are defined may be different depending on n.

We have denoted (and shall denote also below) by E the expectation with respect to P, while by E^n that with respect to P^n in order to make the difference clear.

The relationships among the above three convergence concepts are the following.

Theorem 2.3.5 (i) $X_n \xrightarrow{a.s.} X$ implies $X_n \xrightarrow{\mathsf{P}} X$. The converse is not true in general.

(ii) $X_n \xrightarrow{\mathsf{P}} X$ implies $X_n \stackrel{\mathsf{P}}{\Longrightarrow} \mathcal{L}_X$ in $\mathbb{D}$, where $\mathcal{L}_X$ is the law of X. Although the converse is not always true, the following (iii) holds true.

(iii) When the limit is a deterministic constant c, the two convergences $X_n \xrightarrow{\mathsf{P}^n} c$ and $X_n \stackrel{\mathsf{P}^n}{\Longrightarrow} c$ in $\mathbb{D}$ are equivalent.

The proof of the above theorem is written in most of the standard textbooks for the probability theory.

One of the devices that we most frequently use in our asymptotic statistics would be the following.

Theorem 2.3.6 (Slutsky's theorem) Let $(X_n)_{n=1,2,...}, (Y_n)_{n=1,2,...}$ and X be $\mathbb{R}^p$-valued random variables. Let $c \in \mathbb{R}^p$ be a constant vector. If $X_n \stackrel{\mathsf{P}^n}{\Longrightarrow} X$ in $\mathbb{R}^p$ and $Y_n \xrightarrow{\mathsf{P}^n} c$, then the following (i) and (ii) hold true.

(i) $X_n + Y_n \stackrel{\mathsf{P}^n}{\Longrightarrow} X + c$ in $\mathbb{R}^p$.

(ii) In particular, when Y_n's and c are 1-dimensional, it holds that $Y_n X_n \stackrel{\mathsf{P}^n}{\Longrightarrow} cX$ in $\mathbb{R}^p$.

However, it would be better for readers to regard the above theorem as a special case of the following more general lemma with the help of the continuous mapping theorem stated below.

Lemma 2.3.7 (Slutsky's lemma) Let $(X_n)_{n=1,2,...}$ and X be $\mathbb{R}^p$-valued random variables. Let $(Y_n)_{n=1,2,...}$ be $\mathbb{R}^q$-valued random variables, and $c \in \mathbb{R}^q$ a constant vector. If $X_n \stackrel{\mathsf{P}^n}{\Longrightarrow} X$ in $\mathbb{R}^p$ and $Y_n \xrightarrow{\mathsf{P}^n} c$, then it holds that $(X_n^{\mathrm{tr}}, Y_n^{\mathrm{tr}})^{\mathrm{tr}} \stackrel{\mathsf{P}^n}{\Longrightarrow} (X^{\mathrm{tr}}, c^{\mathrm{tr}})^{\mathrm{tr}}$ in $\mathbb{R}^{p+q}$.

See, e.g., Example 1.4.7 of van der Vaart and Wellner (1996) for a proof of the above lemma in a more general framework.

An important point in the above two devices is that either of the limits of the random sequence $(X_n)_{n=1,2,...}$ or $(Y_n)_{n=1,2,...}$ is a constant. Related to this fact, it should be remarked that:

Discussion 2.3.8 (a) The following claim is **false**. If $X_n \stackrel{\mathsf{P}^n}{\Longrightarrow} X$ in $\mathbb{R}$ and $Y_n \stackrel{\mathsf{P}^n}{\Longrightarrow} Y$ in $\mathbb{R}$, then it holds that $X_n + Y_n \stackrel{\mathsf{P}^n}{\Longrightarrow} X + Y$ in $\mathbb{R}$ and that $X_n Y_n \stackrel{\mathsf{P}^n}{\Longrightarrow} XY$ in $\mathbb{R}$.

(b) The following claim is **true**. If $(X_n,Y_n)^{\mathrm{tr}} \overset{\mathsf{P}^n}{\Longrightarrow} (X,Y)^{\mathrm{tr}}$ in $\mathbb{R}^2$, then it holds that $X_n+Y_n \overset{\mathsf{P}^n}{\Longrightarrow} X+Y$ in $\mathbb{R}$ and that $X_nY_n \overset{\mathsf{P}^n}{\Longrightarrow} XY$ in $\mathbb{R}$.

The first conclusion in (b) of the above discussion follows from either of the following two theorems, while the second one in (b) does from the continuous mapping theorem.

Theorem 2.3.9 (Cramér-Wold's device) Let p-dimensional random vectors $X_n = (X_n^1,...,X_n^p)^{\mathrm{tr}}$ and $X=(X^1,...,X^p)^{\mathrm{tr}}$ be given. A necessary and sufficient condition for $X_n \overset{\mathsf{P}^n}{\Longrightarrow} X$ in $\mathbb{R}^p$ is that it holds for any constant vector $c=(c^1,...,c^p)^{\mathrm{tr}}$ that $c^{\mathrm{tr}}X_n \overset{\mathsf{P}^n}{\Longrightarrow} c^{\mathrm{tr}}X$ in $\mathbb{R}$.

Theorem 2.3.10 (Continuous mapping theorem) Let $(X_n)_{n=1,2,...}$ and X be random variables taking values in a metric space $(\mathbb{D},d)$, and let $(\mathbb{E},e)$ be a metric space. Suppose that a mapping $g:\mathbb{D}\to\mathbb{E}$ is continuous on a set $C\subset\mathbb{D}$ such that $\mathsf{P}(X\in C)=1$.

(i) $X_n \overset{a.s.}{\longrightarrow} X$ implies $g(X_n) \overset{a.s.}{\longrightarrow} g(X)$.

(ii) $X_n \overset{\mathsf{P}}{\longrightarrow} X$ implies $g(X_n) \overset{\mathsf{P}}{\longrightarrow} g(X)$.

(iii) $X_n \overset{\mathsf{P}^n}{\Longrightarrow} X$ in $\mathbb{D}$ implies $g(X_n) \overset{\mathsf{P}^n}{\Longrightarrow} g(X)$ in $\mathbb{E}$.

See, e.g., Theorems 1.9.5 and 1.11.1 of van der Vaart and Wellner (1996) for some proofs of the continuous mapping theorem and its extension.

One of the most important facts in the stochastic convergence theory is that the convergence in law of given sequence of finite-dimensional random variables is characterized by the convergence of their characteristic functions. Hereafter, we denote $i=\sqrt{-1}$.

Theorem 2.3.11 (Lévy's continuity theorem) Let $(X_n)_{n=1,2,...}$ and X be $\mathbb{R}^p$-valued random variables. A necessary and sufficient condition for $X_n \overset{\mathsf{P}^n}{\Longrightarrow} X$ in $\mathbb{R}^p$ is that

$$\lim_{n\to\infty} \mathsf{E}^n[\exp(iz^{\mathrm{tr}}X_n)] = \mathsf{E}[\exp(iz^{\mathrm{tr}}X)], \quad \forall z\in\mathbb{R}^p.$$

If the sequence of the functions $z\mapsto \mathsf{E}^n[\exp(iz^{\mathrm{tr}}X_n)]$ converges pointwise to a function $z\mapsto\phi(z)$ that is continuous at zero, then ϕ is the characteristic function of an $\mathbb{R}^p$-valued random variable X (i.e., $\phi(z)=\mathsf{E}[\exp(iz^{\mathrm{tr}}X)]$) and it holds that $X_n \overset{\mathsf{P}^n}{\Longrightarrow} X$ in $\mathbb{R}^p$.

Remark. The characteristic function $z\mapsto\phi(z)$ of p-dimensional Gaussian distribution with mean vector μ and covariance matrix Σ, namely, $\mathcal{N}_p(\mu,\Sigma)$, is given by

$$\phi(z)=\exp\left(iz^{\mathrm{tr}}\mu-\frac{1}{2}z^{\mathrm{tr}}\Sigma z\right), \quad \forall z\in\mathbb{R}^p.$$

Finally, let us present a sufficient condition for weakly convergent random sequence to deduce the convergence of their moments.

Theorem 2.3.12 Assume $X_n \overset{\mathsf{P}^n}{\Longrightarrow} X$ in $\mathbb{R}$. If the sequence $(X_n)_{n=1,2,\ldots}$ is *asymptotically uniformly integrable*, i.e., if the condition

$$\lim_{K\to\infty} \limsup_{n\to\infty} \mathsf{E}^n \left[|X_n| 1\{|X_n| > K\}\right] = 0 \tag{2.7}$$

is satisfied, then X is integrable and it holds that

$$\lim_{n\to\infty} \mathsf{E}^n[X_n] = \mathsf{E}[X].$$

(See Exercise 2.3.1 for a sufficient condition under which the asymptotically uniform integrability condition (2.7) holds true.)

To close this section, let us make some conventions of stochastic o and O etc.

First, let $(X_n)_{n=1,2,\ldots}$ and $(Y_n)_{n=1,2,\ldots}$ be real-valued random variables and $(R_n)_{n=1,2,\ldots}$ be positive real-valued random variables. All the limit notations appearing here mean to take the limit as $n \to \infty$.

- $X_n = o_{\mathsf{P}^n}(1)$ means that $X_n \xrightarrow{\mathsf{P}^n} 0$.
- $X_n = O_{\mathsf{P}^n}(1)$ means that for any $\varepsilon > 0$ there exists a constant $K > 0$ such that $\limsup_{n\to\infty} \mathsf{P}^n(|X_n| > K) < \varepsilon$. Such a random sequence is said to be *bounded in probability*.
- $X_n = o_{\mathsf{P}^n}(R_n)$ means that $\frac{X_n}{R_n} = o_{\mathsf{P}^n}(1)$.
- $X_n = O_{\mathsf{P}^n}(R_n)$ means that $\frac{X_n}{R_n} = O_{\mathsf{P}^n}(1)$.
- The two sequences $(X_n)_{n=1,2,\ldots}$ and $(Y_n)_{n=1,2,\ldots}$ of random variables are said to be *asymptotically equivalent* if $X_n - Y_n \xrightarrow{\mathsf{P}^n} 0$.

Next, let $(X_n)_{n=1,2,\ldots}$ and $(Y_n)_{n=1,2,\ldots}$ be real-valued random elements and $(R_n)_{n=1,2,\ldots}$ be positive real-valued random elements, all of which are possibly non-measurable. Hereafter, the outer-probability measure of P^n is denoted by P^{n*}.

- $X_n = o_{\mathsf{P}^{n*}}(1)$ means that $X_n \xrightarrow{\mathsf{P}^{n*}} 0$.
- $X_n = O_{\mathsf{P}^{n*}}(1)$ means that for any $\varepsilon > 0$ there exists a constant $K > 0$ such that $\limsup_{n\to\infty} \mathsf{P}^{n*}(|X_n| > K) < \varepsilon$. Such a random sequence is said to be *bounded in outer-probability*.
- $X_n = o_{\mathsf{P}^{n*}}(R_n)$ means that $\frac{X_n}{R_n} = o_{\mathsf{P}^{n*}}(1)$.
- $X_n = O_{\mathsf{P}^{n*}}(R_n)$ means that $\frac{X_n}{R_n} = O_{\mathsf{P}^{n*}}(1)$.

Exercise 2.3.1 A sufficient condition for the asymptotically uniform integrability (2.7) holds true is that there exists a $\delta > 0$ such that $\limsup_{n\to\infty} \mathsf{E}^n[|X_n|^{1+\delta}] < \infty$. Prove this claim.

Exercise 2.3.2 If a real-valued random sequence $(X_n)_{n=1,2,...}$ is uniformly bounded and if $X_n \stackrel{\mathsf{P}^n}{\Longrightarrow} X$ in $\mathbb{R}$, then X is integrable and it holds that $\lim_{n\to\infty} \mathsf{E}^n[X_n] = \mathsf{E}[X]$. Prove this claim.

Exercise 2.3.3 $\lim_{n\to\infty} \mathsf{E}^n[|X_n|] = 0$ implies $X_n = o_{\mathsf{P}^n}(1)$. Prove this claim.

Exercise 2.3.4 $\limsup_{n\to\infty} \mathsf{E}^n[|X_n|] < \infty$ implies $X_n = O_{\mathsf{P}^n}(1)$. Prove this claim.

Exercise 2.3.5 $X_n \stackrel{\mathsf{P}^n}{\Longrightarrow} X$ in $\mathbb{R}$ implies $X_n = O_{\mathsf{P}^n}(1)$. Prove this claim.

Exercise 2.3.6 Let a sequence of $\mathbb{R}^p$-valued random variables $X_n = (X_n^1, ..., X_n^p)^{\mathrm{tr}}$ be given. If $X_n^i \stackrel{\mathsf{P}^n}{\longrightarrow} c^i$ for every $i = 1, ..., p$, where c^i's are constants, then it holds for any continuous function f on $\mathbb{R}^p$ that $f(X_n^1, ..., X_n^p) \stackrel{\mathsf{P}^n}{\longrightarrow} f(c^1, ..., c^p)$. Prove this claim. Prove also that $\max_{1\le i\le p} X_n^i \stackrel{\mathsf{P}^n}{\longrightarrow} \max_{1\le i\le p} c^i$.

Exercise 2.3.7 Prove that if $X_n \stackrel{\mathsf{P}^n}{\Longrightarrow} \mathcal{N}(0,\sigma^2)$ in $\mathbb{R}$ then $cX_n \stackrel{\mathsf{P}^n}{\Longrightarrow} \mathcal{N}(0,c^2\sigma^2)$ in $\mathbb{R}$ for any constant $c \in \mathbb{R}$. More generally, when $X_n \stackrel{\mathsf{P}^n}{\Longrightarrow} \mathcal{N}_p(\mu,\Sigma)$ in $\mathbb{R}^p$ and A is a deterministic $(q \times p)$-matrix, to which limit does AX_n converges in distribution?

3

A Short Introduction to Statistics of Stochastic Processes

This chapter gives a rather informal introduction to statistics of stochastic processes. It starts with explaining a "key point" in statistics. The "key point" is closely related to the essence of martingales, and it will be called the "core of statistics" in this monograph. Although this terminology is the one invented for our discussion, there would be no doubt about the importance and usefulness of the point itself.

The main purpose of this chapter is to give an overview of the martingale theory towards applications to statistics. Although the rigorous descriptions will start formally from the next chapter, the current chapter already includes some explanations of the importance of stochastic integrals and martingale central limit theorems in statistics. The outline of the proofs of asymptotic normality of the maximum likelihood estimators in stochastic process models will be presented, with the emphasis on the role of martingales.

The chapter finishes with exhibiting some concrete examples of counting and diffusion process models.

3.1 The "Core" of Statistics

3.1.1 Two illustrations

In order to explain what is meant by the words "core of statistics", let us start with presenting two illustrative examples; an alternative terminology might be the "orthogonality of noise".

Discussion 3.1.1 (Law of large numbers) Let $X_1, X_2, \ldots$ be an i.i.d. sequence of real-valued random variables. Define the *sample mean* $\overline{X}_n$ of the first n random variables by

$$\overline{X}_n = \frac{1}{n}\sum_{k=1}^{n} X_k.$$

If the mean $\mathsf{E}[X_1] = \mu$ exists, then it holds that $\overline{X}_n$ converges to μ, almost surely. This is called the *strong law of large numbers*, although the proof is not so easy (see, e.g., Theorems 20.2 and 27.5 of Jacod and Protter (2003)). On the other hand, it is very easy to prove the *weak law of large numbers*, i.e., the claim that $\overline{X}_n$ converges in probability to μ, if we strength the assumption up to that the variance of X_1 exists.

DOI: 10.1201/9781315117768-3

Let us show this weaker claim under the stronger assumption for illustration. For any $\varepsilon > 0$, it holds that

$$\begin{aligned}
&\mathsf{P}(|\overline{X}_n - \mu| > \varepsilon) \\
&\quad\le \mathsf{E}\left[\left(\frac{\overline{X}_n - \mu}{\varepsilon}\right)^2 1\{|\overline{X}_n - \mu| > \varepsilon\}\right] \\
&\quad\le \mathsf{E}\left[\left(\frac{\overline{X}_n - \mu}{\varepsilon}\right)^2\right] \\
&\quad= \frac{1}{\varepsilon^2}\left\{\frac{1}{n^2}\sum_{k=1}^{n}\mathsf{E}[(X_k-\mu)^2] + \frac{2}{n^2}\sum_{k<l}\mathsf{E}[(X_k-\mu)(X_l-\mu)]\right\} \\
&\quad=: \frac{1}{\varepsilon^2}\{\alpha_n + \beta_n\} \quad \text{(say)}.
\end{aligned}$$

Notice that the term α_n is computed as

$$\alpha_n = \frac{1}{n^2}\cdot n \cdot \mathrm{Var}[X_1] \to 0.$$

The reason why this term has vanished is that the coefficient $\frac{1}{n^2}$ is multiplied with the summation consisting only of n terms. As for the term β_n, the independence of the random sequence implies that

$$\mathsf{E}[(X_k-\mu)(X_l-\mu)] = \mathsf{E}[X_k-\mu]\mathsf{E}[X_l-\mu] = 0\cdot 0 = 0, \tag{3.1}$$

and thus β_n varnishes not only asymptotically but also for *all* n. The computation (3.1) is the essential point of the proof, and it is the most basic form of the "core of statistics" in our terminologies.

Discussion 3.1.2 (AR (1) model) Let us consider the 1st order autoregressive process model (AR (1) model) given by

$$X_k = \theta X_{k-1} + \varepsilon_k, \qquad k = 1,2,...,$$

with an arbitrary initial value X_0, where ε_k's are i.i.d. random variables with zero mean and finite variance. Our set-up for statistical estimation is the following.

- We are able to observe the realization of $X_0, X_1, ..., X_n$. The noise term ε_k's cannot be observed.
- We assume that θ is a fixed value (say, θ_*), but it is assumed that we do not know the value of θ_*.
- We intend to *estimate* the value of θ_* based on the data $X_0, X_1, \ldots, X_n$. In other words, we want to construct a value, which should be as close to θ_* as possible, by using only $X_0, X_1, ..., X_n$ (not using θ, θ_* and ε_k's!).

One of good methods for this estimation problem is the following. Since $X_k - \theta_* X_{k-1}$ is a value which is close to zero, the value

$$\sum_{k=1}^{n} (X_k - \theta_* X_{k-1})^2$$

should be a small positive value. Now, replacing the unknown value θ_* by a parameter θ, we define

$$\mathbb{M}_n(\theta) := \sum_{k=1}^{n} (X_k - \theta X_{k-1})^2.$$

Then, let us minimize $\mathbb{M}_n(\theta)$ with respect to θ; in practice, by solving the estimating equation $\frac{d}{d\theta}\mathbb{M}_n(\theta) = 0$, we get the estimator

$$\widehat{\theta}_n = \frac{\sum_{k=1}^{n} X_{k-1} X_k}{\sum_{k=1}^{n} X_{k-1}^2}.$$

This is a kind of *least square estimators*.

Among several ways to explain how good this estimator is, let us prove its *consistency* from now on. Substituting $X_k = \theta_* X_{k-1} + \varepsilon_k$, we have

$$\begin{aligned} \widehat{\theta}_n &= \frac{\sum_{k=1}^{n} X_{k-1}(\theta_* X_{k-1} + \varepsilon_k)}{\sum_{k=1}^{n} X_{k-1}^2} \\ &= \theta_* + \frac{\sum_{k=1}^{n} X_{k-1}\varepsilon_k}{\sum_{k=1}^{n} X_{k-1}^2} \\ &= \theta_* + \frac{\frac{1}{n}\sum_{k=1}^{n} X_{k-1}\varepsilon_k}{\frac{1}{n}\sum_{k=1}^{n} X_{k-1}^2}. \end{aligned}$$

It is known that when the true value θ_* satisfies that $|\theta_*| < 1$, the sequence $X_0, X_1, \ldots$ is *ergodic*[1]. Thus, the denominator of the second term on the right-hand side converges almost surely to a positive constant:

$$\frac{1}{n}\sum_{k=1}^{n} X_{k-1}^2 \xrightarrow{a.s.} \int_{\mathbb{R}} x^2 P^\circ(dx),$$

where P° denotes the "invariant measure" for the sequence. Therefore, in order to prove the consistency, it is sufficient to see that the numerator of the second term on the right-hand side converges in probability to zero. This is proved by an argument

[1] It is not easy to define the word "ergodic" in a short way. To read this monograph, it is sufficient to know that, if a given, $\mathcal{X}$-valued, discrete-time stochastic process $(X_n)_{n\in\mathbb{N}_0}$ is *ergodic* with the *invariant measure* P°, then for any P°-integrable function f it holds that $\frac{1}{n}\sum_{k=1}^{n} f(X_{k-1}) \xrightarrow{a.s.} \int_{\mathcal{X}} f(x)P^\circ(dx)$ as $n \to \infty$ (Birkhoff's individual ergodic theorem). There are also some theorems where the almost sure convergence is replaced by L^2-convergence (von Neumann) or by L^p-convergence for $1 < p < \infty$ (Riesz-Yosida-Kakutani).

similar to the proof of weak law of large numbers: observe that

$$\begin{aligned}
&\mathsf{E}\left[\left(\frac{1}{n}\sum_{k=1}^{n} X_{k-1}\varepsilon_k\right)^2\right] \\
&= \frac{1}{n^2}\sum_{k=1}^{n}\mathsf{E}[(X_{k-1}\varepsilon_k)^2] + \frac{2}{n^2}\sum_{k<l}\mathsf{E}[(X_{k-1}\varepsilon_k)(X_{l-1}\varepsilon_l)] \\
&=: \alpha_n + \beta_n \quad \text{(say)}.
\end{aligned}$$

The term α_n is proved to converge to zero under some mild conditions, while the term β_n is analyzed as follows: by using the *tower property* of conditional expectations, we have

$$\begin{aligned}
&\mathsf{E}[(X_{k-1}\varepsilon_k)(X_{l-1}\varepsilon_l)] \\
&= \mathsf{E}[\mathsf{E}[(X_{k-1}\varepsilon_k)(X_{l-1}\varepsilon_l)|\mathcal{F}_{l-1}]] \\
&= \mathsf{E}[X_{k-1}\varepsilon_k X_{l-1}\mathsf{E}[\varepsilon_l|\mathcal{F}_{l-1}]] \\
&= \mathsf{E}[X_{k-1}\varepsilon_k X_{l-1}\cdot 0] \\
&= 0,
\end{aligned}$$

where $\mathcal{F}_{l-1}$ denotes the σ-field generated by $X_0,\dots,X_{l-1}$, that is,

$$\mathcal{F}_k := \sigma(X_0, X_1, \dots, X_k), \quad k = 0,1,2.... \tag{3.2}$$

Thus the term β_n has been proved to converge in probability to zero. We conclude that the least square estimator $\widehat{\theta}_n$ is consistent for θ_*.

Note that the key point of our discussion above has been that

$$\mathsf{E}[X_{k-1}\varepsilon_k|\mathcal{F}_{k-1}] = X_{k-1}\mathsf{E}[\varepsilon_k|\mathcal{F}_{k-1}] = X_{k-1}\cdot 0 = 0, \quad \text{a.s.} \tag{3.3}$$

Generalizing and unifying the computations (3.1) and (3.3) which we have called the *core of statistics*, a random sequence $(\xi_k)_{k=1,2,...}$ satisfying

$$\mathsf{E}[\xi_k|\mathcal{F}_{k-1}] = 0, \quad k = 1,2,\dots, \quad \text{a.s.,}$$

where $\mathcal{F}_{k-1}$ denotes the σ-field generated by the data up to time $k-1$, is called a *martingale difference sequence*.

3.1.2 Filtration, martingale

Let us summarise our discussion in the previous subsection. First, we should introduce an non-decreasing sequence $(\mathcal{F}_k)_{k=0,1,2,...}$ of sub-σ-fields of $\mathcal{F}$, which is called a *filtration*, i.e.,

$$\{\emptyset,\Omega\} \subset \mathcal{F}_0 \subset \mathcal{F}_1 \subset \mathcal{F}_2 \subset \cdots \subset \mathcal{F}_k \subset \cdots \subset \mathcal{F};$$

these are usually taken to be the one like (3.2). Then, stochastic processes of the form

$$M_n = M_0 + \sum_{k=1}^{n} \xi_k, \quad n = 1,2,..., \tag{3.4}$$

where m_k is $\mathcal{F}_k$-measurable and $\mathsf{E}[\xi_k|\mathcal{F}_{k-1}] = 0$, a.s., would play a key role in the analysis of stochastic processes. Equivalently, a stochastic process $(M_n)_{n=0,1,2,...}$ satisfying that each M_n is $\mathcal{F}_n$-measurable and that

$$\mathsf{E}[M_n|\mathcal{F}_{n-1}] = M_{n-1}, \quad \text{a.s.}, \quad n = 1,2,...,$$

is important, and such a stochastic process is called a *martingale*.

We dare to say that *a martingale starting from zero is something like a zero element in the space of stochastic processes*. One may think that a common definition of "zero element" in the space of stochastic processes is $X_n \equiv 0$. However, what is meant by the above statement is that a martingale plays a role of "zero" element *in a weaker sense*, which is often more important and interesting than the trivial zero element. One of the principal messages from this monograph may be this claim, which will be explained in the subsequent sections and chapters.

We close this section with noting that the core of statistics is nothing else than the most important case of the Doob-Meyer decomposition of M^2, that is,

$$M_n^2 = \sum_{k=1}^{n} \mathsf{E}[\xi_k^2|\mathcal{F}_{k-1}] + M'_n,$$

where $(M_n)_{n=0,1,2,...}$ is a martingale given by (3.4) starting from zero, and $(M'_n)_{n=0,1,2,...}$ is another martingale starting from zero. The first term on the right-hand side will be named "predictable quadratic variation" later, and the fact that the expectation of the martingale $(M'_n)_{n=0,1,2,...}$ of the second term is zero has been named as the *core of statistics*[2] (by us).

3.2 A Motivation to Study Stochastic Integrals

3.2.1 Intensity processes of counting processes

In this monograph, the phrase "statistical modelling of counting processes" means that of the *intensities* of those processes. The purpose of this section is to explain what is the intensity of a counting process.

Let $T > 0$ be a fixed time (i.e., a positive constant). We say $X = (X_t)_{t\in[0,T]}$ is a *stochastic process* if it is a collection of real-valued (sometimes, $\mathbb{R}^d$-valued) random

[2]Notice, however, that the core of statistics is not the only successful method in statistics. Indeed, there also exist useful methods, where the expectations of the cross terms are not exactly zero but can be asymptotically negligible, such as the method of *mixing*.

variables X_t, indexed by $t \in [0,T]$, all of which are defined on a common probability space $(\Omega, \mathcal{F}, \mathsf{P})$. (Although some cases where the index set $[0,T]$ is replaced by $[0,\infty)$ or $[0,\infty]$ will be considered later, we shall take it to be a finite time interval for a while.)

Since we may not be able to have any interesting discussion in this vague set-up, let us introduce some more restrictions to our framework to make it more concrete and interesting step by step.

The first restriction is the *adapted* property. When a non-decreasing family $(\mathcal{F}_t)_{t\in[0,T]}$ of sub-σ-fields of $\mathcal{F}$ is introduced, a stochastic process $(X_t)_{t\in[0,T]}$ is said to be *adapted* if for every $t \in [0,T]$, the random variable X_t is not only $\mathcal{F}$-measurable but also $\mathcal{F}_t$-measurable. Throughout this chapter, *all stochastic processes will be assumed to be adapted* although this assumption will not be stated every time. A stochastic process $M = (M_t)_{t\in[0,T]}$ is said to be a *martingale* if it is an adapted process whose almost all paths $t \mapsto M_t(\omega)$ are càdlàg (i.e., right-continuous and having left-hand limit at each time point), such that

$$\mathsf{E}[M_t|\mathcal{F}_s] = M_s \quad \text{a.s.} \quad \text{for every} \quad s \leq t.$$

A stochastic process $N = (N_t)_{t\in[0,T]}$ is said to be a *counting process* if it is an adapted process whose all paths $t \mapsto N_t(\omega)$ are càdlàg, non-decreasing step functions starting from zero (i.e., $N_0(\omega) = 0$) which increase only at jumps with size $+1$. Note that for any given counting process N, there exists a sequence of random times $0 < \tau_1(\omega) < \tau_2(\omega) < \cdots < \tau_{N_T(\omega)}(\omega) \leq T$, where $N_T(\omega) < \infty$, such that the process $(N_t)_{t\in[0,T]}$ can be written as

$$N_t = \sum_k 1\{\tau_k \leq t\}.$$

Now, according to the Doob-Meyer decomposition theorem which will be explained later, there exists a "predictable[3]", increasing process A, uniquely up to an almost sure sense, such that $N - A$ is a martingale; some explanation about the word "predictable" will be added soon. In other words, it is possible to find a unique decomposition

$$\begin{array}{ccccc} N_t & = & A_t & + & \text{martingale}, \qquad \forall t \in [0,\infty), \\ \text{data} & = & \text{trend} & + & \text{noise}. \end{array}$$

Although it may be difficult to find which point is interesting in this fact at first sight, the really important point of this decomposition is that the decomposition is *unique*. Let us add some more explanations for this point. One may think at the first stage of his/her study that even "infinitely many" decompositions into "trend" plus "martingale"; for example, one may think that the "decomposition",

$$\begin{array}{ccccc} N_t & = & N_t & + & 0, \qquad \forall t \in [0,\infty), \\ \text{data} & = & \text{trend} & + & \text{noise}, \end{array}$$

[3]We have not yet given the definition of predictability. For the moment, it is enough to memorize that a left-continuous, adapted process is predictable.

would be possible, too. It should be emphasized that the Doob-Meyer decomposition theorem says that once we assume that the "trend" term is *increasing* and *predictable*, then such an easy decomposition as above is excluded and the decomposition becomes *unique*; note that N is not predictable and thus the above "decomposition" is not the right one. Statisticians may interpret this fact as the message from the theorem saying that *statistical modelling is definitely possible.*

The stochastic process $A = (A_t)_{t\in[0,T]}$ appearing above is called the *predictable compensator* of N. When we construct statistical models of counting processes, it is usually assumed that $t \mapsto A_t(\omega)$ is absolutely continuous with respect to the Lebesgue measure, and in this case we may write

$$A_t = \int_0^t \lambda_s ds.$$

This stochastic process $(\lambda_t)_{t\in[0,T]}$ is called the *intensity process* of N.

A lot of statistical models have been constructed in terms of intensity processes. In such statistical models, if we consider the maximum likelihood estimator (MLE) $\widehat{\theta}_n$ or some similar estimators, then, surprisingly, there exists a sequence of predictable processes $H^n = (H_s^n)_{s\in[0,T_n]}$ such that

$$\sqrt{n}(\widehat{\theta}_n - \theta_*) = \int_0^{T_n} H_s^n(dN_s - \lambda_s ds) + o_P(1),$$

in (not all but) many cases. The first term of the right-hand side is the value of a martingale called *stochastic integral* stopped at T_n, and the asymptotic normality of the MLE is proved based on the martingale central limit theorem for this term. It is therefore important for us to learn and master stochastic integrals and martingale central theorems.

To make the logical structure of this chapter clear, we shall continue our theoretical discussion also in the following sections. However, readers are advised to take a "rest" now, reading Subsection 3.5.1 first to have better understanding what is the intensity process by illustrative examples, and then to come back here again.

3.2.2 Itô integrals and diffusion processes

In Subsection 3.4.2, we will prove the asymptotic normality of the MLEs in a statistical model of 1-dimensional diffusion process given by the stochastic differential equation

$$X_t = X_0 + \int_0^t \alpha(X_s;\theta)ds + \int_0^t \sigma(X_s)dW_s, \tag{3.5}$$

where $(W_s)_{s\in[0,\infty)}$ is a standard Wiener process. Similarly to the cases of counting process models, the MLEs $\widehat{\theta}_T$ based on the observation on the time interval $[0,T]$ can be often written in the form

$$\sqrt{T}(\widehat{\theta}_T - \theta_*) = \int_0^T H_s^T dW_s + o_P(1), \tag{3.6}$$

where $H^T = (H_s^T)_{s\in[0,\infty)}$ is an appropriate predictable process. We therefore should learn the definition and some properties and mathematical tools of the Itô integral

$$\int_0^t H_s dW_s$$

appearing on the right-hand sides of (3.5) and (3.6).

See Subsection 3.5.2 for some concrete models of diffusion processes. Although we continue the description of theoretical issues, readers are advised to read Subsection 3.5.2 first, and then to come back here after getting knowledge of some diffusion process models.

3.3 Square-Integrable Martingales

3.3.1 Predictable quadratic variations

The predictable quadratic variation, which will be explained in this section, is something like a "conditional variance process". On the other hand, the stochastic integral is something like a generalization of the summation with weights of i.i.d. random variables with zero mean and finite variance. It would be well-known to readers that the central limit theorem for independent sequences of random variables requires

"convergence of variance" + "Lindeberg's condition"

to obtain the weak convergence to a Gaussian limit. When this theorem is generalized to the one for stochastic integrals, the conditions are replaced by

"convergence of predictable quadratic variation" + "a Lindeberg type condition".

It is therefore indispensable for understanding the central limit theorems for stochastic integrals to learn what is the predictable quadratic variation.

A martingale $M = (M_t)_{t\in[0,T]}$ is said to be *square-integrable* if the condition

$$\sup_{t\in[0,T]} \mathsf{E}[M_t^2] < \infty$$

is satisfied; since it is proved that $t \mapsto \mathsf{E}[M_t^2]$ is non-decreasing, the above condition is equivalent to $\mathsf{E}[M_T^2] < \infty$. By using the Doob-Meyer decomposition theorem it is proved that for a given square-integrable martingale M there exists a predictable increasing process $\langle M\rangle = (\langle M\rangle_t)_{t\in[0,T]}$, uniquely up to an almost sure sense, such that $M^2 - \langle M\rangle$ is a martingale. This stochastic process $\langle M\rangle$ is called the *predictable quadratic variation* of M.

When $M_0 = 0$, the predictable quadratic variation $t \rightsquigarrow \langle M\rangle_t$ may be regarded as an approximation of $t \rightsquigarrow M_t^2$ by the one that has "poor" measurability. The approximation of $t \rightsquigarrow M_t^2$ by a "stochastic" process with the most poor measurability is the usual variance function $t \mapsto \mathsf{E}[M_t^2]$. The predictable quadratic variation $t \rightsquigarrow \langle M\rangle_t$ is intermediate between the two stochastic processes $t \rightsquigarrow M_t^2$ and $t \mapsto \mathsf{E}[M_t^2]$.

3.3.2 Stochastic integrals

Let $M = (M_t)_{t\in[0,T]}$ be a square-integrable martingale. From now on, for a given predictable process $H = (H_s)_{s\in[0,T]}$ let us define a new stochastic process

$$t \rightsquigarrow \int_0^t H_s dM_s$$

in the following way. First choose a sequence of grids $0 = t_0^n < t_1^n < \cdots < t_n^n = T$ of the interval $[0,T]$, and next for every $t \in [0,T]$ compute

$$X_t^n := \sum_{k:t_{k-1}^n \leq t} H_{t_{k-1}^n}(M_{t_k^n} - M_{t_{k-1}^n});$$

then, letting the grids thinner and thinner as $n \to \infty$, i.e., $\max_k |t_k^n - t_{k-1}^n| \to 0$, "define" the value

$$X_t = \int_0^t H_s dM_s.$$

As a matter of fact, such a limit X is well-defined rigorously if the condition

$$\mathsf{E}\left[\int_0^T H_s^2 d\langle M\rangle_s\right] < \infty$$

is satisfied, and in this case the resulted stochastic process $t \rightsquigarrow X_t = \int_0^t H_s dM_s$ itself is a new square-integrable martingale with the predictable quadratic variation

$$\langle X\rangle_t = \left\langle \int_0^\cdot H_s dM_s \right\rangle_t = \int_0^t H_s^2 d\langle M\rangle_s;$$

it should be emphasized that the last equality is important and very useful for various statistical applications.

We will learn in later chapters that if all paths of $t \rightsquigarrow M_t$ have finite-variation then the stochastic integral $\int_0^t H_s dM_s$ coincides with the value computed as the Lebesgue-Stieltjes integral for every fixed $\omega \in \Omega$, and that if almost all paths of $t \rightsquigarrow M_t$ are continuous then those of $t \rightsquigarrow \int_0^t H_s dM_s$ are also continuous. The counting process considered in Subsection 3.2.1 belongs to the former case, and if we moreover assume that $\mathsf{E}[N_T] = \mathsf{E}[A_T] < \infty$ then $M = N - A$ is a square-integrable martingale with the predictable quadratic variation

$$\begin{aligned}\langle N-A\rangle_t &= A_t \\ &= \int_0^t \lambda_s ds \quad \text{if } A \text{ admits the intensity process,}\end{aligned}$$

and thus it holds for any predictable process H having reasonable integrability assumption that

$$\begin{aligned}\left\langle \int_0^\cdot H_s d(N_s - A_s) \right\rangle_t &= \int_0^t H_s^2 dA_s \\ &= \int_0^t H_s^2 \lambda_s ds \quad \text{if } A \text{ admits the intensity process.}\end{aligned}$$

On the other hand, the standard Wiener process $t \rightsquigarrow W_t$ with an appropriate filtration belongs to the latter case, and it holds that

$$\langle W \rangle_t = t = \int_0^t 1 ds,$$

and the predictable quadratic variation is given by

$$\left\langle \int_0^{\cdot} H_s dW_s \right\rangle_t = \int_0^t H_s^2 d\langle W \rangle_s = \int_0^t H_s^2 ds.$$

3.3.3 Introduction to CLT for square-integrable martingales

Let a sequence of probability spaces with filtration, namely, $(\Omega^n, \mathcal{F}^n, (\mathcal{F}_t^n)_{t\in[0,T_n]}, \mathsf{P}^n)$ be given. We shall present a central limit theorem for sequences of square-integrable martingales $M^n = (M_t^n)_{t\in[0,T_n]}$.

Before stating it, note that the classical CLT for i.i.d. sequences is a special case of the theorem below by setting

$$M_t^n = \sum_{k=1}^{[nt]} \frac{1}{\sqrt{n}} \frac{X_k - \mu}{\sigma}, \qquad \forall t \in [0,1] \qquad \text{and} \qquad T_n \equiv 1.$$

Theorem 3.3.1 Let T_n be a sequence of positive constants; for example, "$T_n \to \infty$ as $n \to \infty$" or "$T_n \equiv T$", or something else. Suppose that a given sequence $M^n = (M_t^n)_{t\in[0,T_n]}$ of square-integrable martingales starting from zero satisfies the following (a) and (b):

(a) $\langle M^n \rangle_{T_n} \xrightarrow{\mathsf{P}} C$, where C appearing as the limit is a constant;

(b) a Lindeberg type condition.

Then, it holds that

$$M_{T_n}^n \stackrel{\mathsf{P}}{\Longrightarrow} \mathcal{N}(0,C) \quad \text{in } \mathbb{R}.$$

Since the Lindeberg type condition is an assumption concerning the jumps of stochastic processes, it is automatically satisfied in the case of continuous martingales. Thus, in such cases, it is sufficient to check the condition (a) only. The following corollary is a useful, special case.

Corollary 3.3.2 Let T_n be a sequence of positive constants. If a sequence $H^n = (H_s^n)_{s\in[0,T_n]}$ of predictable processes satisfies that $\int_0^{T_n} (H_s^n)^2 ds \xrightarrow{\mathsf{P}} C$, where C is a constant, then it holds that

$$\int_0^{T_n} H_s^n dW_s \stackrel{\mathsf{P}}{\Longrightarrow} \mathcal{N}(0,C) \quad \text{in } \mathbb{R}.$$

It is not easy to give a precise description of the Lindeberg type condition in a general, unified way. (We will do it in Theorems 7.1.8 and 7.2.2 later.) In the case of counting processes with intensity processes, however, the condition can be explicitly written in a form which is analogous to that for independent arrays.

Corollary 3.3.3 Let T_n be a sequence of positive constant. Let a sequence $N^n = (N_t^n)_{t\in[0,T_n]}$ of counting processes with intensity processes $(\lambda_s^n)_{s\in[0,T_n]}$ be given. Suppose that a sequence $H^n = (H_s^n)_{s\in[0,T_n]}$ of predictable processes satisfies the following (a) and (b):

(a) $\int_0^{T_n}(H_s^n)^2\lambda_s^n ds \xrightarrow{\mathsf{P}} C$, where C appearing as the limit is a constant;

(b) $\int_0^{T_n}(H_s^n)^2 1\{|H_s^n| > \varepsilon\}\lambda_s^n ds \xrightarrow{\mathsf{P}} 0$ holds for every $\varepsilon > 0$.

Then, it holds that

$$\int_0^{T_n} H_s^n(dN_s^n - \lambda_s^n ds) \overset{\mathsf{P}}{\Longrightarrow} \mathcal{N}(0,C) \quad \text{in } \mathbb{R}.$$

3.4 Asymptotic Normality of MLEs in Stochastic Process Models

3.4.1 Counting process models

In this section, we will see the outline of the proof that the maximum likelihood estimators (MLEs) in a general, simple parametric model of counting processes have the asymptotic normality.

Let $N = (N_t)_{t\in[0,\infty)}$ be a counting process, which is not depending on $n \in \mathbb{N}$. Suppose that this stochastic process is observed on the time interval $[0,T_n]$, where T_n is a sequence of constants that tends to ∞ as $n \to \infty$. Let us consider the parametric model of intensities $\{\lambda^\theta; \theta \in \Theta\}$ of the form

$$\lambda_s^\theta = \alpha(X_s;\theta),$$

where X is a stochastic process taking values in a measurable space $(\mathcal{X},\mathcal{A})$. Suppose that X is ergodic with the invariant measure $P_{\theta_*}^\circ$ under the true parameter value $\theta_* \in \Theta$; then, it holds that for any $P_{\theta_*}^\circ$-integrable function f

$$\frac{1}{T_n}\int_0^{T_n} f(X_s)ds \xrightarrow{\mathsf{P}} \int_{\mathcal{X}} f(x)P_{\theta_*}^\circ(dx).$$

It is known that the log-likelihood function based on the observation $\{N_t, X_t; t \in [0,T_n]\}$ is given by

$$\ell_{T_n}(\theta) = \int_0^{T_n} \log\alpha(X_s;\theta)dN_s - \int_0^{T_n}\alpha(X_s;\theta)ds;$$

see Theorem 6.5.4. From now on, we consider the case where Θ is 1-dimensional, and using the notations like "$\dot{f}(\theta)$" and "$\ddot{f}(\theta)$" for the first and second derivatives of $f(\theta)$ with respect to θ, we have

$$\dot{\ell}_n(\theta) = \int_0^{T_n}\frac{\dot{\alpha}(X_s;\theta)}{\alpha(X_s;\theta)}dN_s - \int_0^{T_n}\dot{\alpha}(X_s;\theta)ds.$$

The MLE $\widehat{\theta}_n$ is defined as a value that satisfies $\dot{\ell}_n(\widehat{\theta}_n) = 0$. Thus it follows from the Taylor expansion that

$$0 = \frac{1}{\sqrt{T_n}}\dot{\ell}_n(\widehat{\theta}_n) = \frac{1}{\sqrt{T_n}}\dot{\ell}_n(\theta_*) + \frac{1}{T_n}\ddot{\ell}_n(\widetilde{\theta}_n)\sqrt{T_n}(\widehat{\theta}_n - \theta_*), \tag{3.7}$$

where $\widetilde{\theta}_n$ is a point on the segment connecting $\widehat{\theta}_n$ and θ_*. Note that the left-hand side of the above formula is zero because of the definition of $\widehat{\theta}_n$.

Here, let us observe two important facts. First, the first term on the right-hand side, i.e., $\dot{\ell}_n(\theta)$ with θ being substituted with θ_*, is the value of a martingale stopped at T_n. Indeed, it holds that

$$\begin{aligned}\dot{\ell}_n(\theta_*) &= \int_0^{T_n} \frac{\dot{\alpha}(X_s;\theta_*)}{\alpha(X_s;\theta_*)} dN_s - \int_0^{T_n} \dot{\alpha}(X_s;\theta_*)ds \\ &= \int_0^{T_n} \frac{\dot{\alpha}(X_s;\theta_*)}{\alpha(X_s;\theta_*)}(dN_s - \alpha(X_s;\theta_*)ds).\end{aligned}$$

We thus are able to apply the martingale CLT to the first term $\frac{1}{\sqrt{T_n}}\dot{\ell}_n(\theta_*)$ of the right-hand side of (3.7) to obtain that

$$\frac{1}{\sqrt{T_n}} \int_0^{T_n} \frac{\dot{\alpha}(Z_s;\theta_*)}{\alpha(Z_s;\theta_*)}(dN_s - \alpha(X_s;\theta_*)ds) \xrightarrow{\mathsf{P}} \mathcal{N}(0, I(\theta_*)) \quad \text{in } \mathbb{R};$$

as a matter of fact, the predictable quadratic variation is computed as

$$\begin{aligned}&\left\langle \frac{1}{\sqrt{T_n}} \int_0^{\cdot} \frac{\dot{\alpha}(X_s;\theta_*)}{\alpha(X_s;\theta_*)}(dN_s - \alpha(X_s;\theta_*)ds) \right\rangle_{T_n} \\ &= \frac{1}{T_n}\int_0^{T_n} \left(\frac{\dot{\alpha}(X_s;\theta_*)}{\alpha(X_s;\theta_*)}\right)^2 \alpha(X_s;\theta_*)ds \\ &\xrightarrow{\mathsf{P}} \int_{\mathcal{X}} \frac{\dot{\alpha}(x;\theta_*)^2}{\alpha(x;\theta_*)} P^{\circ}_{\theta_*}(dx) \\ &=: I(\theta_*),\end{aligned}$$

while checking the Lindeberg type condition is easy.

The second important fact is that the coefficient of the second term on the right-hand side of (3.7), namely, $\frac{1}{T_n}\ddot{\ell}_n(\widetilde{\theta}_n)$, can be proved to converge to $-I(\theta_*)$ in probability. We do not dare to give a proof of this claim because it involves some technical matters and writing it in this introductory chapter may make readers feel boring.

Combining these two facts, we can rewrite (3.7) into

$$\frac{1}{\sqrt{T_n}}\dot{\ell}_n(\theta_*) - I(\theta_*)\sqrt{T_n}(\widehat{\theta}_n - \theta_*) = o_{\mathsf{P}}(1);$$

to be more rigorous, it should be first checked that $\sqrt{n}(\widehat{\theta}_n - \theta_*) = O_{\mathsf{P}}(1)$, and this is indeed possible (see the proof of Theorem 8.2.2 for a rigorous argument). Consequently, we obtain that

$$\begin{aligned}\sqrt{T_n}(\widehat{\theta}_n - \theta_*) &= I(\theta_*)^{-1}\frac{1}{\sqrt{T_n}}\dot{\ell}_n(\theta_*) + o_{\mathsf{P}}(1) \\ &\stackrel{\mathsf{P}}{\Longrightarrow} I(\theta_*)^{-1}\mathcal{N}(0, I(\theta_*)) \quad \text{in } \mathbb{R} \\ &\stackrel{d}{=} \mathcal{N}(0, I(\theta_*)^{-1}).\end{aligned}$$

3.4.2 Diffusion process models

Here, let us show the asymptotic normality of the MLEs for the unknown parameter θ in the simplest model of 1-dimensional diffusion processes

$$X_t = X_0 + \int_0^t \beta(Z_s;\theta)ds + \int_0^t \sigma(Z_s)dW_s,$$

where $s \rightsquigarrow W_s$ is a standard Wiener process and $s \rightsquigarrow Z_s$ is a predictable process taking values in a suitable measurable space $(\mathcal{X}, \mathcal{A})$ which is ergodic. A typical case is that $Z_s = X_s$.

When the stochastic processes X and Z are observed in the time interval $[0, T_n]$, the log-likelihood is given by

$$\ell_n(\theta) = \int_0^{T_n} \frac{\beta(Z_s;\theta)}{\sigma(Z_s)^2}dX_s - \frac{1}{2}\int_0^{T_n} \frac{\beta(Z_s;\theta)^2}{\sigma(Z_s)^2}ds;$$

see Theorem 6.5.4. In the subsequent part, let us assume that θ is a 1-dimensional parameter for simplicity. By using notations $\dot{f}$ and $\ddot{f}$ for the first and the second derivatives of a function $f = f(\theta)$ with respect to θ again, we have

$$\dot{\ell}_n(\theta) = \int_0^{T_n} \frac{\dot{\beta}(Z_s;\theta)}{\sigma(Z_s)^2}dX_s - \int_0^{T_n} \frac{\dot{\beta}(Z_s;\theta)\beta(Z_s;\theta)}{\sigma(Z_s)^2}ds;$$

to be more rigorous, it should be checked that some conditions which guarantee exchanging the order of differentiation and stochastic integration is possible. In the same way as the case of counting processes, the MLE $\widehat{\theta}_n$ is defined as the solution to the estimating equation $\dot{\ell}_n(\widehat{\theta}_n) = 0$. By the Taylor expansion, we have

$$0 = \frac{1}{\sqrt{T_n}}\dot{\ell}_n(\widehat{\theta}_n) = \frac{1}{\sqrt{T_n}}\dot{\ell}_n(\theta_*) + \frac{1}{T_n}\ddot{\ell}_n(\widetilde{\theta}_n)\sqrt{T_n}(\widehat{\theta}_n - \theta_*), \tag{3.8}$$

where $\widetilde{\theta}_n$ is a point on the segment connecting $\widehat{\theta}_n$ and θ_*. Note that the left-hand side of the above formula is zero because of the definition of $\widehat{\theta}_n$.

The rest part of our discussion is exactly the same as the case of counting processes. Let us notice the following two important facts. First, the first term on the right-hand side, i.e., the term $\dot{\ell}_n(\theta)$ with θ being substituted with θ_*, is the value of a martingale stopped at T_n. Indeed,

$$\begin{aligned}\dot{\ell}_n(\theta_*) &= \int_0^{T_n} \frac{\dot{\beta}(Z_s;\theta_*)}{\sigma(Z_s)^2} dX_s - \int_0^{T_n} \frac{\dot{\beta}(Z_s;\theta_*)\beta(Z_s;\theta_*)}{\sigma(Z_s)^2} ds \\ &= \int_0^{T_n} \frac{\dot{\beta}(Z_s;\theta_*)}{\sigma(Z_s)} dW_s.\end{aligned}$$

Thus we can apply the martingale central limit theorem to the first term of the right-hand side of (3.8), namely, $\frac{1}{\sqrt{T_n}}\dot{\ell}_n(\theta_*)$, to obtain that

$$\frac{1}{\sqrt{T_n}} \int_0^{T_n} \frac{\dot{\beta}(Z_s;\theta_*)}{\sigma(Z_s)} dW_s \stackrel{\mathsf{P}}{\Longrightarrow} \mathcal{N}(0, I(\theta_*)) \quad \text{in } \mathbb{R}.$$

As a matter of fact, if the process $t \rightsquigarrow Z_t$ is ergodic with the invariant measure $P^\circ_{\theta_*}$, the predictable quadratic variation is computed as

$$\begin{aligned}&\left\langle \frac{1}{\sqrt{T_n}} \int_0^{\cdot} \frac{\dot{\beta}(Z_s;\theta_*)}{\sigma(Z_s)} dW_s \right\rangle_{T_n} \\ &= \frac{1}{T_n} \int_0^{T_n} \left(\frac{\dot{\beta}(Z_s;\theta_*)}{\sigma(Z_s)} \right)^2 ds \\ &\stackrel{\mathsf{P}}{\longrightarrow} \int_{\mathcal{X}} \left(\frac{\dot{\beta}(z;\theta_*)}{\sigma(z)} \right)^2 P^\circ_{\theta_*}(dz) \\ &=: I(\theta_*).\end{aligned}$$

The other important fact is that the coefficient $\frac{1}{T_n}\ddot{\ell}_n(\widetilde{\theta}_n)$ of the second term of the right-hand side of (3.8) is proved to converge in probability to $-I(\theta_*)$ by using the uniform law of large numbers. The proof of this fact is omitted at the current stage of our study. Combining these two facts, (3.8) is written as

$$\frac{1}{\sqrt{T_n}}\dot{\ell}_n(\theta_*) - I(\theta_*)\sqrt{T_n}(\widehat{\theta}_n - \theta_*) = o_{\mathsf{P}}(1),$$

and consequently, we obtain that

$$\begin{aligned}\sqrt{T_n}(\widehat{\theta}_n - \theta_*) &= I(\theta_d*)^{-1}\frac{1}{\sqrt{T_n}}\dot{\ell}_n(\theta_*) + o_{\mathsf{P}}(1) \\ &\stackrel{\mathsf{P}}{\Longrightarrow} I(\theta_*)^{-1}\mathcal{N}(0, I(\theta_*)) \quad \text{in } \mathbb{R} \\ &\stackrel{d}{=} \mathcal{N}(0, I(\theta_*)^{-1}).\end{aligned}$$

3.4.3 Summary of the approach

As we have seen above, in many cases of statistical parametric models, the Taylor expansion of the log-likelihood

$$
\begin{array}{rcccl}
\frac{1}{\sqrt{n}}\dot{\ell}_n(\widehat{\theta}_n) & = & \frac{1}{\sqrt{n}}\dot{\ell}_n(\theta_*) & + & \frac{1}{n}\ddot{\ell}_n(\widetilde{\theta}_n)\cdot\sqrt{n}(\widehat{\theta}_n-\theta_*), \\
0 & = & \text{(Term I)} & + & \text{(Coefficient II)} \cdot \text{(Target)},
\end{array}
$$

where $\widetilde{\theta}_n$ is a point between $\widehat{\theta}_n$ and θ_*, will play the role of the backbone of the asymptotic theory of MLEs. The left-hand side is zero by the definition of $\widehat{\theta}_n$, and the coefficient of the second term of the right-hand side (namely, Coefficient II) converges in probability to $-I(\theta_*)$. Thus, once the first term of the right-hand side (namely, Term I) is proved to converge in distribution to $\mathcal{N}(0,I(\theta_*))$, then we immediately obtain that

$$
\begin{aligned}
\sqrt{n}(\widehat{\theta}_n-\theta_*) &= -\frac{1}{\text{(Coefficient II)}}\cdot\text{(Term I)} \\
&= I(\theta_*)^{-1}\cdot\frac{1}{\sqrt{n}}\dot{\ell}_n(\theta_*)+o_{\mathsf{P}}(1) \\
&\stackrel{\mathsf{P}}{\Longrightarrow} I(\theta_*)^{-1}\mathcal{N}(0,I(\theta_*)) \quad \text{in } \mathbb{R} \\
&\stackrel{d}{=} \mathcal{N}(0,I(\theta_*)^{-1}).
\end{aligned}
$$

It is the *martingale central limit theorem* applied to Term I that makes this unified method possible. Let us summarise our approach in the following figure.

Asymptotic normality of MLE

$$
\begin{array}{rcccccl}
\frac{1}{\sqrt{n}}\dot{\ell}_n(\widehat{\theta}_n) & = & \frac{1}{\sqrt{n}}\dot{\ell}_n(\theta_*) & + & \frac{1}{n}\ddot{\ell}_n(\widetilde{\theta}_n) & \cdot & \sqrt{n}(\widehat{\theta}_n-\theta_*) \\
 & & & & \downarrow & & \\
0 & = & \text{(martingale)} & + & (-I(\theta_*)) & \cdot & \sqrt{n}(\widehat{\theta}_n-\theta_*) \\
 & & \Downarrow & & & & \\
0 & = & \mathcal{N}(0,I(\theta_*)) & + & (-I(\theta_*)) & \cdot & \sqrt{n}(\widehat{\theta}_n-\theta_*)
\end{array}
$$

Keeping the discussion so far in mind as one of our main motivations, let us start more detailed study from the next chapter.

3.5 Examples

3.5.1 Examples of counting process models

To begin with, let us briefly summarise the "story" of the intensity process for a counting process.

Let $0 < \tau_1 < \tau_2 < \cdots$ be a sequence of random points defined on a probability space. Define the counting process $t \rightsquigarrow N_t$ by

$$N_t = \sum_k 1\{\tau_k \leq t\}, \tag{3.9}$$

and introduce an appropriate filtration. It is known that there always[4] exists a predictable compensator $t \rightsquigarrow A_t$ for N, and we *assume* that it is absolutely continuous with respect to the Lebesgue measure; that is, we assume that A is written in the form

$$A_t(\omega) = \int_0^t \lambda_s(\omega)ds.$$

In this case, the stochastic process $s \rightsquigarrow \lambda_s$ is called the *intensity process* for N.

As a preliminary, let us present the following formula which is helpful to compute the intensity processes in concrete models:

$$\begin{aligned} \lambda_t &= \lim_{s \nearrow t} \lim_{\Delta \downarrow 0} \frac{\mathsf{E}[N_{t+\Delta} - N_t | \mathcal{F}_s]}{\Delta} \quad \forall t > 0 & (3.10) \\ &= \lim_{\Delta \downarrow 0} \frac{\mathsf{E}[N_{t+\Delta} - N_t | \mathcal{F}_{t-}]}{\Delta} & (3.11) \\ &= \lim_{\Delta \downarrow 0} \frac{\mathsf{P}(N_{t+\Delta} - N_t = 1 | \mathcal{F}_{t-})}{\Delta}. & (3.12) \end{aligned}$$

We will actually use the formula (3.10), whose full proof will be given below, while (3.11) and (3.12) have been stated just for intuitive explanations (so, the σ-field "$\mathcal{F}_{t-}$" appearing there, which has not been defined, will not be used in our discussions in the main parts of this monograph).

Choose any $0 \leq s < t \leq t + \Delta$. Since A is the compensator for N, we have that $N - A$ is a martingale and it holds that

$$\mathsf{E}[N_{t+\Delta} - N_t | \mathcal{F}_s] = \mathsf{E}[A_{t+\Delta} - A_t | \mathcal{F}_s] = \mathsf{E}\left[\int_t^{t+\Delta} \lambda_u du \middle| \mathcal{F}_s \right].$$

Divide both sides by Δ, and use the dominated convergence theorem (for the conditional expectation) to have that

$$\lim_{\Delta \downarrow 0} \frac{\mathsf{E}[A_{t+\Delta} - A_t | \mathcal{F}_s]}{\Delta} = \lim_{\Delta \downarrow 0} \mathsf{E}\left[\frac{\int_t^{t+\Delta} \lambda_u du}{\Delta} \middle| \mathcal{F}_s \right]$$

[4]As far as a "good" filtration is introduced, the predictable compensator for a counting process *always* exists.

$$= \mathsf{E}\left[\lim_{\Delta\downarrow 0}\frac{\int_t^{t+\Delta}\lambda_u du}{\Delta}\middle|\mathcal{F}_s\right]$$

$$= \mathsf{E}[\lambda_t|\mathcal{F}_s].$$

These formulas together imply that

$$\lim_{\Delta\downarrow 0}\frac{\mathsf{E}[N_{t+\Delta}-N_t|\mathcal{F}_s]}{\Delta} = \mathsf{E}[\lambda_t|\mathcal{F}_s].$$

Note that this formula holds for any $s < t$. Thus, by choosing a sequence $\{s_n\}$ such that $s_n \uparrow t$ if that is easier to understand, we have that

$$\lim_{s\nearrow t}\lim_{\Delta\downarrow 0}\frac{\mathsf{E}[N_{t+\Delta}-N_t|\mathcal{F}_s]}{\Delta} = \lim_{s\nearrow t}\mathsf{E}[\lambda_t|\mathcal{F}_s] = \mathsf{E}[\lambda_t|\mathcal{F}_{t-}].$$

Since the intensity process λ is defined as a Lebesgue density, it is unique only up to a null set with respect to the Lebesgue measure. Thus, we may assume without loss of generality that $t \rightsquigarrow \lambda_t$ has been taken to be left-continuous, and this convention implies that $\mathsf{E}[\lambda_t|\mathcal{F}_{t-}] = \lambda_t$. Thus the formula (3.10) has been proved.

Example 3.5.1 (Poisson processes) If $\lambda_t(\omega) \equiv \lambda$, a constant, then N is a homogeneous Poisson process with the intensity parameter λ.

More generally, if $\lambda_t(\omega) \equiv \lambda(t)$, a deterministic function, then N is a inhomogeneous Poisson process with the intensity function $\lambda(t)$. Here, notice that the function $\lambda(t)$ is a non-negative, measurable function on $[0,\infty)$ such that $\int_0^t \lambda(s)ds < \infty$ for every $t \in [0,\infty)$.

To prove these claims, our task is to use the assumption that $N_t - \int_0^t \lambda(s)ds$ is a martingale in order to check that all conditions of the definition of (in)homogeneous Poisson process, but this is not so easy. Notice that proving that

"N is an inhomogeneous Poisson process" $\Rightarrow$ "$N_t - \int_0^t \lambda(s)ds$ is a martingale"

is easy (see Proposition 5.1.8), while doing that

"$N_t - \int_0^t \lambda(s)ds$ is a martingale" $\Rightarrow$ "N is an inhomogeneous Poisson process"

is not so easy (see Exercise 6.4.4).

Example 3.5.2 (One-point process) Let us consider the situation where only one point τ_1 may occur. Denote the distribution function of τ_1 by $F(t)$, where F is a probability distribution on $(0,\infty)$ with the Lebesgue density $f(t)$. In this case, the intensity process for the counting process $N_t = 1\{\tau_1 \le t\}$ is given by

$$\lambda_t(\omega) = \alpha(t)1\{t \le \tau_1(\omega)\}, \tag{3.13}$$

where α is the *hazard function* for F defined by

$$\alpha(t) = \frac{f(t)}{1-F(t-)}, \quad \text{which is, in this case, equal to} \quad \frac{f(t)}{1-F(t)}. \tag{3.14}$$

This may be the simplest example where the intensity process is random.

Let us prove the above claim. Since $N_t - \int_0^t \lambda(s)ds$ is a martingale, we shall compute the value $\mathsf{E}[N_{t+\Delta} - N_t|\mathcal{F}_s]$ for every $0 \le s < t \le t+\Delta$. Since

$$N_t - N_s = \begin{cases} 0 & \text{on } \{\tau_1 \le s\}, \\ N_t & \text{on } \{s < \tau_1\}, \end{cases}$$

it follows from the definition of conditional expectation that

$$\mathsf{E}[N_{t+\Delta} - N_t|\mathcal{F}_s] = \begin{cases} 0 & \text{on } \{\tau_1 \le s\}, \\ \dfrac{\mathsf{E}[(N_{t+\Delta} - N_t)1\{s < \tau_1\}]}{\mathsf{P}(s < \tau_1)} & \text{on } \{s < \tau_1\}, \end{cases}$$

where the latter case is computed as

$$\frac{\mathsf{E}[(N_{t+\Delta} - N_t)1\{s < \tau_1\}]}{\mathsf{P}(s < \tau_1)} = \frac{\mathsf{P}(t < \tau_1 \le t+\Delta)}{\mathsf{P}(s < \tau_1)} = \frac{\int_t^{t+\Delta} f(u)du}{1 - F(s)}.$$

Since $\lim_{s \nearrow t}\{\tau_1 \le s\} = \{\tau_1 < t\}$, we have that $\lambda_t = 0$ on the set $\{\tau_1 < t\}$, while it holds on the set $\{t \le \tau_1\}$ that

$$\begin{aligned} \lambda_t &= \lim_{s \nearrow t}\lim_{\Delta \downarrow 0} \frac{\mathsf{E}[N_{t+\Delta} - N_t|\mathcal{F}_s]}{\Delta} \quad \text{(recall (3.10))} \\ &= \lim_{s \nearrow t}\lim_{\Delta \downarrow 0} \frac{\int_t^{t+\Delta} f(u)du/(1-F(s))}{\Delta} \\ &= \lim_{s \nearrow t} \frac{f(t)}{1-F(s)} \\ &= \frac{f(t)}{1-F(t-)}. \end{aligned}$$

The formula (3.13) with (3.14) has been proved.

Example 3.5.3 (Renewal process) Let an i.i.d. sequence $Y_1, Y_2, \dots$ of positive, real-valued random variables with the hazard function $\alpha(t)$ be given. Define

$$\begin{aligned} \tau_0 &= 0 \\ \tau_1 &= Y_1 \\ \tau_2 &= Y_1 + Y_2 \\ \tau_3 &= Y_1 + Y_2 + Y_3 \\ \cdots &= \cdots. \end{aligned}$$

The counting process N defined by (3.9) based on these τ_k's is called a *renewal process*. For $k = 1, 2, \dots$, the one-point process N^k defined by $N_t^k = 1\{\tau_k \le t\}$ starts at τ_{k-1} with the duration time $t - \tau_{k-1}$, and recalling Example 3.5.2 we easily find that the intensity process λ^k for N^k is given by

$$\lambda_t^k = \alpha(t - \tau_{k-1})1\{\tau_{k-1} < t \le \tau_k\}.$$

The intensity process of $N = \sum_k N^k$ is thus given by

$$\lambda_t(\omega) = \sum_k \lambda_t^k(\omega) = \sum_k \alpha(t - \tau_{k-1}(\omega)) 1\{\tau_{k-1}(\omega) < t \le \tau_k(\omega)\};$$

note that only one term on the right-hand side is active (not zero) due to the case study based on the indicator functions of disjoint events $\{\tau_{k-1} < t \le \tau_k\}$'s.

Example 3.5.4 (Failure time data) Let $X_1, X_2, ..., X_n$ be an i.i.d. sequence of positive, real-valued random variables with the distribution, density and hazard functions $F(t)$, $f(t)$ and $\alpha(t)$, respectively. Let $0 < \tau_1 < \tau_2 < \cdots < t_n$ be the order statistics of $X_1, ..., X_n$. Then, the counting process N defined by (3.9) based on τ_k's can be represented by $N = \sum_{k=1}^n N^k$, where N^k denotes the one-point process of X_k defined by $N_t^k = 1\{X_k \le t\}$; that is,

$$N_t = \sum_{k=1}^n 1\{\tau_k \le t\} = \sum_{k=1}^n 1\{X_k \le t\} = \sum_{k=1}^n N_t^k.$$

Since the intensity process for N^k is given by $\lambda_t^k = \alpha(t) 1\{t \le X_k\}$ (recall Example 3.5.2), the intensity process λ for N is given by

$$\lambda_t = \sum_{k=1}^n \lambda_t^k = \sum_{k=1}^n \alpha(t) 1\{t \le X_k\} = \alpha(t) \left\{ n - \sum_{k=1}^n 1\{X_k < t\} \right\} = \alpha(t)\{n - N_{t-}\},$$

and this has the interpretation that

$$\lambda_t = \{ \text{ hazard function } \} \times \{ \text{ the number of individuals surviving at time } t- \},$$

where N_{t-} is the left-hand limit of N at t and the phrase "at time $t-$" may be interpreted as "just before time t".

Statisticians' goal is to estimate F, f and/or α. If all X_k's are observable, then it is unnecessary to introduce τ_k's for this purpose. Under such sampling scheme, the random variables X_k's are more natural and easier to treat than τ_k's. As a matter of fact, it follows from uniform law of large numbers that

$$\sup_{t \in [0,\infty)} \left| \frac{1}{n} \sum_{k=1}^n 1\{X_k \le t\} - F(t) \right| \xrightarrow{\text{a.s.}} 0,$$

and the distribution function F can be estimated well.

On the other hand, what about the situation in the next example?

Example 3.5.5 (Failure time data with censoring) Let $X_1, ..., X_n$ be a failure time data as above. However, in this example, only a part of them are assumed to be observable for statisticians, in the sense that there is another sequence of random variables C_k's that are independent of X_k. Statisticians are supposed to be able to observe the data

$$T_k = X_k \wedge C_k \quad \text{and} \quad \Delta_k = 1\{X_k \le C_k\}.$$

Note that the data which we have in hand is not "X_k's" but "$X_k \wedge C_k$'s", and that the values of "X_k's such that $X_k > C_k$" are missing. Thus, the model consider in this example is one of the *missing data models* whose analysis is an important issue in statistics.

Let us consider the counting process N given by $N_t = \sum_{k=1}^n N_t^k$,

$$N_t^k = 1\{T_k \leq t \text{ and } \Delta_k = 1\},$$

Since C_k's are independent of X_k's, let us introduce the filtration $(\mathcal{F}_t)_{t\in[0,\infty)}$ given by

$$\mathcal{F}_t = \mathcal{F}_0 \vee \sigma(\{T_k \leq s\}; s \in [0,t], k = 1,...,n), \quad \forall t \in [0,\infty),$$

where $\mathcal{F}_0 = \mathcal{C} = \sigma(C_1,...,C_n)$. Then, the intensity process for N with respect to this filtration is given by

$$\lambda_t = \alpha(t)Y_t^n, \tag{3.15}$$

where

$$Y_t^n = n - \sum_{k=1}^n 1\{X_k \wedge C_k < t\} \tag{3.16}$$

denotes the number of individuals at risk, at time t. This model is called the *multiplicative intensity model*; see Aalen (1978).

From now on, let us prove the representation (3.15) for the intensity process. First take any $0 \leq s < t$, and notice that the intensity process for N^k is given by $\lambda_t^k = 0$ on the set $\{T_k < s\}$. On the other hand, observe that it holds on the set $\{s \leq T_k\}$ that

$$\begin{aligned}
&\lim_{\Delta\downarrow 0} \mathsf{E}[N_{t+\Delta}^k - N_t^k | \mathcal{F}_s] \\
&= \lim_{\Delta\downarrow 0} \frac{\mathsf{P}(\{t < T_k \leq t+\Delta\} \cap \{X_k \leq C_k\} \cap \{s \leq T_k\} | \mathcal{C})}{\mathsf{P}(s \leq T_k | \mathcal{C})} \\
&= \lim_{\Delta\downarrow 0} \frac{\mathsf{P}(\{t < X_k \leq t+\Delta\} \cap \{s \leq X_k\})}{\mathsf{P}(s \leq X_k)} \\
&= \lim_{\Delta\downarrow 0} \frac{\mathsf{P}(t < X_k \leq (t+\Delta))}{\mathsf{P}(s \leq X_k)},
\end{aligned}$$

where the fact that C_k's are $\mathcal{F}_0$-measurable is used for deriving the second equality.

The last formula is computed further into

$$= \lim_{\Delta\downarrow 0} \frac{\int_t^{t+\Delta} f(u)du}{1 - F(s-)} = \frac{f(t)}{1 - F(s-)}, \quad \text{on the set } \{s \leq T_k\}.$$

Taking both cases into account, we obtain that

$$\lambda_t = \sum_{k=1}^n \lambda_t^k = \sum_{k=1}^n \lim_{s \nearrow t} \frac{f(t)}{1 - F(s-)} 1\{s \leq T_k\} = \alpha(t)Y_t^n,$$

where

$$Y_t^n = \lim_{s \nearrow t} \left\{ n - \sum_{k=1}^n 1\{T_k \leq s\} \right\} = n - \sum_{k=1}^n 1\{X_k \wedge C_k < t\}.$$

Thus, the formula (3.15) with (3.16) has been proved.

The simple form (3.15) has a big advantage for applications. Since $M_t = N_t - \int_0^t \alpha(s)Y_s^n ds$ is a martingale, taking stochastic integral of $\frac{1\{Y_s^n>0\}}{Y_s^n}$, where $\frac{0}{0}$ should be read as 0, with respect to dM_s, we have that

$$\int_0^t \frac{1\{Y_s^n > 0\}}{Y_s^n}(dN_s - \alpha(s)Y_s^n ds)$$

is a martingale. Thus, we may conclude that

$$\widehat{A}_t^n = \int_0^t \frac{1\{Y_s^n > 0\}}{Y_s^n} dN_s,$$

which is now called the *Nelson-Aalen estimator*, is a good estimator for

$$\int_0^t 1\{Y_s^n > 0\}\alpha(s)ds, \quad \text{which is often equal to} \int_0^t \alpha(s)ds = A(t).$$

Example 3.5.6 (Self-exciting process, ETAS model) A counting process with the intensity process of the form

$$\lambda_t = \alpha + \sum_{k:\tau_k<t} \beta\phi(t-\tau_k), \quad \alpha, \beta > 0,$$

where $\phi(x)$ is a non-negative, non-increasing function on $\mathbb{R}$ (e.g., $\phi(x) = e^{-x}$), is called a *self-exciting process*; see Hawks and Oaks (1974).

Developing this model further, Ogata (1988) introduced the Epidemic Type After Shock model (ETAS model) where the intensity process is given by

$$\lambda_t = \alpha + \sum_{k:\tau_k<t} \frac{\kappa e^{\beta M_k}}{|t-\tau_k+c|^p}, \quad \alpha, \kappa, p > 0.$$

This model has been widely used in seismology, especially for representing the stochastic mechanism of occurrence of small aftershocks following a big earthquake, where τ_k and M_k denote the occurrence time and the magnitude of the k-th aftershock.

Example 3.5.7 (Self-correcting process) A counting process with the intensity process

$$\lambda_t = \alpha\phi(X_t), \quad \alpha > 0,$$

where

$$X_t = \beta t - \gamma N_{t-}, \quad \beta, \gamma > 0,$$

is called a *self-correcting process*. The non-negative function $\phi(x)$ is usually taken to be non-decreasing. This model is to represent the stochastic mechanism of the occurrence times of big earthquakes reflecting the intuitive fact that occurrence of a big earthquake releases the stress of the earth's crust so that there is little possibility of occurrence of the next big earthquake for a while, and that after a long period of the past big earthquake the stress is charged so that the risk of a new big earthquake increases. It is proved that, under some conditions, the process $t \rightsquigarrow X_t$ is ergodic.

3.5.2 Examples of diffusion process models

A Markov process whose almost all paths are continuous is said to be a *diffusion process*. One of the ways to construct diffusion processes is to find a solution to the stochastic differential equation of the form

$$X_t = X_0 + \int_0^t \beta(X_s)ds + \int_0^t \sigma(X_s)dW_s,$$

where $\beta(\cdot)$ and $\sigma(\cdot)$ are suitable functions and $s \rightsquigarrow W_s$ is a standard Wiener process. In this subsection, let us exhibit some concrete examples of such stochastic processes.

Example 3.5.8 (Ornstein-Uhlenbeck process, Vasicek process) A solution to the stochastic differential equation

$$X_t = X_0 - \int_0^t \beta X_s ds + \sigma W_t,$$

where β and σ are some constants, is called an *Ornstein-Uhlenbeck process*.

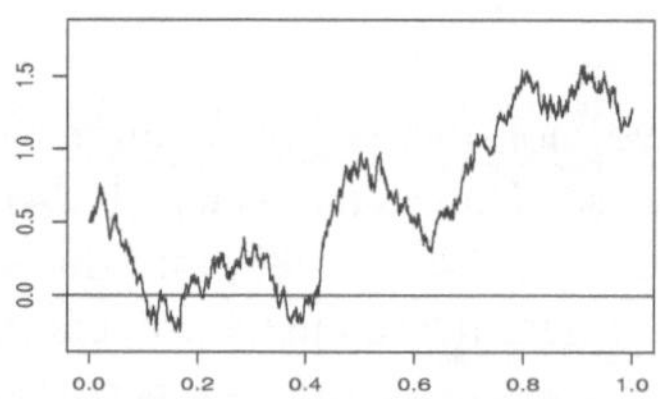

FIGURE 3.1
A path of a Vasicek process, where $\beta_1 = \beta_2 = 1$, $\sigma = 1$, $X_0 = 0.5$. Since we have set $\beta_2 = 1$, the stochastic process is taking values around 1. Compare this with Figure 3.2.

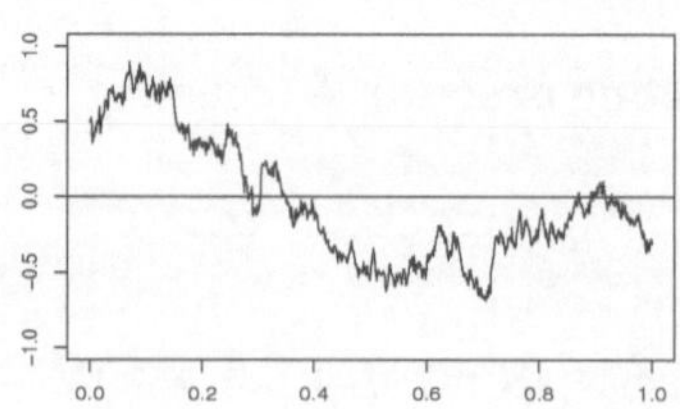

FIGURE 3.2
A path of an Ornstein-Uhlenbeck process, where $\beta = 1$, $\sigma = 1$, $X_0 = 0.5$. Although we took the initial value to be 0.5, it is seen that the stochastic process $t \rightsquigarrow X_t$ is taking values around zero. In fact, the mean of the invariant distribution is zero.

More generally, a solution to the stochastic differential equation

$$X_t = X_0 - \int_0^t \beta_1 (X_s - \beta_2) ds + \sigma W_t,$$

where β_1, β_2 and σ are some constants, is called a *Vasicek process.*

In general, it is rare that the transition density of a diffusion process can be written in an explicit form. The Vasicek process, as well as the Ornstein-Uhlenbeck process as its special case, is one of the examples where an explicit expression of the transition density is possible:

$$\begin{aligned} & p(y,x,t;\beta_1,\beta_2,\sigma) \\ = & \frac{1}{\sqrt{\pi\sigma^2(1-e^{-2\beta_1 t})/\beta_1}} \exp\left(-\frac{(y-e^{-\beta_1 t}x - \beta_2(1-e^{-\beta_1 t}))^2}{\sigma^2(1-e^{-\beta_1 t})/\beta_1} \right), \end{aligned}$$

equivalently, the conditional law of X_t given $X_0 = x$ is

$$\mathcal{N}(e^{-\beta_1 t}x + \beta_2(1-e^{-\beta_1 t}), \sigma^2(1-e^{-\beta_1 t})/(2\beta_1)). \tag{3.17}$$

If $\beta_1 > 0$, then the stochastic process $t \rightsquigarrow X_t$ is ergodic and its invariant distribution is $\mathcal{N}(\beta_2, \sigma^2/2\beta_1)$, which coincides with the distribution obtained by taking formally the "limit" of (3.17) as $t \to \infty$.

Example 3.5.9 (Geometric Brownian motion) A positive real-valued stochastic process

$$X_t = X_0 \exp(\beta t + \sigma W_t),$$

where β and σ are given constants, is called an *geometric Brownian motion.* By Itô's formula which will be explained later, this is rewritten into the form of the stochastic differential equation

$$X_t = X_0 + \int_0^t \left(\beta + \frac{1}{2}\sigma^2 \right) X_s ds + \int_0^t \sigma X_s dW_s,$$

and this model has been widely used in mathematical finance.

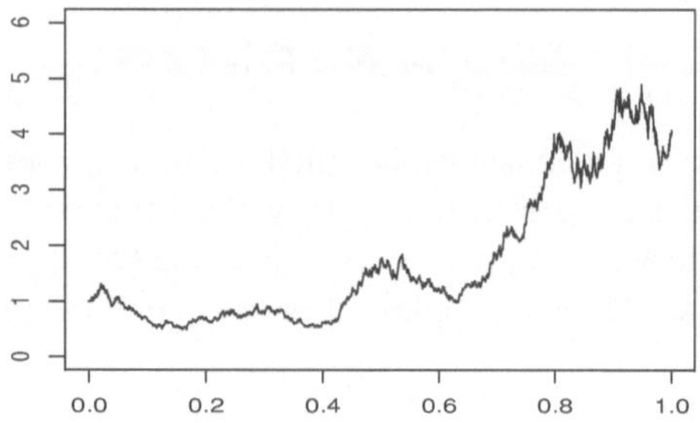

FIGURE 3.3
A path of a geometric Brownian motion, with $\beta = 1$, $\sigma = 1$.

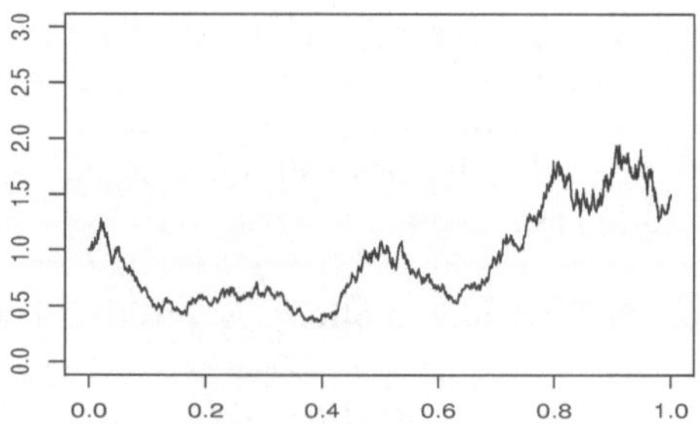

FIGURE 3.4
A path of a geometric Brownian motion, with $\beta = 0$, $\sigma = 1$.

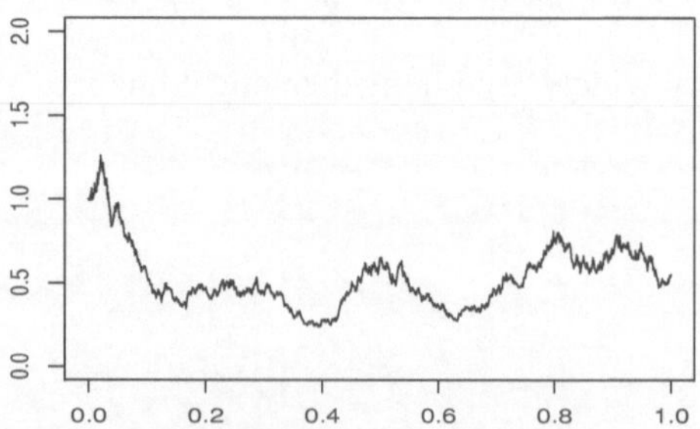

FIGURE 3.5
A path of a geometric Brownian motion, with $\beta = -1$, $\sigma = 1$.

Part II

A User's Guide to Martingale Methods

4

Discrete-Time Martingales

In this chapter, we will learn the definition, some basic facts and the optional sampling theorem for discrete-time martingales, as well as three kinds of useful inequalities (Lenglart's, Bernstein's, and Burkholder's inequalities). All of them will be generalized to the ones for the continuous-time case in subsequent chapters. Among the contents in this chapter, an inconspicuous but important subject is the martingale transformation (Theorem 4.1.3). Although it may look trivial or less interesting at first sight, it is a very important fact which gives the prototypes for the stochastic integrals and the predictable quadratic variation in the theory of continuous-time martingales, and readers are strongly advised to follow its proof.

4.1 Basic Definitions, Prototype for Stochastic Integrals

Let $(\Omega, \mathcal{F})$ be a measurable space. A *discrete-time filtration* is a non-decreasing sequence $\mathbf{F} = (\mathcal{F}_n)_{n \in \mathbb{N}_0}$ of sub-σ-fields of $\mathcal{F}$ with the discrete-time indices $n \in \mathbb{N}_0 = \{0, 1, 2, ...\}$; that is, if $n < n'$ then $\mathcal{F}_n \subset \mathcal{F}_{n'} \subset \mathcal{F}$. When a discrete-time filtration $\mathbf{F}$ and a probability measure P are associated with a measurable space $(\Omega, \mathcal{F})$,

$$\mathbf{B} = (\Omega, \mathcal{F}; \mathbf{F}, \mathsf{P}) = (\Omega, \mathcal{F}; (\mathcal{F}_n)_{n \in \mathbb{N}_0}, \mathsf{P})$$

is called a *discrete-time stochastic basis*. Here and in the sequel, we shall use the notations $\mathbb{N} = \{1, 2, ...\}$ and $\mathbb{N}_0 = \{0, 1, 2, ...\}$.

Definition 4.1.1 Let a discrete-time filtered space $(\Omega, \mathcal{F}; \mathbf{F} = (\mathcal{F}_n)_{n \in \mathbb{N}_0})$ be given.

(i) A discrete-time stochastic process $(X_n)_{n \in \mathbb{N}_0}$ or $(X_n)_{n \in \mathbb{N}}$ is said to be *adapted* (to the filtration $\mathbf{F}$) if X_n is $\mathcal{F}_n$-measurable for every n.

(ii) A discrete-time stochastic process $(X_n)_{n \in \mathbb{N}_0}$ or $(X_n)_{n \in \mathbb{N}}$ is said to be *predictable* (with respect to the filtration $\mathbf{F}$) if X_n is $\mathcal{F}_{(n-1) \vee 0}$-measurable for every n.

Definition 4.1.2 Let a discrete-time stochastic basis $\mathbf{B} = (\Omega, \mathcal{F}; (\mathcal{F}_n)_{n \in \mathbb{N}_0}, \mathsf{P})$ be given.

(i) $(\xi_k)_{k \in \mathbb{N}}$ is said to be a *martingale difference sequence* if it is a real-valued adapted process such that ξ_k is integrable and satisfies that $\mathsf{E}[\xi_k | \mathcal{F}_{k-1}] = 0$ a.s. for every $k \in \mathbb{N}$.

DOI: 10.1201/9781315117768-4

(ii) $(X_n)_{n\in\mathbb{N}_0}$ is said to be a *martingale* if it is a real-valued adapted process such that X_n is integrable for every $n \in \mathbb{N}_0$ and satisfies that $\mathsf{E}[X_n|\mathcal{F}_{n-1}] = X_{n-1}$ a.s. for every $n \in \mathbb{N}$.

Remark. Note that the time index set for a martingale difference sequence is $\mathbb{N} = \{1,2,...\}$, while that for a martingale is $\mathbb{N}_0 = \{0,1,2,...\}$.

When a real-valued, $\mathcal{F}_0$-measurable, integrable random variable M_0 and a martingale difference sequence $(\xi_k)_{k\in\mathbb{N}}$ are given, if we define

$$M_n = M_0 + \sum_{k=1}^{n} \xi_k, \quad n \in \mathbb{N}, \tag{4.1}$$

then $(M_n)_{n\in\mathbb{N}_0}$ is a martingale. Moreover, when a real-valued, $\mathcal{F}_0$-measurable, integrable random variable X_0 and a real-valued, $\mathcal{F}_{k-1}$-measurable random variables H_{k-1} such that $\mathsf{E}[|H_{k-1}\xi_k|] < \infty$ for every $k \in \mathbb{N}$ are also given, if we define

$$X_n = X_0 + \sum_{k=1}^{n} H_{k-1}\xi_k, \quad n \in \mathbb{N}, \tag{4.2}$$

then $(X_n)_{n\in\mathbb{N}_0}$ is a martingale (prove this fact by checking all requirements for being a martingale!), and it is the prototype for stochastic integrals in the continuous-time case which we will learn in Section 6.2. In particular, if we set $H_{k-1} \equiv 1$ in (4.2), then the latter coincides with the former. Since an object called "predictable quadratic variation" will play a very important role in the theory of continuous-time martingales, let us try to understand its idea in advance by presenting the corresponding object in the case of discrete-time martingales here.

Theorem 4.1.3 (Martingale transformation) Let a discrete-time stochastic basis $\mathbf{B}$ be given.

(i) Let us consider (4.1). Assume $\mathsf{E}[M_0^2] < \infty$ and $\mathsf{E}[\xi_k^2] < \infty$ for every $k \in \mathbb{N}$, and define

$$\langle M\rangle_0 = 0, \quad \langle M\rangle_n = \sum_{k=1}^{n} \mathsf{E}[\xi_k^2|\mathcal{F}_{k-1}], \quad n \in \mathbb{N}. \tag{4.3}$$

Then, $((M_n - M_0)^2 - \langle M\rangle_n)_{n\in\mathbb{N}_0}$ is a martingale starting from zero.

(ii) More generally, let us consider (4.2). Assume $\mathsf{E}[X_0^2] < \infty$ and $\mathsf{E}[(H_{k-1}\xi_k)^2] < \infty$ for every $k \in \mathbb{N}$, and define

$$\langle X\rangle_0 = 0, \quad \langle X\rangle_n = \sum_{k=1}^{n} H_{k-1}^2 \mathsf{E}[\xi_k^2|\mathcal{F}_{k-1}], \quad n \in \mathbb{N}. \tag{4.4}$$

Then, $((X_n - X_0)^2 - \langle X\rangle_n)_{n\in\mathbb{N}_0}$ is a martingale starting from zero.

Proof. It suffices to show (ii) only, because (i) is a special case of (ii). For every $n \in \mathbb{N}$, we have

$$\mathsf{E}\left[(X_n - X_0)^2|\mathcal{F}_{n-1}\right]$$

$$
\begin{aligned}
&= \sum_{j=1}^{n}\sum_{k=1}^{n} \mathsf{E}[H_{j-1}\xi_j H_{k-1}\xi_k|\mathcal{F}_{n-1}] \\
&= \sum_{j=1}^{n-1}\sum_{k=1}^{n-1} \mathsf{E}[H_{j-1}\xi_j H_{k-1}\xi_k|\mathcal{F}_{n-1}] \\
&\quad + \sum_{j=1}^{n-1} \mathsf{E}[H_{j-1}\xi_j H_{n-1}\xi_n|\mathcal{F}_{n-1}] + \sum_{k=1}^{n-1} \mathsf{E}[H_{n-1}\xi_n H_{k-1}\xi_k|\mathcal{F}_{n-1}] \\
&\quad + \mathsf{E}[H_{n-1}^2\xi_n^2|\mathcal{F}_{n-1}] \\
&= \sum_{j=1}^{n-1}\sum_{k=1}^{n-1} H_{j-1}\xi_j H_{k-1}\xi_k \\
&\quad + \sum_{j=1}^{n-1} H_{j-1}H_{n-1}\xi_j \mathsf{E}[\xi_n|\mathcal{F}_{n-1}] + \sum_{k=1}^{n-1} H_{k-1}H_{n-1}\xi_k \mathsf{E}[\xi_n|\mathcal{F}_{n-1}] \\
&\quad + H_{n-1}^2 \mathsf{E}[\xi_n^2|\mathcal{F}_{n-1}] \\
&= (X_{n-1}-X_0)^2 + 0 + 0 + H_{n-1}^2 \mathsf{E}[\xi_n^2|\mathcal{F}_{n-1}] \quad \text{a.s.}
\end{aligned}
$$

Thus we have obtained that $\mathsf{E}[(X_n - X_0)^2 - \langle X\rangle_n|\mathcal{F}_{n-1}] = (X_{n-1}-X_0)^2 - \langle X\rangle_{n-1}$ a.s. for every $n \in \mathbb{N}$, which means that $((X_n - X_0)^2 - \langle X\rangle_n)_{n\in\mathbb{N}_0}$ is a martingale starting from zero. □

Discussion 4.1.4 (Prototype for stochastic integral) Since (4.1) implies that $\xi_k = M_k - M_{k-1}$, let us write "$\xi_k = dM_k$". With this notation, (4.2) is written as

$$X_n = X_0 + \sum_{k=1}^{n} H_{k-1} dM_k.$$

Generalizing this up to the continuous-time case, we will be able to define something like

$$X_t = X_0 + \int_0^t H_s dM_s,$$

which will be called "stochastic integral" in Section 6.2.

Discussion 4.1.5 (Prototype for predictable quadratic variation) Since (4.3) implies that $\mathsf{E}[\xi_k^2|\mathcal{F}_{k-1}] = \langle M\rangle_k - \langle M\rangle_{k-1}$, let us write "$\mathsf{E}[\xi_k^2|\mathcal{F}_{k-1}] = d\langle M\rangle_k$". With this notation, (4.4) is written as

$$\langle X\rangle_n = \sum_{k=1}^{n} H_{k-1}^2 d\langle M\rangle_k.$$

Generalizing this up to the continuous-time case, we will be able to define something like

$$\langle X\rangle_t = \int_0^t H_s^2 d\langle M\rangle_s,$$

which will be called "predictable quadratic variation" later.

Exercise 4.1.1 Let Y be a real-valued integrable random variable on a probability space $(\Omega, \mathcal{F}, \mathsf{P})$. Prove that, for any given filtration $\mathbf{F} = (\mathcal{F}_n)_{n\in\mathbb{N}_0}$ on $(\Omega, \mathcal{F})$, if we put $X_n = \mathsf{E}[Y|\mathcal{F}_n]$ for every $n \in \mathbb{N}_0$, then $(X_n)_{n\in\mathbb{N}_0}$ is an $(\mathbf{F}, \mathsf{P})$-martingale.

Exercise 4.1.2 Let $(\xi_k)_{k\in\mathbb{N}}$ be a martingale difference sequence on a discrete-time stochastic basis $\mathbf{B} = (\Omega, \mathcal{F}; (\mathcal{F}_k)_{k\in\mathbb{N}_0}, \mathsf{P})$, such that $\mathsf{E}[|\xi_k|^q] < \infty$ for any $k \in \mathbb{N}$ and any $q \geq 1$. Put $M_0 = 1$ and $M_n = \prod_{k=1}^n (1+\xi_k)$ for every $n \in \mathbb{N}$.

(i) Prove that $(M_n)_{n\in\mathbb{N}_0}$ is a martingale on $\mathbf{B}$ starting from 1.

(ii) Prove that there exists a martingale $(M'_n)_{n\in\mathbb{N}_0}$ on $\mathbf{B}$ starting from zero such that $(M_n - M_0)^2 = \sum_{k=1}^n M_{k-1}^2 \mathsf{E}[\xi_k^2|\mathcal{F}_{k-1}] + M'_n$ a.s. for every $n \in \mathbb{N}$.

(iii) Prove that for any bounded, adapted process $(H_k)_{k\in\mathbb{N}_0}$ on $\mathbf{B}$ there exists a martingale $(M''_n)_{n\in\mathbb{N}_0}$ on $\mathbf{B}$ starting from zero such that $(\sum_{k=1}^n H_{k-1}(M_k - M_{k-1}))^2 = \sum_{k=1}^n H_{k-1}^2 M_{k-1}^2 \mathsf{E}[\xi_k^2|\mathcal{F}_{k-1}] + M''_n$ a.s. for every $n \in \mathbb{N}$.

4.2 Stopping Times, Optional Sampling Theorem

A stopping time is a "random time" that has a "good" measurability in order to develop the theory of martingales based on the concept of filtration. Here, let us give three definitions on stopping times; our set-up here is the situation where a measurable space $(\Omega, \mathcal{F})$ with a discrete-time filtration $\mathbf{F} = (\mathcal{F}_k)_{k\in\mathbb{N}_0}$ (without any probability measure at this moment!) is given.

Definition 4.2.1 Let a discrete-time filtered space $(\Omega, \mathcal{F}; \mathbf{F})$ be given.

(i) T is called a *stopping time* if it is a mapping from Ω to $\mathbb{N}_0 \cup \{\infty\} = \{0, 1, 2, ..., \infty\}$ such that $\{\omega; T(\omega) \leq n\} \in \mathcal{F}_n$ holds for every $n \in \mathbb{N}_0$.

(ii) A stopping time T is said to be *finite* if $T(\omega) < \infty$ for all ω, and *bounded* if there exists a constant c such that $T(\omega) \leq c$ for all ω.

Note that a mapping $T : \Omega \to [0, \infty]$ is a stopping time if and only if the stochastic process $(X_n)_{n\in\mathbb{N}_0}$ defined by $X_n = 1\{T \leq n\}$ is adapted. Notice also that

$$\{\text{bounded stopping times}\} \subset \{\text{finite stopping times}\} \subset \{\text{stopping times}\}. \quad (4.5)$$

Here, we mention the fact that for any given stopping time T we can define a σ-field $\mathcal{F}_T$ in a suitable way; the formal definition is that

$$\mathcal{F}_T = \{A \in \mathcal{F};\ A \cap \{T \leq n\} \in \mathcal{F}_n,\ \forall n \in \mathbb{N}_0\}.$$

Based on this definition, it can be proved that if X is an adapted process then X_T is $\mathcal{F}_T$-measurable for any finite stopping time T, which is very natural and reasonable[1].

[1]However, the corresponding claim in the continuous-time case does not seem true (see Exercise 5.6.1), and we need to introduce the concept of "optional processes" stronger than "adapted processes". Readers do not have to worry too much even if getting an intuitive interpretation of the definition is difficult at the current stage.

Now, let us equip a probability measure P with the filtered space $(\Omega, \mathcal{F}; \mathbf{F})$. If $(X_n)_{n\in\mathbb{N}_0}$ is a martingale and if S and T are bounded stopping times such that $S \leq T$, then both X_T and X_S are integrable and it holds that

$$\mathsf{E}[X_T|\mathcal{F}_S] = X_S, \quad \text{a.s.}$$

This is called the *optional sampling theorem*. In particular, setting $S = 0$ and taking the expectations of the both sides, we have

$$\mathsf{E}[X_T] = \mathsf{E}[\mathsf{E}[X_T|\mathcal{F}_0]] = \mathsf{E}[X_0].$$

We will use the theorem mainly in this form.

4.3 Inequalities for 1-Dimensional Martingales

4.3.1 Lenglart's inequality and its corollaries

Let us prepare a definition.

Definition 4.3.1 When two adapted processes X and Y defined on a discrete-time stochastic basis $\mathbf{B} = (\Omega, \mathcal{F}; (\mathcal{F}_n)_{n\in\mathbb{N}_0}, \mathsf{P})$ are given, we say that X is *L-dominated* by Y if $\mathsf{E}[|X_T|] \leq \mathsf{E}[|Y_T|]$ holds for any bounded stopping time T.

Theorem 4.3.2 (Lenglart's inequality) Let X be a $[0,\infty)$-valued adapted process starting from zero, and A a $[0,\infty)$-valued, predictable, non-decreasing process such that $A_0 \geq 0$ is deterministic, both defined on a discrete-time stochastic basis. If X is L-dominated by A, then it holds for any stopping time T that

$$\mathsf{P}\left(\sup_{n\leq T} X_n \geq \eta\right) \leq \frac{\mathsf{E}[A_T \wedge \delta]}{\eta} + \mathsf{P}(A_T \geq \delta), \quad \forall \eta, \delta > 0,$$

and that

$$\mathsf{E}\left[\sup_{n\leq T}(X_n)^p\right] \leq \left(\frac{2-p}{1-p}\right)\mathsf{E}[(A_T)^p], \quad \forall p \in (0,1),$$

where "A_T" in case of $T(\omega) = \infty$ for some $\omega \in \Omega$ should be read as $A_\infty(\omega) = \lim_{n\to\infty} A_n(\omega)$ which is well-defined for every $\omega \in \Omega$, including the possibility of $A_\infty(\omega) = \infty$, since $n \mapsto A_n(\omega)$ is non-decreasing in n.

Remark. The assumption that A_0 is deterministic cannot be weakened to that it is $\mathcal{F}_0$-measurable. Here is a counter example. Let $a, q \in (0,1)$ be any constants, and put $X_0 = 0$, $X_n = aq$ for all $n \in \mathbb{N}$,

$$A_0 = \begin{cases} a, & \text{with probability } q, \\ 0, & \text{with probability } 1-q, \end{cases}$$

and $A_n = A_0$ for all $n \in \mathbb{N}$. Then, all the assumptions of the theorem, except that A_0 is deterministic, are met. However, if we "apply" the theorem for $T = 1$ then: the first inequality for $\eta = aq$ and $\delta = a^2$ is reduced to

$$1 \leq \frac{a^2 q}{aq} + q = a + q,$$

where the right-hand side can become smaller than 1; the second inequality is reduced to

$$(aq)^p \leq \frac{2-p}{1-p} a^p q, \quad \text{which is equivalent to} \quad 1 \leq \frac{2-p}{1-p} q^{1-p},$$

where the right-hand side, with $p \in (0,1)$ being any fixed constant, can become smaller than 1.

Remark. In the standard textbooks on the martingale theory, it is usually assumed that $A_0 = 0$. The reason way we have slightly extended the inequality up to the above form is that we need a version with a positive, deterministic A_0 in the application of a "stochastic maximal inequality" in Section A1.1.

Although the line of the proof of Theorem 4.3.2 is exactly the same as that for Theorem 6.6.2 for the continuous-time case, we will give a full proof of Theorem 4.3.2 for readers who are interested mainly in the discrete-time case. Before describing it, let us consider an important special case of the theorem.

When $(\xi_k)_{k \in \mathbb{N}}$ is a martingale difference sequence such that $\mathsf{E}[(\xi_k)^2] < \infty$ for all k, if we define $X_\cdot = (\sum_{k=1}^{\cdot} \xi_k)^2$, $A_\cdot = \sum_{k=1}^{\cdot} \mathsf{E}[\xi_k^2 | \mathcal{F}_{k-1}]$, then $X - A$ is a martingale starting from zero as it is seen in Theorem 4.1.3 (i). So we can check all the assumptions in Theorem 4.3.2 using also the optional sampling theorem to obtain that for any stopping time T and every $\varepsilon, \delta > 0$,

$$\begin{aligned} \mathsf{P}\left(\sup_{n \leq T} \left| \sum_{k=1}^{n} \xi_k \right| \geq \varepsilon \right) &= \mathsf{P}\left(\sup_{n \leq T} \left(\sum_{k=1}^{n} \xi_k \right)^2 \geq \varepsilon^2 \right) \\ &\leq \frac{\delta}{\varepsilon^2} + \mathsf{P}\left(\sum_{k=1}^{T} \mathsf{E}[\xi_k^2 | \mathcal{F}_{k-1}] \geq \delta \right). \end{aligned}$$

This observation yields the following corollary.

Corollary 4.3.3 (Corollary to Lenglart's inequality) For every $n \in \mathbb{N}$, let $(\xi_k^n)_{k \in \mathbb{N}}$ be a martingale difference sequence such that $\mathsf{E}^n[(\xi_k^n)^2] < \infty$ for all $k \in \mathbb{N}$, and let T_n be a stopping time, both defined on a discrete-time stochastic basis $\mathbf{B}^n = (\Omega^n, \mathcal{F}^n; (\mathcal{F}_k^n)_{k \in \mathbb{N}_0}, \mathsf{P}^n)$.

(i) As $n \to \infty$,

$$\sum_{k=1}^{T_n} \mathsf{E}^n[(\xi_k^n)^2 | \mathcal{F}_{k-1}^n] = o_{\mathsf{P}^n}(1) \quad \text{implies} \quad \sup_{m \leq T_n} \left| \sum_{k=1}^{m} \xi_k^n \right| = o_{\mathsf{P}^n}(1).$$

(ii) As $n \to \infty$,

$$\sum_{k=1}^{T_n} \mathsf{E}^n[(\xi_k^n)^2 | \mathcal{F}_{k-1}^n] = O_{\mathsf{P}^n}(1) \quad \text{implies} \quad \sup_{m \le T_n} \left| \sum_{k=1}^{m} \xi_k^n \right| = O_{\mathsf{P}^n}(1).$$

Proof of Theorem 4.3.2. To prove the first inequality, notice that

$$\mathsf{P}\left(\sup_{n \le T} X_n \ge \eta\right) = \lim_{m \to \infty} \mathsf{P}\left(\sup_{n \le T} X_n > \eta_m\right), \quad \text{where } \eta_m = \eta - m^{-1}.$$

Thus, it suffices to show the inequality with the left-hand side is replaced by $\mathsf{P}\left(\sup_{n \le T} X_n > \eta\right)$. Next, set $T_m = T \wedge m$ for every $m \in \mathbb{N}$. Since both $\max_{n \le T_m} X_n$ and A_{T_m} are non-decreasing in m, we have

$$\lim_{m \to \infty} \mathsf{P}\left(\max_{n \le T_m} X_n > \eta\right) = \mathsf{P}\left(\sup_{n \le T} X_n > \eta\right),$$

$$\lim_{n \to \infty} \frac{\mathsf{E}[A_{T_n} \wedge \delta]}{\eta} = \frac{\mathsf{E}[A_T \wedge \delta]}{\eta}$$

and

$$\lim_{m \to \infty} \mathsf{P}(A_{T_m} \ge \delta) \le \mathsf{P}(A_T \ge \delta).$$

So, it suffices to show the inequality for each T_m. In other words, we may assume that the stopping time T is bounded. (Recall Discussion 2.1.3 for the necessity of the above arguments.)

The inequality is evident for $\delta \le A_0$ when A_0 is positive, because the second term on the right-hand side is 1. So, let us consider the case $\delta' := \delta - A_0 > 0$, including the case $A_0 = 0$. Set $R = \inf(n : X_n > \eta)$ and $S = \inf(n : A_n - A_0 \ge \delta') = \inf(n : A_n \ge \delta)$. Then we have that $R, S \ge 1$. It is easy to see that R and $S - 1$ are stopping times (check these facts as Exercise 4.3.1). Observing that $\{\max_{n \le T} X_n > \eta\} \subset \{A_T \ge \delta\} \cup \{R \le T < S\}$, we have

$$\mathsf{P}\left(\max_{n \le T} X_n > \eta\right) \le \mathsf{P}(R \le T < S) + \mathsf{P}(A_T \ge \delta).$$

Regarding the first term on the right-hand side, it holds that

$$\begin{aligned}
\mathsf{P}(R \le T < S) &\le \mathsf{P}(R \le T \le (S-1)) \\
&\le \mathsf{P}(X_{R \wedge T \wedge (S-1)} \ge \eta) \\
&\le \frac{1}{\eta} \mathsf{E}[X_{R \wedge T \wedge (S-1)}] \\
&\le \frac{1}{\eta} \mathsf{E}[A_{R \wedge T \wedge (S-1)}],
\end{aligned}$$

where we have used the fact that T is a bounded stopping time to prove the last inequality. So, the first inequality is true because $A_{R \wedge T \wedge (S-1)} \le A_{T \wedge (S-1)} \le A_T \wedge \delta$.

To prove the second inequality, observe that, denoting $X_T^* := \sup_{n\le T} X_n$,

$$\begin{aligned}
\mathsf{E}[(X_T^*)^p] &= \int_0^\infty \mathsf{P}((X_T^*)^p \ge t)dt \\
&= \int_0^\infty \mathsf{P}(X_T^* \ge t^{1/p})dt \\
&\le \int_0^\infty t^{-1/p}\mathsf{E}[A_T \wedge t^{1/p}]dt + \int_0^\infty \mathsf{P}((A_T)^p \ge t)dt \\
&= \mathsf{E}\left[\int_0^{(A_T)^p} dt + \int_{(A_T)^p}^\infty (A_T t^{-1/p})dt + (A_T)^p\right] \\
&= \frac{2-p}{1-p}\mathsf{E}[(A_T)^p].
\end{aligned}$$

The proof is finished. □

Exercise 4.3.1 Prove that R and $S-1$ appearing in the proof of Theorem 4.3.2 are stopping times.

4.3.2 Bernstein's inequality

Theorem 4.3.4 Let $(\xi_k)_{k\in\mathbb{N}}$ be a martingale difference sequence defined on a discrete-time stochastic basis $(\Omega,\mathcal{F};(\mathcal{F}_n)_{n\in\mathbb{N}_0},\mathsf{P})$ such that $|\xi_k| \le a$ for all k, for a constant $a > 0$. Then, it holds for any stopping time T and any $x, v > 0$ that

$$\mathsf{P}\left(\sup_{n\le T}\left|\sum_{k=1}^n \xi_k\right| \ge x,\ \sum_{k=1}^T \mathsf{E}[\xi_k^2|\mathcal{F}_{k-1}] \le v\right) \le 2\exp\left(-\frac{x^2}{2(ax+v)}\right).$$

An important restriction in this inequality is the assumption that ξ_k's are uniformly bounded. See Section 8.2.1 of van de Geer (2000) for an extension to the case where this assumption is replaced by that ξ_k's satisfy some higher-order moment conditions.

4.3.3 Burkholder's inequalities

Theorem 4.3.5 For every $p \ge 1$ there exist some constants $c_p, C_p > 0$ depending only on p, such that for any martingale difference sequence $(\xi_k)_{k\in\mathbb{N}}$ and any stopping time T on a discrete-time stochastic basis $(\Omega,\mathcal{F};(\mathcal{F}_k)_{k\in\mathbb{N}_0},\mathsf{P})$, it holds that

$$c_p\mathsf{E}\left[\left|\sum_{k=1}^T \xi_k^2\right|^{p/2}\right] \le \mathsf{E}\left[\sup_{n\le T}\left|\sum_{k=1}^n \xi_k\right|^p\right] \le C_p\mathsf{E}\left[\left|\sum_{k=1}^T \xi_k^2\right|^{p/2}\right];$$

moreover, it also holds that

$$c_p\mathsf{E}\left[\left|\sum_{k=1}^T \xi_k^2\right|^{p/2}\middle|\mathcal{F}_0\right] \le \mathsf{E}\left[\sup_{n\le T}\left|\sum_{k=1}^n \xi_k\right|^p\middle|\mathcal{F}_0\right] \le C_p\mathsf{E}\left[\left|\sum_{k=1}^T \xi_k^2\right|^{p/2}\middle|\mathcal{F}_0\right], \quad \text{a.s.} \tag{4.6}$$

See Section 2.4 of Hall and Heyde (1980) for a proof of the first displayed inequalities and some variations.

Proof of the second displayed inequalities of Theorem 4.3.5. Choose any $A \in \mathcal{F}_0$. Then, $n \rightsquigarrow M_n^A = (\sum_{k=1}^n \xi_k)1_A$ is a discrete-time martingale, starting from zero, whose quadratic variation is $n \rightsquigarrow [M^A]_n = (\sum_{k=1}^n \xi_k^2)1_A$. Thus, it follows from the first displayed inequalities applied to M^A that

$$c_p\mathsf{E}\left[\left|\sum_{k=1}^T \xi_k^2\right|^{p/2} 1_A\right] \leq \mathsf{E}\left[\sup_{n\leq T}\left|\sum_{k=1}^n \xi_k\right|^p 1_A\right] \leq C_p\mathsf{E}\left[\left|\sum_{k=1}^T \xi_k^2\right|^{p/2} 1_A\right]. \tag{4.7}$$

Here, we define the set N on which the first inequality of (4.6) does not hold by $N = \lim_{m\to\infty} N_m$, where

$$N_m = \left\{ c_p\mathsf{E}\left[\left|\sum_{k=1}^T \xi_k^2\right|^{p/2} \middle| \mathcal{F}_0\right] - \mathsf{E}\left[\sup_{n\leq T}\left|\sum_{k=1}^n \xi_k\right|^p \middle| \mathcal{F}_0\right] \geq m^{-1}\right\}.$$

Let us prove that $\mathsf{P}(N_m) = 0$ for every $m \in \mathbb{N}$. First note that $N_m \in \mathcal{F}_0$. If $\mathsf{P}(N_m)$ were positive, it should hold that

$$\mathsf{E}\left[\left\{ c_p\mathsf{E}\left[\left|\sum_{k=1}^T \xi_k^2\right|^{p/2} \middle| \mathcal{F}_0\right] - \mathsf{E}\left[\sup_{n\leq T}\left|\sum_{k=1}^n \xi_k\right|^p \middle| \mathcal{F}_0\right]\right\} 1_{N_m}\right] \geq m^{-1}\mathsf{P}(N_m) > 0,$$

which contradicts with the first inequality of (4.7). We thus have that $\mathsf{P}(N_m) = 0$ for every $m \in \mathbb{N}$, which implies that $\mathsf{P}(N) = \mathsf{P}(\lim_{m\to\infty} N_m) = \lim_{m\to\infty} \mathsf{P}(N_m) = 0$. The proof of the first inequality of (4.6) is finished.

The second inequality can also be proved in the same way. □

5

Continuous-Time Martingales

Roughly speaking, a martingale is a stochastic process indexed by $[0,\infty)$ whose "conditional trend", analyzed by means of "conditional expectations", is not increasing or decreasing but zero. On the other hand, a stochastic process whose "conditional trend" is increasing is said to be a submartingale.

One of the most fundamental theorems in the martingale theory is the *Doob-Meyer decomposition theorem.* It says that the decomposition

$$\text{"submartingale"} = \text{"predictable increasing process"} + \text{"martingale"}$$

is always possible. Although this claim may not look exciting or interesting at first sight, the true worth of the theorem is that the decomposition is unique. This point is closely related to the fact, which is easy to remember, that any (local) martingale starting from zero which is "predictable" and has "finite-variation (on each compact interval)[1]" is necessarily zero (the degenerate process).

This explanation has been chosen as an introduction to this chapter in order to announce at this stage that two of the important concepts in the martingale theory are "predictability" and "finite-variation". With these two properties in hands, we will be able to build up a lot of important objects in the theory, including the "predictable quadratic (co)-variation" of a square-integrable martingale.

The chapter finishes with an explanation about a deep theory concerning the decomposition of local martingales.

5.1 Basic Definitions, Fundamental Facts

Let $(\Omega,\mathcal{F})$ be a measurable space. A *filtration* is a family $\mathbf{F}=(\mathcal{F}_t)_{t\in[0,\infty)}$ of sub-σ-fields of $\mathcal{F}$ which is non-decreasing and right-continuous, in the sense that

$$\mathcal{F}_s \subset \mathcal{F}_t \subset \mathcal{F} \quad \text{for every} \quad 0 \le s \le t < \infty, \quad \text{(non-decreasing)}$$

and

$$\mathcal{F}_t = \bigcap_{s\in(t,\infty)} \mathcal{F}_s, \quad \forall t \in [0,\infty), \quad \text{(right-continuous).}$$

When a filtration $\mathbf{F}=(\mathcal{F}_t)_{t\in[0,\infty)}$ is associated with a measurable space $(\Omega,\mathcal{F})$, we shall call

$$(\Omega,\mathcal{F};\mathbf{F}) = (\Omega,\mathcal{F};(\mathcal{F}_t)_{t\in[0,\infty)})$$

[1] Any increasing process has finite-variation on any compact interval.

DOI: 10.1201/9781315117768-5

a *filtered space*. When a filtration $\mathbf{F} = (\mathcal{F}_t)_{t\in[0,\infty)}$ and a probability measure P are associated with a measurable space $(\Omega, \mathcal{F})$, we shall call

$$\mathbf{B} = (\Omega, \mathcal{F}; \mathbf{F}, \mathsf{P}) = (\Omega, \mathcal{F}; (\mathcal{F}_t)_{t\in[0,\infty)}, \mathsf{P})$$

a *stochastic basis*.

A stochastic basis $(\Omega, \mathcal{F}; (\mathcal{F}_t)_{t\in[0,\infty)}, \mathsf{P})$ is said to be *complete* if $\mathcal{F}$ is P-complete and $\mathcal{F}_0$ contains all P-null sets, where $\mathcal{F}$ is said to be P-*complete* if $A \subset N \in \mathcal{F}$ and $\mathsf{P}(N) = 0$ imply $A \in \mathcal{F}$. See Definition I.1.3 and a subsequent remark in Jacod and Shiryaev (2003).

A stochastic process $(X_t)_{t\in[0,\infty)}$ is said to be *adapted* to the filtration $(\mathcal{F}_t)_{t\in[0,\infty)}$ if X_t is $\mathcal{F}_t$-measurable for every $t \in [0,\infty)$. This property means that X_t is a random variable determined only by the information up to time t.

Now, let us suspend our discussion on stochastic processes for a while. An $\mathbb{R}^d$-valued function $t \mapsto x(t)$ defined on $[0,\infty)$ is said to be *càdlàg*[2] if it is right-continuous and has left-hand limits at all points, that is, $x(t) = \lim_{s\searrow t} x(s)$ for every $t \in [0,\infty)$ and $\lim_{s\nearrow t} x(s)$ exists for every $t \in (0,\infty)$. We denote the left-hand limit at t by $x(t-)$ for every $t \in (0,\infty)$, and set formally $x(0-) = x(0)$; moreover, we denote $\Delta x(t) = x(t) - x(t-)$, which means the jump of x at t.

Turning back to our discussion on stochastics, let us recall some of the notations and conventions given at the beginning of this monograph. When an $\mathbb{R}^d$-valued stochastic process $(X_t)_{t\in[0,\infty)}$ is given, we may regard $X_t(\omega)$ as a function of t and ω. When we fix a $t \in [0,\infty)$, the function $\omega \mapsto X_t(\omega)$ may be regarded as an $\mathbb{R}^d$-valued random variable. When we fix a $\omega \in \Omega$, the function $t \mapsto X_t(\omega)$ is called a *path*.

We mean by *all paths* the totality of paths $t \mapsto X_t(\omega)$, $\omega \in \Omega$, while by *almost all paths* a collection of paths $t \mapsto X_t(\omega)$, $\omega \in \Omega \setminus N$, where N is a P-*null set* in the sense that $N \in \mathcal{F}$ and $\mathsf{P}(N) = 0$. Two stochastic processes X and Y are said to be *indistinguishable* if there exists a P-null set N such that $X_t(\omega) = Y_t(\omega)$ holds for all $t \in [0,\infty)$, for every $\omega \in \Omega \setminus N$, in other words, almost all paths are exactly the same. A stochastic process Y is said to be a *version* of X if for every $t \in [0,\infty)$ there exists a P-null set $N = N_t$ such that $X_t(\omega) = Y_t(\omega)$ for any $\omega \in \Omega \setminus N$; note that the P-null sets $N = N_t$ in the latter definition may be chosen depending on $t \in [0,\infty)$. Hence, if X and Y are indistinguishable then X and Y are versions of each other. Although the converse is not true in general, it is easy to show the following.

Exercise 5.1.1 If X and Y are stochastic processes whose almost all paths are right-continuous, and if X and Y are versions of each other, then X and Y are indistinguishable. Prove this claim.

Exercise 5.1.2 Construct some examples of stochastic processes X and Y such that X and Y are versions of each other (i.e., $\mathsf{P}(X_t = Y_t) = 1$ for all $t \in [0,\infty)$) and that X and Y are not indistinguishable (i.e., $\mathsf{P}^*(X_t = Y_t, \forall t \in [0,\infty)) < 1$, where P^* denotes the outer probability measure defined by $\mathsf{P}^*(A) := \inf(\mathsf{P}(B) : A \subset B \in \mathcal{F})$ for any $A \subset \Omega$ that may not be measurable). **[Comment:** It is possible even to construct an example such that $\mathsf{P}(X_t = Y_t) = 1$ for all $t \in [0,\infty)$ and that $\mathsf{P}(X_t = Y_t, \forall t \in [0,\infty)) = 0$.]

[2] This is an abbreviation of the phrase "continu *à d*roite avec des *l*imite *à g*auche" in French.

Definition 5.1.1 (Càdlàg process, etc.) Let a measurable space $(\Omega, \mathcal{F})$ be given; at this moment, we do not associate any probability measure to this space.

(i) An $\mathbb{R}^d$-valued stochastic process $t \rightsquigarrow X_t$ is said to be a *càdlàg* process if *all* paths $t \mapsto X_t(\omega)$ are càdlàg. The definitions of *continuous process, right-continuous process, left-continuous process* and *process with left-hand limits* are given in the same ways.

(ii) For a given $\mathbb{R}^d$-valued process X with left-hand limits, denote the left-hand limit at t by X_{t-} for every $t \in (0, \infty)$, and set formally $X_{0-} = X_0$; moreover, put $\Delta X_t = X_t - X_{t-}$.

Definition 5.1.2 (Martingale, etc.) Let $X = (X_t)_{t \in [0,\infty)}$ be a real-valued adapted process on a stochastic basis $(\Omega, \mathcal{F}; \mathbf{F} = (\mathcal{F}_t)_{t \in [0,\infty)}, \mathsf{P})$ such that its *almost all* paths are càdlàg and that $\mathsf{E}[|X_t|] < \infty$ for every $t \in [0, \infty)$. Consider the following three properties:

$$X_s = \mathsf{E}[X_t | \mathcal{F}_s] \quad \text{a.s.} \quad \text{for every} \quad 0 \le s \le t < \infty; \tag{5.1}$$

$$X_s \le \mathsf{E}[X_t | \mathcal{F}_s] \quad \text{a.s.} \quad \text{for every} \quad 0 \le s \le t < \infty; \tag{5.2}$$

$$X_s \ge \mathsf{E}[X_t | \mathcal{F}_s] \quad \text{a.s.} \quad \text{for every} \quad 0 \le s \le t < \infty. \tag{5.3}$$

The process X is said to be a *martingale*, *submartingale* or *supermartingale*, respectively, if (5.1), (5.2), or (5.3) is satisfied, respectively[3].

Remark. When it is necessary to emphasize the choice of the filtration $\mathbf{F}$ or that of the pair $(\mathbf{F}, \mathsf{P})$ of the filtration and the probability measure, the expression "$\mathbf{F}$-martingale" or "$(\mathbf{F}, \mathsf{P})$-martingale" is used instead of "martingale". Similar remarks are given also for "submartingale", "supermartingale", and all other objects based on a filtration (and a probability measure) which will appear in subsequent parts of this monograph.

Note that martingales, submartingales and supermartingales are not "càdlàg processes" in the sense of Definition 5.1.1 that requires the càdlàg property for all paths. On the other hand, "semimartingale" which will be defined in Section 6.1, are assumed to be a process whose all paths are càdlàg; see also a remark after Definition 5.4.4 below. Nevertheless, here is an important result due to J.L. Doob, which is true when the filtration is right-continuous as we assume throughout this monograph.

Theorem 5.1.3 (Regularization) Let a (not necessarily complete) stochastic basis be given. For any given real-valued adapted process X satisfying (5.1), which may not be càdlàg even almost surely, there exists a martingale $\widetilde{X}$ whose *all* paths are càdlàg such that $\widetilde{X}$ is a version of X. In particular, if X is a martingale (thus, almost all paths X are càdlàg), then there exists a càdlàg martingale $\widetilde{X}$ (thus, *all* paths are càdlàg) such that X and $\widetilde{X}$ are indistinguishable. In summary, *any martingale on a stochastic basis whose filtration is right-continuous has a "càdlàg modification",*

[3] An "$\mathbb{R}^d$-valued martingale" is also defined in an obvious way, while we avoid using the terminologies "$\mathbb{R}^d$-valued submartingale" and "$\mathbb{R}^d$-valued supermartingale" since the inequalities for $\mathbb{R}^d$-valued functions with $d \ge 2$ may cause confusion.

i.e., a martingale defined on the same stochastic basis whose all paths are càdlàg, which is indistinguishable from the original one.

Proof. See Theorem 7.27 of Kallenberg (2002) for a proof of the former claim. The latter follows from the former using also Exercise 5.1.1. □

Exercise 5.1.3 (Regularization with localization) Prove that if X is a local martingale (thus, almost all paths are càdlàg), then there exists a càdlàg local martingale $\widetilde{X}$ (thus, *all* paths are càdlàg) such that X and $\widetilde{X}$ are indistinguishable. Such a process $\widetilde{X}$ is called a "*càdlàg modification*" of X. **[Comment:** Try to solve this exercise after learning the definition of "local martingale", i.e., Definition 5.7.1. **]**

Some interesting examples will be given soon, and here we first give some simple ones for an illustration. When $t \rightsquigarrow M_t$ is a martingale,

$$\begin{aligned} X_t &= t + M_t \quad \text{is a submartingale,} \\ X_t &= -t + M_t \quad \text{is a supermartingale, but} \\ X_t &= \sin t + M_t \quad \text{is neither of them.} \end{aligned}$$

Including also the last one, all of the above examples are *semimartingales*. Although the precise definition of semimartingale will be given later, at this stage readers may think that a semimartingale is a real-valued càdlàg adapted process $t \rightsquigarrow X_t$ of an additive form

$$X_t = \text{“adapted process with finite-variation”} + \text{“martingale”}.$$

Now, let us learn two important examples. The first one is standard Wiener processes.

Definition 5.1.4 (Wiener process) A real-valued stochastic process $W = (W_t)_{t\in[0,\infty)}$ defined on a probability space $(\Omega, \mathcal{F}, \mathsf{P})$ (with no filtration!) is said to be a *standard Wiener process* or a *standard Brownian motion* if the following properties (a) and (b) are satisfied.

(a) $W_0(\omega) = 0$ for *almost all* ω, and *almost all* paths $t \mapsto W_t(\omega)$ are continuous.

(b) For any $n \in \mathbb{N}$ and any $0 = t_0 < t_1 < \cdots < t_n$, the random variables $W_{t_k} - W_{t_{k-1}}$, $k = 1, ..., n$, are independent and they are distributed as $\mathcal{N}(0, t_k - t_{k-1})$, $k = 1, ..., n$, respectively.

Remark. It is not so easy to see that standard Wiener processes do exist, in other words, to see that it is possible to construct a probability space $(\Omega, \mathcal{F}, \mathsf{P})$ on which a stochastic process W satisfying the requirements (a) and (b) in Definition 5.1.4 is defined. However, it is well-known that such a construction is indeed possible; see, e.g., Section 37 of Billingsley (1995). Compare this problem for standard Wiener processes with that for Poisson processes which we will discuss after Definitions 5.1.6 and 5.1.7 below.

Here, we give two examples of martingales based on a standard Wiener process. Before doing it, notice that, generally speaking, in order to discuss whether a given stochastic process $X = (X_t)_{t\in[0,\infty)}$ is a martingale or not, first we have to introduce a filtration to which X is adapted. One of the most typical methods is to introduce the *filtration generated by an independent*[4] *pair of a right-continuous process X taking values in $\mathbb{R}^d$ and a sub-σ-field $\mathcal{H}$ of $\mathcal{F}$*, which is defined by $\mathbf{F}^{X,\mathcal{H}} = (\mathcal{F}_t^{X,\mathcal{H}})_{t\in[0,\infty)}$, where

$$\mathcal{F}_t^{X,\mathcal{H}} = \mathcal{H} \vee \bigcap_{s\in(t,\infty)} \sigma(\{\omega : X_r(\omega) \in B\} : r \in [0,s],\, B \in \mathfrak{B}(\mathbb{R}^d)), \quad \forall t \in [0,\infty).$$

Proposition 5.1.5 (Martingales related to Wiener process) Let $W = (W_t)_{t\in[0,\infty)}$ be a standard Wiener process defined on a probability space $(\Omega, \mathcal{F}, \mathsf{P})$. Introduce the filtration $\mathbf{F}^{W,\mathcal{H}}$ generated by W and a sub-σ-field $\mathcal{H}$ of $\mathcal{F}$ that is independent of W. Then the following claims hold true.

(i) The standard Wiener process $(W_t)_{t\in[0,\infty)}$ is an $\mathbf{F}^{W,\mathcal{H}}$-martingale.

(ii) The stochastic process $(X_t)_{t\in[0,\infty)}$ defined by $X_t = W_t^2 - t$ for every $t \in [0,\infty)$ is an $\mathbf{F}^{W,\mathcal{H}}$-martingale.

As far as proving this proposition, it is unnecessary to use high-level tools in the stochastic analysis. However, one of the purposes of this monograph is to give a clear, unified survey of the theory of martingales. Thus, the above proposition will be rewritten in a more general form that is easy to remember; see Example 5.9.2 (i) and Exercise 5.9.3.

The second example is Poisson processes.

Definition 5.1.6 (Poisson process) Let $\lambda > 0$ be a constant. A real-valued stochastic process $(N_t)_{t\in[0,\infty)}$ defined on a probability space $(\Omega, \mathcal{F}, \mathsf{P})$ is said to be a *(homogeneous) Poisson process* with the intensity parameter λ if the following two properties (a) and (b) are satisfied:

(a) $N_0(\omega) = 0$ for *all* $\omega \in \Omega$, and *all* paths $t \mapsto N_t(\omega)$ are càdlàg.

(b) For any $n \in \mathbb{N}$ and any $0 = t_0 < t_1 < \cdots < t_n$, the random variables $N_{t_k} - N_{t_{k-1}}$, $k = 1, ..., n$, are independent and they are distributed as Poisson distributions with mean $\lambda(t_k - t_{k-1})$, $k = 1, ..., n$, respectively.

Definition 5.1.7 (Inhomogeneous Poisson process) Let a $[0,\infty)$-valued, measurable function $\lambda = (\lambda(t))_{t\in[0,\infty)}$ such that $\int_0^t \lambda(s)ds < \infty$ for every $t \in [0,\infty)$ be given. A real-valued stochastic process $(N_t)_{t\in[0,\infty)}$ defined on a probability space $(\Omega, \mathcal{F}, \mathsf{P})$ is said to be an *inhomogeneous Poisson process* with the intensity function λ if the following properties (a) and (b) are satisfied.

(a) $N_0(\omega) = 0$ for *all* $\omega \in \Omega$, and *all* paths $t \mapsto N_t(\omega)$ are càdlàg.

(b) For any $n \in \mathbb{N}$ and any $0 = t_0 < t_1 < \cdots < t_n$, the random variables $N_{t_k} - N_{t_{k-1}}$, $k = 1, ..., n$, are independent and they are distributed as Poisson distributions with mean $\int_{t_{k-1}}^{t_k} \lambda(s)ds$, $k = 1, ..., n$, respectively.

[4] Two sub-σ-fields $\mathcal{G}$ and $\mathcal{H}$ of $\mathcal{F}$ are said to be independent if $\mathsf{P}(A \cap B) = \mathsf{P}(A)\mathsf{P}(B)$ holds for any $A \in \mathcal{G}$ and $B \in \mathcal{H}$. A stochastic process $(X_t)_{t\in[0,\infty)}$ and a sub-σ-field $\mathcal{H}$ are said to be independent if $\sigma(X_t : t \in [0,\infty))$ and $\mathcal{H}$ are independent.

Remark. Usually, the definition of (in)homogeneous Poisson processes demands the property (a) concerning paths for *all* $\omega \in \Omega$, while that of standard Wiener processes does it only for *almost all* ω. Some reasons why we adopt this definition will be explained in a remark after Definition 5.4.4 below.

Remark. It is not difficult to see that homogeneous Poisson processes do exit. Indeed, first introduce an i.i.d. sequence $(X_k)_{k=1,2,\ldots}$ of $(0,\infty)$-valued, exponential random variables with mean $1/\lambda$; such a sequence does exist. Next, define $N_0 = 0$ and $N_t = \max(n : \sum_{k=1}^{n} X_k \le t)$ for every $t \in (0,\infty)$. Then, it can be proved that this stochastic process $(N_t)_{t\in[0,\infty)}$ satisfies the requirements (a) and (b) in Definition 5.1.7.

Exercise 5.1.4 (i) Prove that the stochastic process $(N_t)_{t\in[0,\infty)}$ constructed in the above remark indeed satisfies all requirements for being a homogeneous Poisson process with the intensity parameter λ.

(ii) Find a way to construct inhomogeneous Poisson processes.

Proposition 5.1.8 (Martingales related to Poisson process) Let $N = (N_t)_{t\in[0,\infty)}$ be an inhomogeneous Poisson process with the intensity function $(\lambda(t))_{t\in[0,\infty)}$, defined on a probability space. Introduce the filtration $\mathbf{F}^{N,\mathcal{H}}$ generated by N and a sub-σ-field $\mathcal{H}$ that is independent of N.

(i) The stochastic process $(X_t)_{t\in[0,\infty)}$ defined by $X_t = N_t - \int_0^t \lambda(s)ds$ is an $\mathbf{F}^{N,\mathcal{H}}$-martingale.

(ii) The stochastic process $(X_t)_{t\in[0,\infty)}$ defined by

$$X_t = \left(N_t - \int_0^t \lambda(s)ds\right)^2 - \int_0^t \lambda(s)ds$$

is an $\mathbf{F}^{N,\mathcal{H}}$-martingale.

See Example 5.9.2 (ii) and Exercise 5.9.3 for how to memorize the results of this proposition and their proofs, respectively.

5.2 Discre-Time Stochastic Processes in Continuous-Time

For a given discrete-time stochastic process $X = (X_k)_{k\in\mathbb{N}_0}$ on a stochastic basis $(\Omega,\mathcal{F};\mathbf{F} = (\mathcal{F}_k)_{k\in\mathbb{N}_0},\mathsf{P})$, there are two methods to treat it in the framework of continuous-time stochastic processes.

(The 1st method.) Introduce the filtration $\mathbf{F}^c = (\mathcal{F}_t^c)_{t\in[0,\infty)}$ by

$$\mathcal{F}_t^c := \mathcal{F}_k, \quad t \in [k,k+1), \quad k \in \mathbb{N}_0$$

and extend the definition of X to $(X_t)_{t\in[0,\infty)}$ by

$$X_t := X_k, \quad t \in [k,k+1), \quad k \in \mathbb{N}_0.$$

Then, all paths of $t \rightsquigarrow X_t$ are càdlàg. Moreover, if the original discrete-time stochastic process $X = (X_k)_{k\in\mathbb{N}_0}$ is adapted to $\mathbf{F}$, then the extended stochastic process $X = (X_t)_{t\in[0,\infty)}$ is adapted to $\mathbf{F}^c$.

(The 2nd method.) Let $m_n \in \mathbb{N}$ be given. Introduce the filtration $\widetilde{\mathbf{F}}^{c,n} = (\widetilde{\mathcal{F}}_u^{c,n})_{u\in[0,\infty)}$ (or, $(\widetilde{\mathcal{F}}_u^{c,n})_{u\in[0,1]}$) by

$$\widetilde{\mathcal{F}}_u^{c,n} := \mathcal{F}_k, \quad u \in [k/m_n, (k+1)/m_n), \quad k \in \mathbb{N}_0 \text{ (or, } \{0,1,...,m_n\}),$$

and define the stochastic process $X^n = (X_u^n)_{u\in[0,\infty)}$ (or, $(X_u^n)_{u\in[0,1]}$) by

$$\widetilde{X}_u^n := X_k, \quad u \in [k/m_n, (k+1)/m_n), \quad k \in \mathbb{N}_0 \text{ (or, } \{0,1,...,m_n\}).$$

Also in this case, all paths of $u \rightsquigarrow \widetilde{X}_u^n$ are càdlàg. Moreover, if the original discrete-time stochastic process X is adapted to $\mathbf{F}$, then the new stochastic process $\widetilde{X}^n$ is adapted to $\mathbf{F}^{c,n}$.

When we put $m_n \equiv 1$, the 2nd method is reduced to the 1st one. It is often to put $m_n = n$, which is actually used, e.g., for Donsker's theorem (Corollary 7.2.4).

5.3 $\varphi(M)$ Is a Submartingale

The following theorem will be used, for example, for the construction of "predictable quadratic co-variation" in Theorem 5.9.1 by setting $\varphi(x) = x^2$.

Proposition 5.3.1 Let $\varphi : \mathbb{R} \to \mathbb{R}$ be a convex function, and let M be a martingale defined on a stochastic basis $(\Omega, \mathcal{F}; (\mathcal{F}_t)_{t\in[0,\infty)}, \mathsf{P})$. If $\varphi(M_t)$ is integrable for every $t \in [0,\infty)$, then $\varphi(M)$ is a submartingale.

Proof. Since any convex function is continuous, it is clear that: almost all paths of $\varphi(M)$ are càdlàg; since M_t is $\mathcal{F}_t$-measurable for every $t \in [0,\infty)$, $\varphi(M_t)$ is $\mathcal{F}_t$-measurable for every $t \in [0,\infty)$; so $\varphi(M)$ is an adapted process. Now, it follows from Jensen's inequality for conditional expectation that if $0 \le s \le t < \infty$ then

$$\varphi(\mathsf{E}[M_t|\mathcal{F}_s]) \le \mathsf{E}[\varphi(M_t)|\mathcal{F}_s] \quad \text{a.s.}$$

Since M is a martingale, the inside of the function $\varphi(\cdot)$ on the left-hand side is M_s a.s. So we have

$$\varphi(M_s) \le \mathsf{E}[\varphi(M_t)|\mathcal{F}_s] \quad \text{a.s.}$$

The proof is complete. □

5.4 "Predictable" and "Finite-Variation"

There are two properties which play a key role to build up a lot of important objects in the martingale theory, namely, "predictability" and "having finite-variation". We would be able to have a real understanding of the message from the Doob-Meyer decomposition theorem only after we master these two properties.

5.4.1 Predictable and optional processes

Throughout this subsection, let a filtered space $(\Omega, \mathcal{F}; \mathbf{F})$, with no probability measure, be given.

Definition 5.4.1 (Predictable σ-field) The predictable σ-field, denoted by $\mathcal{P}$, is a σ-field on $\Omega \times [0,\infty)$ which is generated by all real-valued left-continuous adapted processes.

Here, let us read the above definition step-by-step. First take a real-valued left-continuous adapted process $(X_t)_{t\in[0,\infty)}$ and regard it as a real-valued function $(\omega,t) \mapsto X_t(\omega)$ defined on $\Omega \times [0,\infty)$. Next define the class $\mathcal{A}_X$ of subsets of $\Omega \times [0,\infty)$ by

$$\mathcal{A}_X = (\{(\omega,t) : X_t(\omega) \in B\} : B \in \mathfrak{B}(\mathbb{R})).$$

Doing this operation for all real-valued left-continuous adapted processes X and define $\mathcal{P}$ as the smallest σ-filed including all of $\mathcal{A}_X$'s:

$$\mathcal{P} = \sigma(\mathcal{A}_X : X \in \{\text{real-valued left-continuous adapted processes}\}).$$

As it is clear from the definition, the predictable σ-field $\mathcal{P}$ is constructed by using adapted processes, so it can be defined only when a filtration $\mathbf{F} = (\mathcal{F}_t)_{t\in[0,\infty)}$ has been introduced. Notice also that it is defined *not* depending on any probability measure P. Therefore, the predictable σ-field is defined for $(\Omega, \mathcal{F}; \mathbf{F})$ before the family $\{\mathsf{P}_\theta : \theta \in \Theta\}$ of probability measures is introduced to build up a "statistical model $(\Omega, \mathcal{F}; (\mathcal{F}_t)_{t\in[0,\infty)}, \{\mathsf{P}_\theta; \theta \in \Theta\})$".

Definition 5.4.2 (Predictable process) A stochastic process $(X_t)_{t\in[0,\infty)}$ is said to be *predictable* if the mapping $(\omega,t) \mapsto X_t(\omega)$ is $\mathcal{P}$-measurable as a real-valued function on $\Omega \times [0,\infty)$.

In practice, it would be enough to remember the following two facts, both clear from the definition of predictable processes.

(i) Any adapted process whose *all* paths are left-continuous is predictable. Thus, any adapted process whose *all* paths are continuous is predictable[5].

[5]When the stochastic basis is complete, the words "*all* paths" in the claim (i) may be replaced by "*almost all* paths".

(ii) Any deterministic function (that is, a function not depending on ω) which is Borel measurable as a function on $[0,\infty)$ is predictable (with respect to any filtration) as a stochastic process.

To close this subsection, let us introduce another concept concerning the measurability of stochastic processes called "optionality". The importance of this concept will be well understood when readers observe, e.g., Lemma 5.6.2 and Exercise 5.6.1 below.

Definition 5.4.3 (Optional σ-field, optional process) (i) The optional σ-field, denoted by $\mathcal{O}$, is a σ-field on $\Omega \times [0,\infty)$ which is generated by all real-valued càdlàg adapted processes.

(ii) A stochastic process X is said to be *optional* if it is $\mathcal{O}$-measurable when it is regarded as a real-valued function on $\Omega \times [0,\infty)$.

By the definition, any càdlàg adapted process is optional. It can be proved that, more generally, any adapted process whose *all* paths are left-continuous is optional, and that any adapted process whose *all* paths are right-continuous is optional (see Proposition I.1.24 and Remark I.1.26 of Jacod and Shiryaev (2003)), and the former of these facts implies that $\mathcal{P} \subset \mathcal{O}$. Hence, any predictable process is optional. Moreover, it is proved also that any optional process is adapted (see Proposition I.1.21 (a) of Jacod and Shiryaev (2003)). In summary, we may conclude that:

$$\{\text{predictable processes}\} \subset \{\text{optional processes}\} \subset \{\text{adapted processes}\};$$

$$\{\text{right-continuous adapted processes}\} \subset \{\text{optional processes}\}.$$

5.4.2 Processes with finite-variation

The concept of "having finite-variation" is originally introduced not for stochastic processes with randomness but for real-valued functions on the real line. Omitting to write the formal definition here, let us recall that a necessary and sufficient condition for a function $F : [0,T] \to \mathbb{R}$ starting from zero to have finite-variation is that it has a unique decomposition $F = F^a - F^b$ where both F^a and F^b are non-decreasing function starting from zero. Recall also that the Lebesgue-Stieltjes integral of a measurable function h on $[0,T]$ with respect to a function with finite-variation F on $[0,T]$ is defined by $\int_0^T h(t)dF(t) = \int_0^T h(t)dF^a(t) - \int_0^T h(t)dF^b(t)$, where the two terms on the right-hand side are the Lebesgue integrals of h with respect to the measures $dF^a(t) = \mu^a(dt)$ and $dF^b(t) = \mu^b(dt)$ constructed by $\mu^a((s,t]) = F^a(t) - F^a(s)$ and $\mu^b((s,t]) = F^b(t) - F^b(s)$, $0 \le s < t \le T$, respectively. This definition makes sense if at least one of the two terms on the right-hand side is finite.

Now, let us turn to the world of stochastics.

Definition 5.4.4 (Increasing process, Process with finite-variation) Let a filtered space $(\Omega, \mathcal{F}; \mathbf{F})$ be given.

(i) A stochastic process $A = (A_t)_{t\in[0,\infty)}$ is said to be an *increasing process* if it is an adapted process such that $A_0(\omega) = 0$ for *all* $\omega \in \Omega$ and that *all* paths $t \mapsto A_t(\omega)$ are càdlàg and non-decreasing.

(ii) A stochastic process $A = (A_t)_{t\in[0,\infty)}$ is said to be a *process with finite-variation* if it is an adapted process such that $A_0(\omega) = 0$ for *all* $\omega \in \Omega$ and that *all* paths $t \mapsto A_t(\omega)$ are càdlàg and have finite-variation on each compact interval $[0,T]$.

Remark. All the definitions of "càdlàg process", "continuous process", "right-continuous process", "left-continuous process" and "process with left-hand limits" given in Definition 5.1.1, as well as "*increasing process*" and "*process with finite-variation*" given in Definition 5.4.4 strictly demand the properties that are concerned are satisfied for *all* paths. If the requirements for all $\omega \in \Omega$ in these definitions were replaced by those only for almost all ω, some inconsistency would occur in our discussion on the martingale theory. For example, defining "predictable process" based on left-continuous adapted processes would need a more difficult discussion involving a probability measure, and such a definition is usually not adopted. Also, in our study we will often intend to introduce a stopping time as a good "first hitting time" of an "increasing process" via (5.5) below (see Theorems 5.5.2 and 5.5.4), but if we adopted a weaker (incorrect, in our sense) definition of "increasing process", then the resulted first hitting time, which is suitably defined only for almost all ω, might be neither a stopping time nor even a measurable random variable any more[6].

In contrast, in the definitions of martingales, submartingales and supermartingales, we have demanded only that *almost all* paths are càdlàg, because of the reasons including that, otherwise, standard Wiener processes would be excluded in our discussion, and that, otherwise, the Doob-Meyer decomposition theorem would not be able to be well formulated, and so on....

Remark. Readers should not confuse an "*increasing process*", which is defined on a filtered space, with a "non-decreasing process" defined on a measurable space $(\Omega, \mathcal{F})$. In this monograph, the phrase "*increasing process*" is a special terminology defined in Definition 5.4.4 (i), while an "non-decreasing process" is simply a process whose all paths are non-decreasing.

It is clear from the corresponding fact for the deterministic function with finite-variation that a process with finite-variation has a unique decomposition $A = A^a - A^b$, where all paths of both A^a and A^b start from zero and are non-decreasing. Moreover, it is possible to prove the following more important facts; see Proposition I.3.3 of Jacod and Shiryaev (2003) for a proof.

Lemma 5.4.5 Let A be a stochastic process, defined on a filtered space $(\Omega, \mathcal{F}; \mathbf{F})$, whose all paths have finite-variation, and denote its unique decomposition into the difference of two non-decreasing processes starting from zero by $A = A^a - A^b$.

---

[6] An alternative method to make things caused by some problems discussed here run smoothly is to introduce a filtration and a probability measure at this stage and to assume that the stochastic basis is complete.

(i) If A is an adapted process, then A^a and A^b are also adapted processes.
(ii) If A is a predictable process, then A^a and A^b are also predictable processes.

Definition 5.4.6 (Integrable processes) Let a stochastic basis $\mathbf{B}$ be given.
(i) An increasing process is said to be *integrable* if $\mathsf{E}[A_\infty] < \infty$, where $A_\infty(\omega)$ is defined as the limit of $A_t(\omega)$ as $t \to \infty$ for every $\omega \in \Omega$.
(ii) A *process with integrable-variation* A is a process with finite-variation such that the increasing processes A^a and A^b appearing in the unique decomposition $A = A^a - A^b$ are integrable.

We remark that, for any increasing process $A = (A_t)_{t\in[0,\infty)}$, the $[0,\infty]$-valued random variable A_∞ is well-defined[7] as we have already used this fact in the above definition.

Finally, let us define the Stieltjes integral process of a real-valued predictable (or, more generally, optional) process $H = (H_t)_{t\in[0,\infty)}$ with respect to an adapted process $A = (A_t)_{t\in[\infty)}$ with finite-variation. Since $t \mapsto H_t(\omega)$ is Borel measurable as a function on $[0,\infty)$ for every $\omega \in \Omega$, due to Fubini's theorem we can formally set

$$\int_0^t H_s(\omega)dA_s(\omega) := \int_0^t H_s(\omega)dA_s^a(\omega) - \int_0^t H_s(\omega)dA_s^b(\omega), \quad \forall t \in [0,\infty), \quad (5.4)$$

for every $\omega \in \Omega$. This definition makes sense if at least one of the two terms on the right-hand side is finite.

Theorem 5.4.7 (The Stieltjes integral process) For a real-valued optional process H and an adapted process A with finite-variation defined on a filtered space $(\Omega,\mathcal{F};\mathbf{F})$, suppose that the value $\int_0^t H_s dA_s$ computed by (5.4) are finite for every $t \in [0,\infty)$, for *all* $\omega \in \Omega$. Then, $t \rightsquigarrow \int_0^t H_s dA_s$ is an adapted process with finite-variation; this stochastic process is called the *Stieltjes integral process* of H with respect to A.

If, moreover, H and A are predictable processes, then the Stieltjes integral process $t \rightsquigarrow \int_0^t H_s dA_s$ is also predictable.

See Proposition I.3.5 of Jacod and Shiryaev (2003) for a proof.

Warning! A continuous function does not necessary have finite-variation. For example, it is well-known that almost all paths of a standard Wiener process $t \rightsquigarrow W_t$ do not have finite-variation. Therefore, it is not possible to define a stochastic integral with respect to a standard Wiener process in an easy way like

$$\int_0^t H_s(\omega)dW_s(\omega) := \int_0^t H_s(\omega)dW_s^a(\omega) - \int_0^t H_s(\omega)dW_s^b(\omega), \quad \forall \omega.$$

This is why defining the Itô integral in the L^2-sense, which we will learn later, is necessary.

[7] In this monograph, the time parameter t of a stochastic process $t \rightsquigarrow X_t$ ranges only over $[0,\infty)$ in principle. The only exceptions where we define a random variable "X_∞" are the following two cases: the case where X is a non-decreasing process; the case where X is a uniformly integrable martingale, including the case where X is a square-integrable martingale as a special case. See Theorems 5.7.3 and 5.7.4 for the latter.

5.4.3 A role of the two properties

Let us try to understand a role of the predictability and the property of having finite-variation in order to have a good perspective of the martingale theory at this stage. As a corollary to the Doob-Meyer decomposition theorem, it will be proved that if a martingale starting from zero is a predictable process as well as a process with finite-variation, then it is necessarily zero (the degenerate process). There are many examples of predictable martingales that do not have finite-variation, including standard Wiener processes adapted to the filtration on a complete stochastic basis, while there also exist plenty of examples of martingales with finite-variation that are not predictable, including the compensated martingales of Poisson processes. However, the only martingale that has both of the two properties is a trivial stochastic process that is degenerate. This fact is closely related to the "uniqueness" of several important objects in the martingale theory.

5.5 Stopping Times, First Hitting Times

As it was so in the discrete-time case, the concept of stopping time, which is a "random time" that has a nice measurability related to a given filtration, is indispensable to develop the theory of continuous-time martingales.

Definition 5.5.1 (Stopping time) Let a filtered space $(\Omega, \mathcal{F}; \mathbf{F} = (\mathcal{F}_t)_{t\in[0,\infty)})$ be given.

(i_1) A stopping time is a mapping $T : \Omega \to [0,\infty]$ such that $\{T \leq t\} \in \mathcal{F}_t$ holds for every $t \in [0,\infty)$.

(i_2) For a given stopping time T, define

$$\mathcal{F}_T = \{A \in \mathcal{F} : A \cap \{T \leq t\} \in \mathcal{F}_t \text{ for all } t \in [0,\infty)\},$$

which becomes a sub-σ-field of $\mathcal{F}$ (prove this claim!).

(ii) A stopping time T is said to be *finite* if $T(\omega) < \infty$ for all ω, and *bounded* if there exists a constant $c > 0$ such that $T(\omega) \leq c$ for all ω.

Note that a mapping $T : \Omega \to [0,\infty]$ is a stopping time if and only if the stochastic process $(X_t)_{t\in[0,\infty)}$ defined by $X_t = 1\{T \leq t\}$ is adapted. It is clear that the relationships (4.5) hold true also for the continuous-time case. On the other hand, it may be difficult to understand the meaning of the definition of $\mathcal{F}_T$ intuitively. At this stage, readers are advised to memorize the fact that, if X is an optional process, then X_T is $\mathcal{F}_T$-measurable for any finite stopping time T (see Proposition I.1.21 of Jacod and Shiryaev (2003)); this claim is not always true if X is merely an adapted process. Since the σ-field $\mathcal{F}_T$ will appear only at some restricted places where we apply the optional sampling theorem as something like an "automatic machine", let us go ahead not worrying about its interpretation too much!

An important special case of stopping times is *predictable times*. Before proceeding with the study for predictable times, let us first consider the "first hitting time" defined by

$$T = \inf(t : X_t \in B), \tag{5.5}$$

where X is a certain $\mathbb{R}^d$-valued stochastic process and B is a Borel subset of $\mathbb{R}^d$. The "first hitting time" plays an important role in the proofs of various theorems in our study, especially when it is a predictable time that can approximated by a sequence of stopping times called an "announcing sequence". Here, the meaning of the quotation mark for the words "first hitting time" above is that the $[0,\infty]$-valued mapping T defined by (5.5) is *not* always even a stopping time. So, we shall present some sufficient conditions under which a given "first hitting time" becomes a stopping time or a predictable time.

Theorem 5.5.2 (When does a first hitting time become a stopping time?) (i) Consider the case where a stochastic basis, which may not be complete, is given.

(i_1) If X is an $\mathbb{R}^d$-valued optional process[8] and if B is a Borel subset of $\mathbb{R}^d$, then there exists a stopping time $\widetilde{T}$ such that $\widetilde{T} = T$, P^*-almost surely[9], where $T = \inf(t : X_t \in B)$; as a matter of fact, T itself may be neither a stopping time nor even an $\mathcal{F}$-measurable random variable.

(i_2) If X is an $\mathbb{R}^d$-valued adapted process whose *all* paths are right-continuous, and if B is an open subset of $\mathbb{R}^d$, then $T = \inf(t : X_t \in B)$ is a stopping time.

(i_3) If X is a real-valued adapted process whose *all* paths are right-continuous and non-decreasing and if $c \in \mathbb{R}$, then $T = \inf(t : X_t \geq c)$ is a stopping time.

(ii) When the stochastic basis is complete, the $[0,\infty]$-valued mapping T given in (i_1) is a stopping time for itself, and the words "all paths" in the claims (i_2) and (i_3) may be replaced by "almost all paths".

Proof. (i_1) and the first claim of (ii). These claims are well-known but difficult to prove; see Theorem IV.50 of Dellacherie and Meyer (1978) for a proof, which is based on their Theorem III.44.

(i_2). Using the right-continuity of all paths of X and the assumption that B is open, we have

$$\{T < t\} = \bigcup_{s \in [0,t) \cap \mathbb{Q}} \{X_s \in B\}.$$

Since X is adapted, the right-hand side is in $\mathcal{F}_t$. So the claim (i_2) follows from Exercise 5.5.1 (ii) below.

(i_3). When X is a stochastic process whose all paths are right-continuous and non-decreasing, it holds that $\{T \leq t\} = \{X_t \geq c\}$, which belongs to $\mathcal{F}_t$ because X is adapted. So the definition of stopping time has been checked.

The remaining claims in (ii) are proved similarly to (i_2) and (i_3). □

---

[8] This claim is true in a more general situation where X is an $\mathbb{R}^d$-valued "progressively measurable" process.

[9] The meaning of the above claim "$\widetilde{T} = T$, P^*-almost surely" is that there exists an $\mathcal{F}$-measurable, P-null set N such that $\{\omega : \widetilde{T}(\omega) \neq T(\omega)\} \subset N$.

Now let us proceed to a discussion on "predictable time".

Definition 5.5.3 (Predictable time) Let a filtered space $(\Omega, \mathcal{F}; \mathbf{F})$ be given. A *predictable time* is a mapping $T : \Omega \to [0,\infty]$ such that $\{(\omega,t) : 0 \le t < T(\omega)\} \in \mathcal{P}$ where $\mathcal{P}$ is the predictable σ-field with respect to the filtration $\mathbf{F}$.

A predictable time with respect to a filtration is a stopping time with respect to the same filtration (Exercise 5.5.1 (iv) below).

Theorem 5.5.4 (Predictable time, Announcing sequence) (i) Consider the case where a stochastic basis, which may not be complete, is given.

(i_1) If X is a real-valued, predictable process whose *all* paths are right-continuous and non-decreasing and if $c \in \mathbb{R}$, then $T = \inf(t : X_t \ge c)$ is a predictable time.

(i_2) If T is a predictable time, then there exists an increasing sequence (T_n) of stopping times, such that $T_n < T$ a.s. on $\{T > 0\}$ and $\lim_{(n)} T_n = T$ a.s. The sequence (T_n) is called an *announcing sequence* for T.

(ii) When the stochastic basis is complete, the words "all paths" in the claim (i_1) may be replaced by "almost all paths", and the announcing sequence (T_n) in the claim (i_2) can be chosen such that $T_n < T$ on $\{T > 0\}$ and $\lim_{(n)} T_n = T$.

Proof. Since $\{(\omega,t) : T(\omega) = t\} \subset \{(\omega,t) : T(\omega) \le t\} = \{(\omega,t) : X_t(\omega) \ge c\} \in \mathcal{P}$, the claim ($\mathrm{i}_1$) follows from Proposition I.2.13 of Jacod and Shiryaev (2003). The first claim in (ii) is proved in a similar way.

The claim (i_2) and the second claim in (ii) are well-known but difficult to prove; see Theorem IV.76 of Dellacherie and Meyer (1978). □

We often encounter some situations where we would like to introduce a stopping time T satisfying that $X_t \le c$ for all $t \in [0,T]$ a.s. for a given, increasing process $t \rightsquigarrow X_t$. However, when X may have jumps, constructing such a stopping time T is not easy: the claim "$X_T \le c$ a.s." is not true even if we define $T = \inf(t : X_t \ge c)$, or $T = \inf(t : X_t \ge c - \varepsilon)$, or etc. However, in the case where X is a real-valued *predictable* process whose all paths are non-decreasing, if we introduce the predictable time $T = \inf(t : X_t \ge c)$ and an announcing sequence (T_n) for T, then we have that $X_t \le X_{T_n} \le c$ for all $t \le T_n < T$ a.s. on $\{T > 0\}$. In this way, we are able to have a "good" $X_{T_n} \le c$ a.s. that "converges" to X_T. (However, it is *wrong* to argue that "by letting $n \to \infty$, we get $X_{T_n} \uparrow X_T$, so $X_T \le c$ a.s.". This argument is true if X is left-continuous.)

To close this section, we summarise some operations which we can use for stopping times and predictable times, in the form of exercise.

Exercise 5.5.1 (Operations on stopping times) Prove the following claims.

(i) If T is a stopping time and if $c \ge 0$, then $T + c$ is a stopping time. [**Remark:** This is not always true if c is negative.]

(ii) A mapping $T : \Omega \to [0,\infty]$ is a stopping time if and only if $\{T < t\} \in \mathcal{F}_t$ for all $t \in [0,\infty)$. [**Hint:** Use the right-continuity of the filtration.]

(iii) If (T_n) is a sequence of stopping times, then $\bigwedge_{(n)} T_n$ and $\bigvee_{(n)} T_n$ are stopping times.

(iv) Any predictable time is a stopping time.

(v) If (T_n) is a sequence of predictable times, then $T = \bigvee_{(n)} T_n$ is a predictable time. Although $S = \bigwedge_{(n)} T_n$ is not a predictable time in general, it is a predictable time if $\bigcup_{(n)} \{S = T_n\} = \Omega$.

5.6 Localizing Procedure

Some readers who have already studied the martingale theory to some extent may have been bothered with the procedure to construct a new class of stochastic processes, which is named like "local ABC", where the name of a given class of stochastic processes is inserted to the part "ABC". The procedure is based on some stopping times in a rather abstract way. In this section, let us explain why such a procedure is necessary. We start with stating the definition of the procedure; for a given stochastic process X and a stopping time T, the notation for the stopped process

$$X^T = (X_t^T)_{t\in[0,\infty)}, \quad \text{where} \quad X_t^T := X_{t\wedge T}, \ \forall t \in [0,\infty),$$

will be often used in the subsequent part of this monograph.

Definition 5.6.1 (Localizing procedure) For a given class $\mathcal{C}$ of stochastic processes defined on a stochastic basis, the *localized class* of $\mathcal{C}$, which we denote by $\mathcal{C}_{\text{loc}}$, is defined as follows. A stochastic process X belongs to $\mathcal{C}_{\text{loc}}$ if there exists a non-decreasing sequence (T_n) of stopping times such that $T_n(\omega) \uparrow \infty$ as $n \to \infty$ for almost all ω and that every stopped processes X^{T_n} belongs to $\mathcal{C}$.

Here is a preliminary lemma concerning some measurability issues on stopped processes.

Lemma 5.6.2 Let a stochastic process $X = (X_t)_{t\in[0,\infty)}$ and a stopping time T defined on a filtered space be given.

(i) If X is an adapted process, and if T takes values in a countable set $\{t_i; i \in \mathbb{N}\} \subset [0,\infty]$, then X^T is adapted; in particular, if T is a deterministic time (like $T = n$) then X^T is adapted.

(ii) If X is an optional process, then the process X^T is also optional; in particular, X^T is adapted.

(iii) If X is a predictable process, then the process X^T is also predictable; in particular, X^T is optional and adapted.

Proof. To prove (i), for any $B \in \mathfrak{B}(\mathbb{R})$ and any $t \in [0,\infty)$, observe that

$$\begin{aligned}
\{X_t^T \in B\} &= \{X_{t\wedge T} \in B\} \\
&= (\{X_T \in B\} \cap \{T \le t\}) \cup (\{X_t \in B\} \cap \{T > t\}) \\
&= \left(\bigcup_{i: t_i \le t} \{X_{t_i} \in B\} \cap \{T = t_i\} \right) \cup (\{X_t \in B\} \cap \{T > t\}),
\end{aligned}$$

which belongs to $\mathcal{F}_t$, because all $\{T \le t\}$, $\{T < t\}$ and $\{T = t\}$ belong to $\mathcal{F}_t$ (see Exercise 5.5.1 (ii)). Thus we have proved that X^T is adapted.

See Propositions I.1.21 and I.2.4 of Jacod and Shiryaev (2003) for the proofs of (ii) and (iii), respectively. □

Exercise 5.6.1 Regarding (i) of Lemma 5.6.2, prove or disprove the following claim: "If X is an adapted process and if T is a (general) stopping time, then X^T is adapted." [**Commnet:** This claim is probably false, although the author does not know any counter example. However, it follows from (ii) of the lemma that: for any stopping time T and any càdlàg adapted process X, the stopped process X^T is adapted; this claim probably turns out to be false if we replace the "càdlàg" assumption on the adapted process X by the "a.s. càdlàg" one.]

Now, let us try to get a better understanding of the localizing procedure with the following illustrative example.

Example 5.6.3 According to Definition 5.4.4, a homogeneous Poisson process $N = (N_t)_{t\in[0,\infty)}$ with the intensity parameter λ is not integrable: $\mathsf{E}[N_\infty] = \infty$. So we would like to introduce a concept of "local integrability". Some readers might think that a natural definition is that

$$\mathsf{E}[N_t] < \infty, \quad \forall t \in [0,\infty).$$

In the case of a homogeneous Poisson process N, it indeed holds that $\mathsf{E}[N_t] = \lambda t < \infty$, hence this way of definition may look natural and reasonable at first sight. However, in some cases where the intensity λ is not a constant but a stochastic process $(\lambda_t)_{t\in[0,\infty)}$ (that is, in the case of counting process that we will consider in Example 5.8.7), it is not clear whether $\mathsf{E}[N_t] = \mathsf{E}\left[\int_0^t \lambda_s ds\right]$ is finite or not.

However, once the definition of "local ABC" is introduced as above, we can easily check that a number of non-decreasing processes X, including a homogeneous Poisson process N, are locally integrable by introducing the localizing sequence of stopping times (T_n) given by

$$T_n = \inf(t : X_t \ge n).$$

If X is a $[0,\infty)$-valued right-continuous adapted process with non-decreasing paths, and if $\Delta X \le a$ a.s. for a constant $a > 0$ (these are true for inhomogeneous Poisson processes), then it follows from Theorem 5.5.2 that T_n is actually a stopping time, and

$$X_\infty^{T_n} = X_{\infty\wedge T_n} \le \sup_{t\in[0,T_n]} X_t \le \sup_{t\in[0,T_n]} \{X_{t-} + \Delta X_t\} \le n + a, \quad \text{a.s.},$$

which implies that $\mathsf{E}[X_\infty^{T_n}] \le n + a < \infty$. Therefore, X is locally integrable in the sense of Definition 5.6.1.

In conclusion, we may think that the procedure has been invented to generalize the property that X_t satisfies the property "ABC" for every t to the one that $X_{t\wedge T_n}$ satisfies the property "ABC". It is clear that the latter is much weaker than the former.

5.7 Integrability of Martigales, Optional Sampling Theorem

The purpose of this section is to explain how to use the "optional sampling theorem", which is a tool for computing the values like $\mathsf{E}[X_T]$, where X is a martingale and T is a stopping time, under certain conditions on X and/or T. The theorem will be applied to prove the fundamental fact that if $t \rightsquigarrow X_t$ is a martingale and if T is a stopping time, then $t \rightsquigarrow X_{t\wedge T}$ is also a martingale (in many cases) at the end of this section.

First, let us introduce some definitions concerning special or more generalized classes of martingale, as well as the "Class (*D*)" of general stochastic processes. The Class (*D*) is a concept that is necessary to describe the Doob-Meyer decomposition theorem that will appear later.

Definition 5.7.1 Let a stochastic basis be given.

(i) A real-valued stochastic process X is said to be a *uniformly integrable martingale* if it is a martingale satisfying that

$$\lim_{K\to\infty} \sup_{t\in[0,\infty)} \mathsf{E}\left[|X_t| 1\{|X_t| > K\}\right] = 0.$$

The class of all uniformly integrable martingales is denoted by $\mathcal{M}$. The localized class of $\mathcal{M}$ is denoted by $\mathcal{M}_{\mathrm{loc}}$. Each element of $\mathcal{M}_{\mathrm{loc}}$ is said to be a *local martingale*[10].

(ii) A real-valued stochastic process X is a *square-integrable martingale* if it is a martingale satisfying that

$$\sup_{t\in[0,\infty)} \mathsf{E}[X_t^2] < \infty.$$

The class of all square-integrable martingales is denoted by $\mathcal{M}^2$. The localized class of $\mathcal{M}^2$ is denoted by $\mathcal{M}^2_{\mathrm{loc}}$. Each element of $\mathcal{M}^2_{\mathrm{loc}}$ is said to be a *locally square-integrable martingale*.

(iii) A real-valued stochastic process X is said to belong to *Class (D)* if the class $\{X_T; T \in \mathcal{T}\}$ of real-valued random variables, where $\mathcal{T}$ is the set of all finite stopping times, is uniformly integrable, that is,

$$\lim_{K\to\infty} \sup_{T\in\mathcal{T}} \mathsf{E}\left[|X_T| 1\{|X_T| > K\}\right] = 0.$$

Not small number of readers may have questions something like "Why do we have to introduce such definitions? While the definition (ii) looks natural, I cannot understand the purpose to introduce the concept (i) of uniformly integrable martingale". Here, let us state *an* answer to this question. When X is a martingale, of course it holds, by definition, that if $s \le t$ then

$$X_s = \mathsf{E}[X_t|\mathcal{F}_s] \quad \text{a.s.}$$

[10]One may think that the terminology "locally uniformly integrable martingale" might be better. The reason why the terminology "local martingales" is used for $\mathcal{M}_{\mathrm{loc}}$ is that the class $\mathcal{M}_{\mathrm{loc}}$ coincides with the localized class of all martingales, namely, $\{\text{martingales}\}_{\mathrm{loc}}$. This fact will be proved in Theorem 5.7.2 (iii).

In applications, we often encounter some situations where we would be happy if we could replace s,t above with any stopping times S,T such that $S \le T$ to have the formula

$$X_S = \mathsf{E}[X_T|\mathcal{F}_S] \quad \text{a.s.}$$

Actually, we will need this kind of formula in the proofs of martingale central limit theorems, Girsanov's theorems, and many others. Unfortunately, the latter formula is not always true if X is merely a martingale. However, it is true if X is a *uniformly integrable* martingale; this fact is a part of the "optional sampling theorem". Even only this reason would be sufficient to make us find the importance of the uniform integrability.

Now, let us state some claims concerning the relationships among the classes of stochastic processes introduced above. This course of explanation may not look logically of the right order at first sight, because some theorems that will appear later are needed to prove those claims. However, we dare to announce them here for the sake of convenience. Of course, no logical flaw will remain after all; the claims of Theorem 5.7.2 will be proved at the end of this section after all necessary tools are prepared.

Theorem 5.7.2 (i) $\mathcal{M}^2 \subset \mathcal{M}$. $\mathcal{M}^2_{\text{loc}} \subset \mathcal{M}_{\text{loc}}$.

(ii) $\mathcal{M} \subset$ Class (D).

(iii) $\mathcal{M} \subset \{\text{martingales}\} \subset \mathcal{M}_{\text{loc}}$. Thus the class $\mathcal{M}_{\text{loc}}$ coincides with the class obtained by localizing the class of all martingales.

(iv) $\mathcal{M} = \mathcal{M}_{\text{loc}} \cap$ Class (D).

(v) If X is an optional process such that $X \in \mathcal{M}_{\text{loc}}$ and that $|\Delta X| \le a$ for a constant $a \ge 0$, then $X \in \mathcal{M}^2_{\text{loc}}$. In particular, any càdlàg local martingale X such that $|\Delta X| \le a$ for a constant $a \ge 0$, including any continuous local martingale X, belongs to $\mathcal{M}^2_{\text{loc}}$.

As it was stated above, in practice we often encounter some situations where we would like to use the formula for a martingale X and some stopping times S,T such that $S \le T$ of the form

$$X_S = \mathsf{E}[X_T|\mathcal{F}_S] \quad \text{a.s.}$$

Although this equation is not always true, it is indeed true if X and S,T satisfy certain conditions. This type of identities are called the *optional sampling theorems*. First we present a version of the theorems which is easy to remember.

Theorem 5.7.3 (Optional sampling theorem, I) Let X be a right-continuous adapted process defined on a stochastic basis $\mathbf{B} = (\Omega, \mathcal{F}; (\mathcal{F}_t)_{t\in[0,\infty)}, \mathsf{P})$.

(i) When X is a submartingale, for any *bounded* stopping times S,T such that $S \le T$ it holds that X_T is integrable and

$$X_S \le \mathsf{E}[X_T|\mathcal{F}_S] \quad \text{a.s.}$$

(i') In particular, when X is a martingale, for any *bounded* stopping times S,T such that $S \le T$ it holds that X_T is integrable and

$$X_S = \mathsf{E}[X_T|\mathcal{F}_S] \quad \text{a.s.}$$

(ii) When X is a *uniformly integrable* martingale, there exists an integrable random variable X_∞ such that $X_\infty = \lim_{t\to\infty} X_t$ a.s. and that for any stopping times S,T such that $S \leq T$ it holds that

$$X_S = \mathsf{E}[X_T|\mathcal{F}_S] = \mathsf{E}[X_\infty|\mathcal{F}_S] \quad \text{a.s.}$$

(iii) When X is merely a local martingale, we cannot use such a convenient formula in general.

Remark. Under either (i') or (ii), it follows from the obtained formula that

$$\mathsf{E}[X_T] = \mathsf{E}[X_0].$$

To see this, just set $S = 0$ and take the expectations of the both sides.

Remark. See Theorem 7.29 of Kallenberg (2002) for a proof of (i). The assertion (i') is a special case of (i). In fact, assume that X is a martingale; then since both of X and $-X$ are submartingales, the assertion (i) yields that $X_S \leq \mathsf{E}[X_T|\mathcal{F}_S]$ a.s. and that $-X_S \leq \mathsf{E}[-X_T|\mathcal{F}_S]$ a.s., which implies $X_S = \mathsf{E}[X_T|\mathcal{F}_S]$ a.s.

Remark. The assertion (ii) above is a special case of Theorem 5.7.4 (ii). We also mention that (i') can be viewed also as a special case of (ii); this claim is almost immediate from the fact, which is worth remembering for itself, that *any martingale with the time parameter t varying only over a compact interval* $[0,c]$ *is a uniformly integrable martingale*. A more clear description of the latter fact is the following; when a $(\mathcal{G}_t)_{t\in[0,c]}$-martingale $(Y_t)_{t\in[0,c]}$ is given, if we formally set $Y_t := Y_c$ and $\mathcal{G}_t := \mathcal{G}_c$ for any $t \in (c,\infty)$, then it follows from "(a) $\Rightarrow$ (c)" of Theorem 5.7.4 that the extended stochastic process $(Y_t)_{t\in[0,\infty)}$ is a uniformly integrable martingale with respect to the extended filtration $(\mathcal{G}_t)_{t\in[0,\infty)}$.

Now let us present the theorem in a more detailed form.

Theorem 5.7.4 (Optional sampling theorem, II) Let a stochastic basis $\mathbf{B}$ be given.

(i) For a given martingale X, the following three conditions are equivalent.

(a) There exists an integrable random variable X_∞ such that $\lim_{t\to\infty} \mathsf{E}[|X_t - X_\infty|] = 0$.

(b) There exists an integrable random variable X_∞ such that $X_t = \mathsf{E}[X_\infty|\mathcal{F}_t]$ a.s. for any $t \in [0,\infty)$.

(c) $X \in \mathcal{M}$.

Under one (thus, all) of these conditions it also holds that $X_\infty = \lim_{t\to\infty} X_t$ a.s.

(ii) For any *right-continuous* uniformly integrable martingale X, it holds that X_T is integrable for any stopping time T, and that for any stopping times S,T such that $S \leq T$,

$$X_S = \mathsf{E}[X_T|\mathcal{F}_S] = \mathsf{E}[X_\infty|\mathcal{F}_S] \quad \text{a.s.}$$

See Theorems 7.21 and 7.29 of Kallenberg (2002) for the proofs of (i) and (ii), respectively.

Using the above theorem, we can provide a convenient criterion for the martingale property. Indeed, when we would like to show that a given stochastic process is a martingale, it is often wiser to use the following method than to check directly the conditions in the definition of martingale.

Theorem 5.7.5 (Criterion for martingale property) For any given càdlàg adapted process X, a necessary and sufficient condition for being a martingale is that for any bounded stopping time T, the random variable X_T is integrable and $\mathsf{E}[X_T] = \mathsf{E}[X_0]$ holds true.

Proof. The necessity is immediate from the optional sampling theorem.

To show the sufficiency, choose any $0 \le s < t$ and any $A \in \mathcal{F}_s$, and define $T = t1_{A^c} + s1_A$. Then, it is easy to see that T is a bounded, and thus it holds that

$$\mathsf{E}[X_0] = \mathsf{E}[X_T] = \mathsf{E}[X_t 1_{A^c}] + \mathsf{E}[X_s 1_A].$$

On the other hand, since t itself is also a bounded stopping time, it holds that

$$\mathsf{E}[X_0] = \mathsf{E}[X_t] = \mathsf{E}[X_t 1_{A^c}] + \mathsf{E}[X_t 1_A].$$

Comparing these we obtain $\mathsf{E}[X_s 1_A] = \mathsf{E}[X_t 1_A]$, which means $X_s = \mathsf{E}[X_t|\mathcal{F}_s]$ a.s. Therefore, X is a martingale. □

Using this criterion, we can prove that the stopped process of a martingale is also a martingale in many cases. Recall that: for a given adapted process X and a stopping time T, a sufficient condition for X^T to be adapted is that *all* paths of $t \rightsquigarrow X_t$ are right-continuous; another sufficient condition is that T is a deterministic time (like $T = n$). See Lemma 5.6.2 for more details.

Theorem 5.7.6 Let T be a stopping time, and let X be a right-continuous adapted process.

(i) If X is a martingale, then X^T is also a martingale.
(ii) $X \in \mathcal{M}$ implies $X^T \in \mathcal{M}$.
(iii) $X \in \mathcal{M}^2$ implies $X^T \in \mathcal{M}^2$.
(iv) $X \in \mathcal{M}_{\text{loc}}$ implies $X^T \in \mathcal{M}_{\text{loc}}$.
(v) $X \in \mathcal{M}^2_{\text{loc}}$ implies $X^T \in \mathcal{M}^2_{\text{loc}}$.

Proof. Since X is an optional process by the assumptions, for any stopping time T', the stopped process $X^{T'}$ is optional (and thus, adapted). Almost all paths of $X^{T'}$ are càdlàg under any of (i) – (v) of the current theorem.

To prove (i), choose a càdlàg modification $\widetilde{X}$ of X (see Theorem 5.1.3). If S is a bounded stopping time, then $S \wedge T$ is so, too. It follows from the necessity of Theorem 5.7.5 (or directly from the optional sampling theorem) that

$$\mathsf{E}[\widetilde{X}^T_S] = \mathsf{E}[\widetilde{X}_{S\wedge T}] = \mathsf{E}[\widetilde{X}_0].$$

So, using the sufficiency of Theorem 5.7.5 we have that $\widetilde{X}^T$ is a martingale. Since X is indistinguishable from $\widetilde{X}$, for any $0 \le s \le t < \infty$ it holds that

$$\begin{aligned} \mathsf{E}[X^T_t|\mathcal{F}_s] &= \mathsf{E}[\widetilde{X}^T_t|\mathcal{F}_s] \quad \text{a.s.} \\ &= \widetilde{X}^T_s \quad \text{a.s.} \\ &= X^T_s \quad \text{a.s.} \end{aligned}$$

The proof of (i) is finished.

The claims (ii) and (iii) are immediate from (i); indeed, in either case the "integrability" of X implies that of X^T. The claims (iv) and (v) are easy consequences from (ii) and (iii), respectively. □

Remark. Once the above results are presented, one may think that in order for X belonging to a localized class $\mathcal{C}_{\text{loc}}$ of some adapted processes $\mathcal{C}$, like the set of martingales, it must be almost always necessary to assume that X is a right-continuous adapted (or, optional) process, because, otherwise, X^T may not be even an adapted process. However, this worry is melancholy. For example, the statement "$X \in \mathcal{M}_{\text{loc}}$" *demands* that there exists a localizing sequence (T_n) of stopping times such that $X^{T_n} \in \mathcal{M}$, and the condition that X^{T_n} is an adapted process is *included* in this demand. As another example, notice also that the localizing sequence can sometimes be chosen as $T_n = n$, and in this case X^{T_n} is adapted for any adapted process X that may not be right-continuous.

We are now ready to prove the claims of Theorem 5.7.2 that was announced several pages before.

Proof of Theorem 5.7.2. (i) To prove the former, notice that

$$\sup_{t\in[0,\infty)} \mathsf{E}[|X_t|1\{|X_t|>K\}] \leq \sup_{t\in[0,\infty)} \mathsf{E}\left[\frac{(X_t)^2}{K}1\{|X_t|>K\}\right] \leq \frac{\sup_{t\in[0,\infty)}\mathsf{E}[(X_t)^2]}{K},$$

and let $K \to \infty$. To prove the latter, assume $X \in \mathcal{M}^2_{\text{loc}}$ and let (T_n) be a sequence of stopping times for the localization. Since $X^{T_n} \in \mathcal{M}^2 \subset \mathcal{M}$, the sequence (T_n) plays the role of a localizing sequence of stopping times to check that $X \in \mathcal{M}_{\text{loc}}$.

(ii) Let $\widetilde{X}$ be a càdlàg modification of X (see Theorem 5.1.3). It follows from Theorem 5.7.3 (ii) that there exists an integrable random variable $\widetilde{X}_\infty$ such that $\widetilde{X}_T = \mathsf{E}[\widetilde{X}_\infty|\mathcal{F}_T]$, a.s., for any stopping time T, and thus Lemma A1.2.1 implies that $\widetilde{X} \in$ Class (D). Since $\widetilde{X}$ is indistinguishable from X, we have $X \in$ Class (D).

(iii) The former inclusion is clear. On the other hand, if X is a martingale, then it follows from Theorem 5.7.6 (i) that X^{T_n} is a martingale for the stopping time $T_n = n$. So, using "(a) $\Rightarrow$ (c)" of Theorem 5.7.4 we have $X^{T_n} \in \mathcal{M}$, and therefore $X \in \mathcal{M}_{\text{loc}}$. Finally, noting the general fact that $\mathcal{C} \subset \mathcal{C}'$ implies $\mathcal{C}_{\text{loc}} \subset \mathcal{C}'_{\text{loc}}$, we have that $\mathcal{M}_{\text{loc}} \subset \{\text{martingales}\}_{\text{loc}}$. To prove that $\{\text{martingales}\}_{\text{loc}} \subset \mathcal{M}_{\text{loc}}$, for a given $X \in \{\text{martingales}\}_{\text{loc}}$, choose a localizing sequence (T_n) so that X^{T_n} is a martingale; then $X^{T_n \wedge n} \in \mathcal{M}$ by using Theorem 5.7.6 (i) and "(a) $\Rightarrow$ (c)" of Theorem 5.7.4, and thus $(T_n \wedge n)$ plays the role of a localizing sequence of stopping times to establish that $X \in \mathcal{M}_{\text{loc}}$.

(iv) "$\subset$" is clear from (ii) and (iii). To prove "$\supset$", choose any $X \in \mathcal{M}_{\text{loc}} \cap$ Class (D), and its càdlàg modification $\widetilde{X}$ (see Exercise 5.1.3). Let (T_n) be a localizing sequence for $\widetilde{X}$. Then, for any bounded stopping time T it holds that

$$\mathsf{E}[\widetilde{X}_{T\wedge T_n}] = \mathsf{E}[\widetilde{X}^{T_n}_T] = \mathsf{E}[\widetilde{X}^{T_n}_0] = \mathsf{E}[\widetilde{X}_0].$$

Since $\{\widetilde{X}_{T\wedge T_n}\}_{n\in\mathbb{N}}$ is uniformly integrable and $\widetilde{X}_{T\wedge T_n} \xrightarrow{a.s.} \widetilde{X}_T$ as $n\to\infty$, Theorem 2.3.12 yields that $\lim_{n\to\infty}\mathsf{E}[\widetilde{X}_{T\wedge T_n}] = \mathsf{E}[\widetilde{X}_T]$. Hence $\mathsf{E}[\widetilde{X}_T] = \mathsf{E}[\widetilde{X}_0]$ and thus $\widetilde{X}$ is a martingale by Theorem 5.7.5. Therefore, X is also a martingale, and it is uniformly integrable because $X \in \text{Class}(D)$. We have proved that $X \in \mathcal{M}$.

(v) For a given $X \in \mathcal{M}_{\text{loc}}$, choose a càdlàg modification $\widetilde{X}$ (see Exercise 5.1.3). For every $n \in \mathbb{N}$, define the stopping time $T_n = \inf(t : |\widetilde{X}_t| > n)$. Then it holds that

$$\sup_{t\in[0,\infty)} |\widetilde{X}_t^{T_n}| \leq \sup_{t\in[0,T_n]} \{|\widetilde{X}_{t-}| + |\Delta\widetilde{X}_t|\} \leq n+a.$$

Thus it holds that $\sup_{t\in[0,\infty)}\mathsf{E}[(\widetilde{X}_t^{T_n})^2] \leq (n+a)^2 < \infty$, and thus $\widetilde{X}^{T_n} \in \mathcal{M}^2$. Since we have assumed that X is optional, X^{T_n} is an adapted process. Moreover, since X is indistinguishable from $\widetilde{X}$, we may conclude that $X^{T_n} \in \mathcal{M}^2$; thus $X \in \mathcal{M}^2_{\text{loc}}$. □

Exercise 5.7.1 When X and T belong to one of the sub-classes of right-continuous local martingales and stopping times, respectively, listed in the following table, classify the claim $\mathsf{E}[X_T] = \mathsf{E}[X_0]$ into "always true, with both sides being finite values" or "not". Answer to this question by filling "Yes" or "No" in the table, where "stopping time" is abbreviated to "S.T.".

	(general) S.T.	finite S.T.	bounded S.T.	$T = t$
martingale				
$\mathcal{M}$				
$\mathcal{M}^2$				
$\mathcal{M}_{\text{loc}}$				
$\mathcal{M}^2_{\text{loc}}$				

5.8 Doob-Meyer Decomposition Theorem

In order to describe Itô's formula, it is necessary to define the "predictable quadratic co-variation" for locally square-integrable martingales. The Doob-Meyer decomposition theorem, together with Doob's inequality, provides the existence and the uniqueness of the predictable quadratic co-variation, as it will be shown in Theorem 5.9.1. Throughout this section, let a stochastic basis $(\Omega, \mathcal{F}; (\mathcal{F}_t)_{t\in[0,\infty)}, \mathsf{P})$ be given.

5.8.1 Doob's inequality

Theorem 5.8.1 (Doob's inequality) If $X \in \mathcal{M}^2$, then it holds that

$$\mathsf{E}\left[\sup_{t\in[0,\infty)} X_t^2\right] \leq 4 \sup_{t\in[0,\infty)} \mathsf{E}[X_t^2] = 4\mathsf{E}[X_\infty^2],$$

where X_∞ is the random variable appearing in Theorem 5.7.4 (i). Moreover, it holds that $\lim_{t\to\infty}\mathsf{E}[(X_t - X_\infty)^2] = 0$.

Notice that $\sup_t$ on the left-hand side is taken in the inside of the probability. The outstanding point in Doob's inequality is that such a quantity that is difficult to compute is bounded by the quantity (multiplied by a universal constant) that is easy to handle on the right-hand side.

Proof of Theorem 5.8.1. Noting also that X admits a càdlàg modification (see Theorem 5.1.3), the former inequality is essentially a special case of Theorem II.1.7 of Revuz and Yor (1999) which provides an L^p-inequality for any $p \geq 1$.

To prove the latter equality, first notice that $t \mapsto \mathsf{E}[X_t^2]$ is non-decreasing (this is easy to prove), and then replace "$\sup_{t\in[0,\infty)}$" with "$\lim_{t\to\infty}$" to deduce the claim from Lebesgue's convergence theorem with the help of the former inequality and the fact that $X_t \xrightarrow{\text{a.s.}} X_\infty$ which was given in Theorem 5.7.4 (i).

Since $\sup_{t\in[0,\infty)}(X_t - X_\infty)^2$ is integrable, the last claim follows from the Lebesgue's convergence theorem. □

5.8.2 Doob-Meyer decomposition theorem

Before reading this section, recall the definitions of "increasing process", "process with finite-variation", "integrable, increasing process" and "process with integrable-variation" given by Definitions 5.4.4 and 5.4.6. Notice especially that all these definitions require that all paths of the stochastic process under consideration are càdlàg, starting from zero at $t = 0$ and having finite-variation.

Lemma 5.8.2 (Properties of increasing processes) (i) Any increasing process is a process with finite-variation.

(ii) Suppose that X is a locally integrable, increasing process. Then X belongs to the class obtained by localizing the class of all submartingales belonging to Class (D); that is, $X \in (\{\text{submartingales}\} \cap \text{Class } (D))_{\text{loc}}$. In particular, X is a local submartingale.

(iii) If X is a predictable increasing process, then X is locally integrable and all of the conclusions for X in (ii) are true.

Proof. The claim (i) is evident.

To show (ii), by localization, it is enough to prove that any adapted, integrable, increasing process X is a submartingale belonging to Class (D). Being a submartingale is evident. On the other hand, denoting the set of all finite stopping times by $\mathcal{T}$ we have

$$\sup_{T\in\mathcal{T}} \mathsf{E}[|X_T|1\{|X_T| > K\}] \leq \sup_{T\in\mathcal{T}} \mathsf{E}[X_\infty 1\{X_\infty > K\}] = \mathsf{E}[X_\infty 1\{X_\infty > K\}],$$

and the right-hand side converges to zero as $K \to \infty$ because X_∞ is integrable. Thus X belongs to Class (D), and this implies the claim (ii).

To show (iii), for a given predictable, increasing process X, define $T_n = \inf(t : X \geq n)$. Then T_n is a predictable time (see Theorem 5.5.4 (i)), and therefore there exists an announcing sequence $(T_{n,m})_{m=1,2,\ldots}$ for T_n, that is, we have that $T_{n,m} < T_n$ a.s. on $\{T_n > 0\}$ and that $T_{n,m} \uparrow T_n$ a.s. (see Theorem 5.5.4 (ii)). Since $X_{T_{n,m}} \leq n$ a.s. on $\{T_n > 0\}$ and $X_{T_n} = X_0 = 0$ a.s. on $\{T_n = 0\}$, it holds that $\mathsf{E}[X_{T_{n,m} \wedge T_n}] \leq \mathsf{E}[X_{T_{n,m}} 1\{T_n > 0\}] + \mathsf{E}[X_{T_n} 1\{T_n = 0\}] \leq n$. Letting $m \to \infty$, we deduce from the monotone convergence theorem that $\mathsf{E}[X_{T_n}] \leq n$, which means that X^{T_n} is an integrable process. We therefore have proved that X is locally integrable. □

Theorem 5.8.3 (The Doob-Meyer decomposition) If X is a right-continuous submartingale belonging to Class (D), then there exists a unique (up to indistinguishability) predictable integrable increasing process A such that $X - A \in \mathcal{M}$. (Here, the phrase "unique up to indistinguishability" means that if A and A' are two processes satisfying the required properties then there exists a P-null set $N \in \mathcal{F}$ such that "$A_t(\omega) = A'_t(\omega)$ for all $t \in [0,\infty)$" holds for every $\omega \in \Omega \setminus N$.)

See Lemmas 25.7–25.11 of Kallenberg (2002) for a proof of the theorem. See also Section III.3 of Protter (2005) for some results under somewhat different settings.

Here, recall that an optional process is an adapted process, that an adapted process may not be an optional process in general, and that an adapted process whose *all* paths are right-continuous is an optional process.

Corollary 5.8.4 (The Doob-Meyer decomposition with localization) For a given right-continuous adapted process X, the following conditions (a) and (b) are equivalent:

(a) X belongs to the class obtained by localizing the class of all submartingales belonging to Class (D); that is, $X \in (\{\text{submartingales}\} \cap \text{Class } (D))_{\text{loc}}$;

(b) X admits a unique (up to indistinguishability) decomposition $X = A + M$, where A is a predictable increasing process and $M \in \mathcal{M}_{\text{loc}}$.

Moreover, under one (and thus both) of these conditions, the predictable increasing process A appearing in (b) is locally integrable.

Proof. To show that (a) $\Rightarrow$ (b), first let us introduce a sequence (T_n) of stopping times that makes each X^{T_n} to be a submartingale belonging to Class (D). We deduce from Theorem 5.8.3 that a unique decomposition $X^{T_n} = A^n + M^n$ exists, where A^n is an integrable, predictable increasing process such that $M^n \in \mathcal{M}$. The uniqueness implies $(A^{n+1})^{T_n} = A^n$ and $(M^{n+1})^{T_n} = M^n$. So, setting $A_t = \sum_{(n)}(A^n_{t\wedge T_n} - A^n_{t\wedge T_{n-1}})$ and $M_t = M_0 + \sum_{(n)}(M^n_{t\wedge T_n} - M^n_{t\wedge T_{n-1}})$ for every $t \in [0,\infty)$, where $T_0 := 0$, we construct a predictable increasing process $A = (A_t)_{t\in[0,\infty)}$ and a local martingale $M = (M_t)_{t\in[0,\infty)}$, satisfying that $X = A + M$. Therefore, A and M satisfy all of the requested properties, except for the uniqueness of decomposition.

It is clear that the pair (A,M) has been constructed uniquely, after a localizing sequence (T_n) was introduced at the beginning of the proof. Let (T'_n) be any other localizing sequence, and construct a unique pair (A',M') using this sequence. Now define $T''_n = T_n \wedge T'_n$, then the uniqueness implies that $(A_{t\wedge T''_n}, M_{t\wedge T''_n}) = (A'_{t\wedge T''_n}, M'_{t\wedge T''_n})$ for all $t \leq T''_n$, a.s. Let $n \to \infty$ to have that $(A,M) = (A',M')$ up to indistinguishability. The claim that (a) $\Rightarrow$ (b) has been established.

To show that (b) $\Rightarrow$ (a), first choose a càdlàg modification $\widetilde{M}$ of M, and put $X_t(\omega) = A_t(\omega) + \widetilde{M}_t(\omega)$ for all t, ω; then X is a right-continuous adapted process. Choose a localizing sequences (T_n) for A^{T_n} being an integrable, predictable increasing process (recall Lemma 5.8.2 (iii)), and (T_n') for $\widetilde{M}^{T_n'}$ being a uniformly integrable martingale, respectively. Next, define $T_n'' = T_n \wedge T_n'$. Then $A^{T_n''}$ and $\widetilde{M}^{T_n''}$ satisfy the same properties as A^{T_n} and $\widetilde{M}^{T_n'}$, respectively (since M itself is merely adapted, $M^{T_n''}$ may not be an adapted process). This (T_n'') plays the role of a localizing sequence for X corresponding to the condition (a). The claim that (b) $\Rightarrow$ (a) has been proved.

The last assertion of the corollary is immediate from Lemma 5.8.2 (iii). □

Roughly speaking, the Doob-Meyer decomposition theorem says that

$$\text{submartingale} = \text{predictable increasing "trend"} + \text{martingale}.$$

While decomposing a submartingale into an increasing trend plus a martingale is not unique in general, the key point of the theorem is that once we require the two properties of "predictability" and "having finite-variation" for the "trend" term, the decomposition becomes unique. This point would become more evident if we see the following corollary; recall the discussion in Section 5.4.3.

Corollary 5.8.5 If X is a local martingale starting from zero and if it is a predictable process with finite-variation, then $X_t = 0$ for all $t \in [0, \infty)$, a.s.

Proof. Due to Lemma 5.4.5 (ii), we can write $X = A - B$ where A and B are predictable increasing processes. With the help of Lemma 5.8.2 (iii) we apply Corollary 5.8.4 to deduce that there exists an increasing, predictable process A' such that $A - A' \in \mathcal{M}_{\text{loc}}$.

Now, observe that both A and B satisfy the conditions which are requested for A'. Since the existence of A' is unique, we have $A = B$, which implies $X = 0$. □

Example 5.8.6 (Locally integrable, increasing process) Let X be a locally integrable, increasing process. With the help of Lemma 5.8.2 (ii) we apply Corollary 5.8.4 to obtain that there exists a unique (up to indistinguishability) predictable increasing process X^p such that $X - X^p \in \mathcal{M}_{\text{loc}}$. This stochastic process X^p is called the *predictable compensator* for X, and it is always locally integrable by Lemma 5.8.2 (iii) or by the last claim of Corollary 5.8.4.

More generally, when a process X with locally integrable variation is given, decomposing it into the difference of two locally integrable, increasing processes and applying the above argument to each of the two processes, it is proved that there exists a unique (up to indistinguishability) predictable process X^p with finite-variation such that $X - X^p \in \mathcal{M}_{\text{loc}}$. This process X^p, which is proved to be locally integrable, is also called the *predictable compensator* for X.

Let X be a process with locally integrable-variation, and let T be a stopping time. Since $X - X^p$ is merely a local martingale, it is wrong to apply the optional sampling theorem to deduce $\mathsf{E}[(X - X^p)_T] = \mathsf{E}[(X - X^p)_0] = 0$ directly. However, the following claim is true.

Exercise 5.8.1 For a process X with locally integrable-variation and a stopping time T, if $\mathsf{E}[|X_T^p|] < \infty$ then $\mathsf{E}[X_T] = \mathsf{E}[X_T^p]$. Prove this claim.

Example 5.8.7 (Counting process) A stochastic process $N = (N_t)_{t\in[0,\infty)}$ is said to be a *counting process* if it is an $\mathbb{N}_0$-valued adapted process whose *all* paths $t \mapsto N_t(\omega)$ are càdlàg and non-decreasing, and satisfy that $N_0(\omega) = 0$ and $\Delta N_t(\omega) = N_t(\omega) - N_{t-1}(\omega) \leq 1$ for all $t \in [0,\infty)$. Any counting process N is locally integrable with the localizing sequence (T_n) of stopping times given by $T_n = \inf(t : N_t \geq n)$; recall Theorem 5.5.2 (ii) and note that $N_{T_n}(\omega) \leq n$ for all ω. So, by the argument in Example 5.8.6 there exists a unique (up to indistinguishability) predictable increasing process A such that $N - A \in \mathcal{M}_{\mathrm{loc}}$. This stochastic process A is called the *predictable compensator* for N, and it is always locally integrable. When A is absolutely continuous with respect to the Lebesgue measure almost surely, that is, when it can be written as

$$A_t(\omega) = \int_0^t \lambda_s(\omega)ds, \quad \forall t \in [0,\infty), \quad \text{for almost all } \omega,$$

the stochastic process $t \rightsquigarrow \lambda_t$ is called the *intensity process* for N.

5.9 Predictable Quadratic Co-Variations

Using the Doob-Meyer decomposition together with Jensen's and Doob's inequalities we can prove the existence and the uniqueness of the quadratic co-variation for locally square-integrable martingales.

Theorem 5.9.1 (Predictable quadratic co-variation) (i) If $X \in \mathcal{M}^2_{\mathrm{loc}}$, then there exists a unique (up to indistinguishability) predictable increasing process, denoted by $\langle X,X\rangle$ or $\langle X\rangle$, such that $X^2 - \langle X\rangle \in \mathcal{M}_{\mathrm{loc}}$. In particular, if $X \in \mathcal{M}^2$, then $\langle X\rangle$ is an integrable predictable increasing process, and $X^2 - \langle X\rangle \in \mathcal{M}$.

(ii) If $X,Y \in \mathcal{M}^2_{\mathrm{loc}}$, then there exists a unique (up to indistinguishability) predictable process with finite-variation, denoted by $\langle X,Y\rangle$, such that $XY - \langle X,Y\rangle \in \mathcal{M}_{\mathrm{loc}}$. Moreover, it holds that

$$\langle X,Y\rangle = \frac{1}{4}\{\langle X+Y\rangle - \langle X-Y\rangle\}.$$

In particular, if $X,Y \in \mathcal{M}^2$, then $\langle X,Y\rangle$ is a predictable process with integrable-variation, and $XY - \langle X,Y\rangle \in \mathcal{M}$.

We call $\langle X\rangle$ the *predictable quadratic variation* for X, and $\langle X,Y\rangle$ the *predictable quadratic co-variation* for X,Y.

Proof. If $X \in \mathcal{M}^2$, choose a càdlàg modification $\widetilde{X}$ of X. Then it follows from Proposition 5.3.1 that $\widetilde{X}^2$ is a right-continuous submartingale, and moreover, it is immediate from Doob's inequality that $\widetilde{X}^2$ belongs to Class (D) (check the latter fact as an Exercise 5.9.1). So we can apply Theorem 5.8.3 to show the existence and the uniqueness of $\langle\widetilde{X}\rangle$. Due to the uniqueness, we may safely denote this by $\langle X\rangle$, because it holds for any other modification $\widetilde{X}'$ of X that $\langle\widetilde{X}'\rangle$ is indistinguishable from $\langle\widetilde{X}\rangle$.

If $X \in \mathcal{M}^2_{\mathrm{loc}}$, choosing a càdlàg modification $\widetilde{X}$ of X, by localization we can prove that $\widetilde{X}^2$ is a right-continuous adapted process belonging to $(\{\text{submartingales}\} \cap \text{Class } (D))_{\mathrm{loc}}$. So the assertion follows from Corollary 5.8.4 instead of Theorem 5.8.3.

The claims in (ii) are immediate from the fact that $XY = \frac{1}{4}\{(X+Y)^2 - (X-Y)^2\}$ and (i). □

Exercise 5.9.1 Prove that $X \in \mathcal{M}^2$ implies $X^2 \in$ Class (D).

In order to state an intuitive explanation for the predictable quadratic co-variation, let first us recall the interpretation of the expectation and the conditional expectation of a real-valued, $\mathcal{F}$-measurable, integrable random variable X. When a sub-σ-field $\mathcal{G}$ of $\mathcal{F}$, namely,

$$\{\emptyset, \Omega\} \subset \mathcal{G} \subset \mathcal{F},$$

is given, we may have the following interpretation.

$\mathsf{E}[X]$	$\leftarrow$--	$\mathsf{E}[X\vert\mathcal{G}]$	$\leftarrow$--	X
$\{\emptyset, \Omega\}$-measurable		$\mathcal{G}$-measurable		$\mathcal{F}$-measurable

Here, the notation "$\widetilde{A} \leftarrow\text{--}\ A$" may be read as "$\widetilde{A}$ is an object obtained by reducing the information that A has"; recall the explanation in Section 2.2. In the case of a locally square-integrable martingale M starting from zero, we may have a similar interpretation concerning the variance function and the "conditional variance process", the latter of which has been formally named the *predictable quadratic co-variation*.

$t \mapsto \mathsf{E}[X_t^2]$	$\leftarrow$--	$t \leadsto \langle X\rangle_t$	$\leftarrow$--	$t \leadsto X_t^2$
deterministic function		predictable process with finite-variation		adapted process

When a locally square-integrable martingale X starting from zero is given, $X^2 - \langle X\rangle$ is merely a local martingale in general. It is therefore wrong to apply the optional sampling theorem to deduce that $\mathsf{E}[(X^2 - \langle X\rangle)_T] = \mathsf{E}[(X^2 - \langle X\rangle)_0] = 0$ for a stopping time T; this argument is wrong even if T is a bounded stopping time or a fixed time $T = t$. However, the following claims, which are similar to Exercise 5.8.1, hold true.

Exercise 5.9.2 Prove that, for any stopping time T, if $X \in \mathcal{M}^2_{\mathrm{loc}}$ and $\mathsf{E}[\langle X\rangle_T] < \infty$, then it holds that $\mathsf{E}[\sup_{t\in[0,T]} X_t^2] < \infty$.

It is wise to memorize the facts announced in Propositions 5.1.5 and 5.1.8 in the following way. The proofs for the claims in this example are left for readers as Exercise 5.9.3. See also Exercises 6.4.3 and 6.4.4.

Example 5.9.2 (i) Let W be a standard Wiener process, and introduce the filtration $\mathbf{F}^{W,\mathcal{H}}$ generated by W and the σ-field $\mathcal{H}$ which is independent of W. Then, W is an $\mathbf{F}^{W,\mathcal{H}}$-martingale as well as a locally squire-integrable $\mathbf{F}^{W,\mathcal{H}}$-martingale with the predictable quadratic variation given by

$$\langle W \rangle_t = t, \quad \forall t \in [0, \infty).$$

(ii-a) Let N be a homogeneous Poisson process with the intensity parameter $\lambda > 0$, and introduce the filtration $\mathbf{F}^{N,\mathcal{H}}$ generated by N and the σ-field $\mathcal{H}$ that is independent of N. Then, the stochastic process $X = (X_t)_{t \in [0,\infty)}$ defined by $X_t = N_t - \lambda t$ is an $\mathbf{F}^{N,\mathcal{H}}$-martingale as well as a locally square-integrable $\mathbf{F}^{N,\mathcal{H}}$-martingale with the predictable quadratic variation given by

$$\langle X \rangle_t = \lambda t, \quad \forall t \in [0, \infty).$$

(ii-b) Let N^k, $k = 1, ..., m$, be counting processes with respect to a common filtration, and denote the predictable compensator for N^k for by A^k; we do not assume even that A^k's are continuous processes. Then, the stochastic processes $N^k - A^k$'s are not only local martingales but also locally square-integrable martingales. Moreover, if N^k's have no simultaneous jump, then it holds that

$$\langle N^k - A^k, N^{k'} - A^{k'} \rangle = \begin{cases} A^k, & \text{if } k = k', \\ 0, & \text{otherwise.} \end{cases}$$

Exercise 5.9.3 Prove the claims in Example 5.9.2. [**Suggestion and Hint:** As for (i) and (ii-a), try to find some proofs by elementary computation going back to the definition of martingales, not using any high-level stochastic calculus. As for (ii-b), use the formula for integration by parts; see Definition 6.3.1 and (6.3).]

Exercise 5.9.4 Let T be a stopping time, and let X, Y be right-continuous adapted processes. Prove the following claims.

(i) If $X \in \mathcal{M}^2_{\text{loc}}$ then $\langle X^T \rangle_t = \langle X \rangle_{t \wedge T}$ for all $t \in [0, \infty)$, a.s.

(ii) If $X, Y \in \mathcal{M}^2_{\text{loc}}$ then $\langle X^T, Y^T \rangle_t = \langle X, Y \rangle_{t \wedge T}$ for all $t \in [0, \infty)$, a.s.

5.10 Decompositions of Local Martingales

In the general theory of Hilbert spaces, when a closed subset H_0 of a Hilbert space $\mathbb{H}$ with the inner product $(\cdot, \cdot)_{\mathbb{H}}$ is given, its orthogonal complement $H_0^{\perp} := \{h : (h, h_0)_{\mathbb{H}} = 0, \forall h_0 \in H_0\}$ is also a closed subspace of $\mathbb{H}$, and moreover, any $h \in \mathbb{H}$ admits a unique decomposition $h = h_0 + h_1$ where $h_0 \in H_0$ and $h_1 \in H_0^{\perp}$.

The goal of this section is to prove a deep result that *any local martingale admits a unique (up to indistinguishability) decomposition into the sum of a continuous local martingale and a "purely discontinuous" local martingale*. This result is based on the following facts: the space of local martingale is (not exactly, but) something like a "Hilbert space"; the space of continuous local martingales is something like a "closed subspace" of the space of local martingales; the space of "purely discontinuous" local martingales is something like an "orthogonal complement" of the space of continuous local martingales.

Let us first introduce some notations and the definition of "purely discontinuous" local martingales.

Definition 5.10.1 (i) $\mathcal{L}$ denotes the class of all $M \in \mathcal{M}_{\text{loc}}$ such that $M_0 = 0$ a.s.

(ii) $\mathcal{L}^c$ denotes the class of all $M \in \mathcal{L}$ which are continuous.

(iii) $\mathcal{L}^d$ denotes the class of all $M \in \mathcal{L}$ that is purely discontinuous. Here, an element $M \in \mathcal{L}$ is said to be *purely discontinuous* if M is *orthogonal* to any $N \in \mathcal{L}^c$, in the sense that MN belongs to $\mathcal{L}$ for any $N \in \mathcal{L}^c$.

Note that $\mathcal{L}^c \subset \mathcal{M}^2_{\text{loc}} \subset \mathcal{M}_{\text{loc}}$; see Theorem 5.7.2 (v). On the other hand, note that $\mathcal{L}^d \subset \mathcal{M}_{\text{loc}}$ by definition, but $\mathcal{L}^d \not\subset \mathcal{M}^2_{\text{loc}}$.

Here are two well-known facts concerning (a.s.) continuous local martingales and purely discontinuous local martingales.

Lemma 5.10.2 (i_1) For a given local martingale, if it is a predictable process, then its almost all paths are continuous. The converse is true if the stochastic basis is complete.

(i_2) It is true that $\mathcal{L} \cap \{\text{continuous processes}\} = \mathcal{L}^c$ by definition, but it is *not* always true that "$\mathcal{L} \cap \{\text{predictable processes}\} = \mathcal{L}^c$" even when the stochastic basis is complete.

(ii) Any local martingale M with finite-variation such that $M_0 = 0$ a.s. is purely discontinuous. The converse is not true: there are many examples of purely discontinuous local martingales that do not have finite-variation.

Proof. See Proposition 25.16 of Kallenberg (2002) for a proof of (i_1). The claim (i_2) is due to the fact that we demand that *all* paths of an element of $\mathcal{L}^c$ are continuous in the definition. See Lemma I.4.14 of Jacod and Shiryaev (2003) for a proof of (ii). □

Now, let us proceed with our discussion on the decomposition problem. Our starting point is the following.

Lemma 5.10.3 If $M \in \mathcal{L}^c \cap \mathcal{L}^d$, then $M_t = 0$ for all $t \in [0,\infty)$, a.s.

Proof. Since $M \in \mathcal{M}^2_{\text{loc}}$, it has a (unique) quadratic predictable variation $\langle M \rangle$. On the other hand, M, which is an element of $\mathcal{L}^d$, has to be orthogonal to itself because M is continuous by assumption. Hence $M^2 \in \mathcal{L}$, and this implies that $\langle M \rangle = 0$. It then follow from Exercise 5.9.2 (iii) that $M = M_0 = 0$ a.s. □

The following lemma will provide the core part of our argument for the most general result, Theorem 5.10.6.

Lemma 5.10.4 (Direct sum decomposition of $\mathcal{M}^2_{\text{loc}}$) Any $M \in \mathcal{M}^2_{\text{loc}}$ admits a unique (up to indistinguishability) decomposition $M = M_0 + M^c + M^d$, where
(i) $M^c \in \mathcal{L}^c \cap \mathcal{M}^2_{\text{loc}}$ (actually, $\mathcal{L}^c \subset \mathcal{M}^2_{\text{loc}}$), and
(ii) $M^d \in \mathcal{L}^d \cap \mathcal{M}^2_{\text{loc}}$.
Moreover, it holds that

$$\langle M \rangle = \langle M^c + M^d \rangle = \langle M^c \rangle + 2\langle M^c, M^d \rangle + \langle M^d \rangle = \langle M^c \rangle + \langle M^d \rangle.$$

The above lemma can be proved based on the facts that $\mathcal{M}^2$ is a Hilbert space if we equip it with the inner product $(M,N)_{\mathcal{M}^2} := E[M_\infty N_\infty]$, where M_∞, N_∞ are random variables corresponding to $M, N \in \mathcal{M}^2$, appearing in Theorem 5.7.4, and that the set of all continuous elements of $\mathcal{M}^2$ is a closed subset of $\mathcal{M}^2$, using also some localization procedure. See Page 39 of Jacod and Shiryaev (2003) for the details of these facts; among them, the fact that $\mathcal{M}^2$ is complete will be proved in Lemma 6.2.2 (i) below.

We prepare another lemma that gives a preliminary decomposition in the proof of Theorem 5.10.6, and it is useful not only here but also for the construction of stochastic integrals with respect to semimartingales in Subsection 6.2.3, and for the proofs of Rebolledo's central limit theorems (Theorems 7.1.8 and 7.2.2).

Lemma 5.10.5 (A non-unique decomposition of $\mathcal{M}_{\text{loc}}$) For a given constant $a > 0$, any $M \in \mathcal{M}_{\text{loc}}$ admits a (not unique, actually depending on the constant a) decomposition $M = M_0 + M' + M''$, where
(i) $M' \in \mathcal{L}$, and M' is a process with finite-variation, and
(ii) $M'' \in \mathcal{L}$ and $|\Delta M''| \leq a$ (and thus, $M'' \in \mathcal{M}^2_{\text{loc}}$).

This lemma is proved in a constructive way as follows: first define a local martingale $M' := A - A^p$ that contains "all big jumps", where $A := \sum_{s \leq \cdot} \Delta M_s 1\{|\Delta M_s| > a/2\}$ and A^p is the predictable compensator of A, and then define a local martingale $M'' = M - M'$ that has "only small jumps". Although it may not be so evident, for example, that $|\Delta M''| \leq a$, it can be eventually proved that all of the required properties for M' and M'' are satisfied; see Proposition I.4.17 of Jacod and Shiryaev (2003) for the details.

Theorem 5.10.6 (Canonical decomposition of $\mathcal{M}_{\text{loc}}$) Any $M \in \mathcal{M}_{\text{loc}}$ admits a unique (up to indistinguishability) decomposition $M = M_0 + M^c + M^d$, where $M^c \in \mathcal{L}^c$ and $M^d \in \mathcal{L}^d$.

Proof. To show the uniqueness, choose two decompositions $M = M_0 + M^{c,1} + M^{d,1} = M_0 + M^{c,2} + M^{d,2}$, and observe that $M^{c,1} - M^{c,2} = M^{d,2} - M^{d,1}$. Apply Lemma 5.10.3 to show that both sides are zero almost surely.

We cannot directly apply Lemma 5.10.4 to show the existence, because M may not belong to $\mathcal{M}^2_{\text{loc}}$. Thus, first apply Lemma 5.10.5 to have $M = M_0 + M' + M''$, and then apply Lemma 5.10.4 to decompose M'', which has a càdlàg modification belonging to $\mathcal{M}^2_{\text{loc}}$, into $M'' = M'''^{,c} + M'''^{,d}$, a.s. Then, setting $M^c := M'''^{,c}$ and $M^d := M' + M'''^{,d}$ we obtain a desired decomposition. □

6

Tools of Semimartingales

The chapter starts with defining stochastic integrals and investigating their properties. We first give the definition of the stochastic integral of a predictable process with respect to a locally square-integrable martingale. A formula for the quadratic variation of stochastic integral is also provided. This formula will be frequently used as a fundamental tool in the latter part of this monograph to study the statistical analysis of stochastic processes.

In the middle part of this chapter, an intuitive explanation of Itô's formula will be given. A (special) semimartingale is a stochastic process that can be written in the form of the sum of a predictable process with finite-variation and a local martingale. The main reason why semimartingales are so important is that they are given in the additive form of two good processes. On the other hand, some stochastic processes appearing in applications are not of an additive form, and the treatment for such processes may look difficult at first sight. Itô's formula is a powerful tool to transform a smooth functional of semimartingales into an additive form that is easy to analyze.

The chapter finishes with presenting Girsanov's theorem, which provides the likelihood ratio processes for semimartingales.

6.1 Semimartingales

Let us introduce the notion of "semimartingale"; readers will soon find that it is an important class of stochastic processes as integrators of stochastic integrals through our discussion in Section 6.2.

Definition 6.1.1 (Semimartingale) (i) Let a stochastic basis $(\Omega, \mathcal{F}; (\mathcal{F}_t)_{t\in[0,\infty)}, \mathsf{P})$ be given. A stochastic process X is said to be a (real-valued) *semimartingale* if it is a càdlàg process of the form

$$X_t = X_0 + A_t + M_t, \quad \forall t \in [0,\infty), \tag{6.1}$$

where X_0 is an $\mathcal{F}_0$-measurable random variable, A is an adapted process with finite-variation and $M \in \mathcal{L}$; this decomposition is *not* unique.

(ii) X is said to be a (real-valued) *special semimartingale* if it is a semimartingale in the sense of (i) where A is *predictable*; the decomposition (6.1) is unique (up to indistinguishability). Furthermore, a special semimartingale X can be expressed by a unique (up to indistinguishability) form

$$X_t = X_0 + A_t + M_t^c + M_t^d, \quad \forall t \in [0,\infty),$$

DOI: 10.1201/9781315117768-6

where X_0 is an $\mathcal{F}_0$-measurable random variable, A is a predictable process with finite-variation, $M^c \in \mathcal{L}^c$ and $M^d \in \mathcal{L}^d$; we have used the unique decomposition of a local martingale given by Theorem 5.10.6, namely, $M = M^c + M^d$. This decomposition is called the *canonical decomposition* of a special semimartingale.

Remark. It is required that *all* paths of a semimartingale are càdlàg. When the objects on the right-hand side of (6.1) are given, the second term $t \rightsquigarrow A_t$ is required to be càdlàg by the definition of a "process with finite-variation", while the last term $t \rightsquigarrow M_t$, a local martingale, is required to have càdlàg paths only almost surely. However, due to Theorem 5.1.3 with localization (Exercise 5.1.3) it is possible to find a càdlàg local martingale $t \rightsquigarrow M_t$ to make $t \rightsquigarrow X_t$ defined by (6.1) to be càdlàg. One of the reasons why we emphasize this demand for *all* paths of a semimartingale to be càdlàg is that the integrands of some stochastic integrals appearing in, e.g., the formula for the integration by parts (Definition 6.3.1) and Itô's formula (Theorem 6.4.1), have to be predictable processes; if the paths of the process $t \rightsquigarrow X_t$ were càdlàg merely almost surely, then the process $t \rightsquigarrow X_{t-}$ would not be a predictable process in general[1].

Remark. The "canonical decomposition" of a special semimartingale should not be confused with the "canonical representation" of a (not necessarily special) semimartingale, given by using Lemma 5.10.5, which is unique only after a "truncation function for jumps" is introduced; see Section II.2c of Jacod and Shiryaev (2003).

Example 6.1.2 (i) A counting process N with the predictable compensator A is a special semimartingale. In fact, it holds that $N = A + M$, where A is a predictable increasing process and $M = N - A \in \mathcal{L}^d \subset \mathcal{L}$.

(ii) Let β, σ be measurable functions on $\mathbb{R}$ that satisfy some appropriate properties. Then, for a given standard Wiener process W and an initial value X_0, there exists a special semimartingale X with respect to the filtration $\mathbf{F}^{W,X_0}$ generated by W and X_0 such that

$$X_t = X_0 + \int_0^t \beta(X_s)ds + \int_0^t \sigma(X_s)dW_s, \quad \forall t \in [0, \infty),$$

where the definition of the last term, namely, the Itô integral process, will be given in the next section. The function $\beta(\cdot)$ is called a *drift coefficient* and $\sigma(\cdot)$ is called a *diffusion coefficient*. We will study this example in more detail in Subsection 6.5.3.

[1] Another possible way to avoid this problem is to assume that the stochastic basis is complete.

6.2 Stochastic Integrals

6.2.1 Starting point of constructing stochastic integrals

In this monograph, we treat three types of stochastic integrals of the form

$$\int_0^t H_s dX_s, \quad \forall t \in [0,\infty),$$

where H is a predictable (sometimes, more generally, optional) process that satisfies certain properties. The properties required for H depend on the stochastic process X. All stochastic integrals appearing in this monograph belong to either of the following three cases.

Case I. $X = A$ is a process with finite-variation; in this case, H is an optional (or predictable) process such that the value $\int_0^t H_s(\omega)dA_s(\omega)$ defined as the usual Lebesgue-Stieltjes integral for every $t \in [0,\infty)$ is finite for all ω's. This case has been already studied in Theorem 5.4.7, and we refer it as the *Stieltjes integral process*.

Case II. $X = M$ is a locally square-integrable martingale; in this case, H can be a predictable process such that the process $\int_0^\cdot H_s^2 d\langle M\rangle_s$, defined as the Stieltjes integral process, is locally integrable. We refer this stochastic integral as the *Itô integral process*, named after Kiyosi Itô who created the most important part of stochastic integrals.

Case III. X is a (general) semimartingale; in this case, H should be a predictable process that is locally bounded[2], and this requirement is more restrictive than that for H in Case II.

The starting point of our construction in both of the Cases II and III is to introduce a class $\mathcal{E}$ of "simple" predictable processes H and to give a "natural" definition of stochastic integrals in such a case.

Definition 6.2.1 (Stochastic integral of simple process) Let a measurable space $(\Omega,\mathcal{F})$ with a filtration $(\mathcal{F}_t)_{t\in[0,\infty)}$ be given. A *simple process* is a stochastic process H that has the following form: there exists $n \in \mathbb{N}$, time points $0 = t_0 < t_1 < \cdots < t_n < t_{n+1} = \infty$, and bounded $\mathcal{F}_{t_k}$-measurable random variables Y_k, $k = 0,1,...,n$, such that

$$H_t = \begin{cases} Y_0, & t = 0, \\ Y_k, & t \in (t_{k-1}, t_k],\ k = 1,2,..,n, \\ Y_n, & t \in (t_n,\infty). \end{cases}$$

Note that any simple process is predictable. We denote by $\mathcal{E}$ the class of all simple

[2] A stochastic process $X = (X_t)_{t\in[0,\infty)}$ is said to be *bounded* if there exists a constant $K > 0$ such that $\sup_{t\in[0,\infty)} |X_t(\omega)| \le K$ for all $\omega \in \Omega$. A stochastic process is said to be *locally bounded* if it belongs to a localized class of all bounded processes. Any left-continuous process is locally bounded.

processes. For a given $H \in \mathcal{E}$ and a given semimartingale X, define a new semimartingale $(\int_0^t H_s dX_s)_{t\in[0,\infty)}$ by

$$\int_0^t H_s dX_s := \sum_{k=1}^{n+1} Y_{k-1}(X_{t\wedge t_k} - X_{t_{k-1}})1\{t_{k-1} < t\}, \quad \forall t \in [0,\infty).$$

In order to build up rich classes of stochastic integrals, we will extend the class $\mathcal{E}$ of predictable processes H's to more general classes of predictable processes, corresponding to the Cases II and III in Subsections 6.2.2 and 6.2.3, respectively.

6.2.2 Stochastic integral w.r.t. locally square-integrable martingale

In this section, we consider the case where $X = M$ is a locally square-integrable martingale. We denote by $L^2(M)$ the class of predictable processes H such that $t \leadsto \int_0^t H_s^2 d\langle M\rangle_s$, which is defined as the Stieltjes integral process, is integrable. The localized class of $L^2(M)$ is denoted by $L^2_{\text{loc}}(M)$.

Now, we prepare two lemmas.

Lemma 6.2.2 (i) $\mathcal{M}^2$ is complete with respect to the metric defined by the L_2-type norm $||M|| := \sqrt{\mathsf{E}[M_\infty^2]}$, where $M_\infty = \lim_{t\to\infty} M_t$ a.s.; see Theorem 5.7.4.

(ii) For every $M \in \mathcal{M}^2$, the class $\mathcal{E}$ introduced in Definition 6.2.1 is dense in $L^2(M)$ with respect to the metric defined by the L^2-type norm $|H| := \sqrt{\mathsf{E}[\int_0^\infty H_s^2 d\langle M\rangle_s]}$.

Proof. (i) Fix any Cauchy sequence (M^n) in $\mathcal{M}^2$. The sequence (M_∞^n) is then Cauchy in $L^2 = L^2(\Omega, \mathcal{F}_\infty, \mathsf{P})$, where $\mathcal{F}_\infty = \bigvee_{t\in[0,\infty)} \mathcal{F}_t$, and thus converges to an element $\xi \in L^2$. Define $M \in \mathcal{M}^2$ by $M_t = \mathsf{E}[\xi|\mathcal{F}_t]$ for every $t \in [0,\infty)$, and note that $M_\infty = \xi$ a.s. because ξ is $\mathcal{F}_\infty$-measurable. Hence

$$||M^n - M|| = \sqrt{\mathsf{E}[(M_\infty^n - M_\infty)^2]} = \sqrt{\mathsf{E}[(M_\infty^n - \xi)^2]} \to 0.$$

(ii) See Lemma II.1.1 and Proposition I.5.1 of Ikeda and Watanabe (1989). □

Lemma 6.2.3 For every $M \in \mathcal{M}^2$ and every $H \in \mathcal{E}$, the stochastic process $X = (X_t)_{t\in[0,\infty)}$, where $X_t = \int_0^t H_s dM_s$ is constructed by the method in Definition 6.2.1 is a square-integrable martingale with the predictable quadratic variation $\langle X\rangle_t = \int_0^t H_s^2 d\langle M\rangle_s$.

Proof. Do the same computation as that in the proof of Theorem 4.1.3 for discrete-time martingales. (Not difficult! Readers are strongly advised to do this computation as an exercise.) □

Using these lemmas, we shall define the stochastic integral $X = (X_t)_{t\in[0,\infty)}$, where $X_t = \int_0^t H_s dM_s$, with respect to a given $M \in \mathcal{M}^2$, not only for $H \in \mathcal{E}$ but also for $H \in L^2(M)$.

Let $H \in L^2(M)$ be given. First, due to (iii) of Lemma 6.2.2, we can choose a sequence (H^n) of elements of $\mathcal{E}$ such that $|H^n - H| \to 0$. Next, define $X^n := \int_0^t H_s^n dM_s$ as in Definition 6.2.1. Then, it follows from Lemma 6.2.3 that $||X^n - X^m||^2 = \mathsf{E}[\int_0^\infty (H_s^n - H_s^m)^2 d\langle M\rangle_s] = |H^n - H^m|^2$. Since it holds that $|H^n - H^m| \leq |H^n - H| + |H - H^m| \to 0$, we have $||X^n - X^m|| \to 0$ and thus (X^n) is a Cauchy sequence in $\mathcal{M}^2$. Due to the fact that $\mathcal{M}^2$ is complete (Lemma 6.2.2 (i)), there exists a "unique" limit X in $\mathcal{M}^2$ (up to indistinguishability). Let us show that this "unique" limit X, which may look depending on the choice of the sequence (H^n) at first sight, is indeed unique and well-defined. If X and $\widetilde{X}$ are the limits constructed by (H^n) and $(\widetilde{H}^n)$, respectively, then it holds that $||X - \widetilde{X}|| = \lim_{(n)} ||X^n - \widetilde{X}^n|| = \lim_{(n)} |H^n - \widetilde{H}^n| \leq \lim_{(n)}\{|H^n - H| + |\widetilde{H}^n - H|\} = 0$.

Finally, let us discuss the localization for this construction. Thanks to Doob's regularization, when $M \in \mathcal{M}^2_{\text{loc}}$ is given, we may assume that M is a càdlàg process. Introduce a localizing sequence (T_n) which makes $M^{T_n} \in \mathcal{M}^2$. Construct $X_t^n = \int_0^t H_s dM_s^{T_n}$ for every $n \in \mathbb{N}$, and define $X_t := \sum_{(n)}(X^n_{t\wedge T_n} - X^n_{t\wedge T_{n-1}})$ for every $t \in [0,\infty)$. Then, $X = (X_t)_{t\in[0,\infty)}$ is a locally square-integrable martingale with the localizing sequence (T_n). To see that this construction does not depend on the choice of localizing sequence, take another localizing sequence (T_n'), and construct $X' \in \mathcal{M}^2_{\text{loc}}$ as above. Then, it holds that $X_{t\wedge T_n''} = X'_{t\wedge T_n''}$ for all $t \leq T_n'' := T_n \wedge T_n'$, and thus for all $t \in [0,\infty)$, almost surely. Let $n \to \infty$ to conclude that $X = X'$ up to indistinguishability.

Theorem 6.2.4 (The Itô integral process) For any $M \in \mathcal{M}^2_{\text{loc}}$ and any $H \in L^2_{\text{loc}}(M)$, the above procedure defines a unique (up to indistinguishability) locally square-integrable martingale $(\int_0^t H_s dM_s)_{t\in[0,\infty)}$ satisfying the following properties.

(i) It starts from zero at $t = 0$, a.s.

(ii) $\int_0^\cdot H_s dM_s \in \mathcal{M}^2$ if and only if $H \in L^2(M)$.

(iii) $\int_0^t (aH_s + bK_s)dM_s = a\int_0^t H_s dM_s + b\int_0^t K_s dM_s$, for all $t \in [0,\infty)$, a.s., for any $K \in L^2_{\text{loc}}(M)$ and any constants $a, b \in \mathbb{R}$.

(iv) If M is continuous a.s., then $\int_0^\cdot H_s dM_s$ is also continuous a.s.

(v) If M is a process with finite-variation, then $\int_0^\cdot H_s dM_s$ is also a process with finite-variation and it coincides almost surely with the Stieltjes integral process defined by fixing each $\omega \in \Omega$.

(vi) $\Delta(\int_0^\cdot H_s dM_s)_t = H_t \Delta M_t$ for all $t \in [0,\infty)$, a.s.

(vii) $\langle \int_0^\cdot H_s dM, \int_0^\cdot K dN\rangle_t = \int_0^t H_s K_s d\langle M,N\rangle_s$ for all $t \in [0,\infty)$, a.s., for any $N \in \mathcal{M}^2_{\text{loc}}$ and any $K \in L^2_{\text{loc}}(N)$.

Most of these properties are clear from the construction; among them, the important property (vii) is proved by recalling Lemma 6.2.3 in the construction, and its roots exists in the computation that we presented as Theorem 4.1.3 for discrete time martingales. The proofs for the properties that are not so evident are found in standard books, including Jacod and Shiryaev (2003), Kallenberg (2002), Revuz and Yor (1999). In the above construction, we followed the same line as that of the fundamental paper of Kunita and Watanabe (1967), which inherits the original idea of Kiyosi Itô; see the authoritative book of Ikeda and Watanabe (1989) for more details.

Example 6.2.5 Let $\{H^i\}_{i=1,...,d}$ be predictable processes (satisfying some integrability conditions) on a stochastic basis.

(i) Let W be a standard Wiener process. Supposing that W is a locally square-integrable martingale with respect to the given filtration, we consider the Itô integral processes

$$X_t^i = \int_0^t H_s^i dW_s, \quad \forall t \in [0,\infty), \quad i = 1,...,d.$$

If we regard $M = W$, then by Example 5.9.2 (i) we have

$$\langle M \rangle_t = \langle W \rangle_t = t = \int_0^t ds, \quad \text{thus} \quad d\langle M \rangle_s = ds.$$

Hence it holds for every $(i,j) \in \{1,...,d\}^2$ that

$$\langle X^i, X^j \rangle_t = \int_0^t H_s^i H_s^j d\langle M \rangle_s = \int_0^t H_s^i H_s^j ds, \quad \forall t \in [0,\infty).$$

(ii) Let N^k, $k = 1,...,m$, be counting processes that have no simultaneous jumps, and suppose that each N^k admits an intensity process λ^k. Let us consider

$$X_t^i = \sum_{k=1}^m \int_0^t H_s^i (dN_s^k - \lambda_s^k ds), \quad \forall t \in [0,\infty), \quad i = 1,...,d.$$

If we regard $M^k = N^k - \int_0^\cdot \lambda_s^k ds$, then by Example 5.9.2 (ii-b) we have

$$\langle M^k, M^{k'} \rangle_t = \begin{cases} \int_0^t \lambda_s ds, & \text{if } k = k', \\ 0, & \text{otherwise.} \end{cases}$$

Hence it holds for every $(i,j) \in \{1,...,d\}^2$ that

$$\langle X^i, X^j \rangle_t = \sum_{k=1}^m \sum_{k'=1}^m \int_0^t H_s^i H_s^j d\langle M^k, M^{k'} \rangle_s = \sum_{k=1}^m \int_0^t H_s^i H_s^j \lambda_s^k ds, \quad \forall t \in [0,\infty).$$

6.2.3 Stochastic integral w.r.t. semimartingale

Let a semimartingale X and a locally bounded measurable function H are given. In order to construct the stochastic integrable "$\int_0^\cdot H_s dX_s$" in Case III, first decompose X into

$$\begin{aligned} X_t &= X_0 + A_t + M_t \\ &= X_0 + A_t + M_t' + M_t'', \quad \forall t \in [0,\infty), \end{aligned} \tag{6.2}$$

where the decomposition $M = M' + M''$ is the one given in Lemma 5.10.5 for any fixed constant $a > 0$. Based on this representation, we define the stochastic integrals of H with respect to the second, third, and fourth terms on the right-hand side, namely,

$$\int_0^\cdot H_s dA_s, \quad \int_0^\cdot H_s dM_s' \quad \text{and} \quad \int_0^\cdot H_s dM_s'',$$

by the methods of Cases I, I, and II, respectively.

Let us discuss this construction in details. Since A and M' are processes with finite-variation, the first two stochastic integrals above are well-defined uniquely, because the Lebesgue-Stieltjes integral $\int_0^t H_s(\omega)dY_s(\omega)$ of a locally bounded, predictable process H, for all ω's, is actually finite for *any* process Y with finite-variation; in order to define $\int_0^t H_s dA_s$ and $\int_0^t H_s dM_s'$ as finite values in any case (i.e., no matter what A and M' are), we have been content with restricting H to be a process that is locally bounded. On the other hand, since M'' is a locally square-integrable martingale, the third stochastic integral $\int_0^\cdot H_s dM_s''$ can be defined uniquely, too. However, the constructions of the second and third stochastic integrals depend on the choice of the decomposition $M = M' + M''$ in (6.2), and this might give readers an impression that the construction might be depending on the choice of the constant $a > 0$ in Lemma 5.10.5. To see this impression is a needless fear, consider two possible decompositions $M = M' + M'' = \widetilde{M}' + \widetilde{M}''$ and observe that

$$\begin{aligned}
\int_0^\cdot H_s dM_s' + \int_0^\cdot H_s dM_s'' &= \int_0^\cdot H_s d(\widetilde{M}' - M'' + \widetilde{M}'')_s + \int_0^\cdot H_s dM_s'' \\
&= \int_0^\cdot H_s d\widetilde{M}' - \int_0^\cdot H_s d(M'' - \widetilde{M}'')_s + \int_0^\cdot H_s dM_s'' \\
&= \int_0^\cdot H_s d\widetilde{M}' - \int_0^\cdot H_s dM'' + \int_0^\cdot H_s d\widetilde{M}'' + \int_0^\cdot H_s dM_s'' \\
&= \int_0^\cdot H_s d\widetilde{M}' + \int_0^\cdot H_s d\widetilde{M}'',
\end{aligned}$$

where the second and third equalities are due to the uniqueness of the Stieltjes integral process and the Itô integral process, respectively. Hence, this construction is actually "well-defined" in the sense that the resulting stochastic integral "$\int_0 H_s dX_s$" does not depend on the choice of the decomposition $M = M' + M''$ in (6.2).

Most of the properties of the stochastic integrals for $\int_0^\cdot H_s dX_s$, such as "càdlàg", "adapted", "linear" and so on, are inherited from the ones of each term defined as the Stieltjes integral process or the Itô integral process. Among them, recall the remark after Definition 6.1.1 for the reason why it is possible to find a modification whose all paths are càdlàg. They are summarised in the following theorem.

Theorem 6.2.6 (Stochastic integral w.r.t. semimartingale) For any semimartingale X and any real-valued predictable process H that is locally bounded, the above procedure defines a unique (up to indistinguishability) semimartingale $(\int_0^t H_s dX_s)_{t\in[0,\infty)}$ satisfying the following properties.

(i) It starts from zero at $t = 0$, a.s.

(ii) $\int_0^t (aH_s + bK_s)dX_s = a\int_0^t H_s dX_s + b\int_0^t K_s dX_s$ for all $t \in [0,\infty)$, a.s., for any constants a and b, where K is any real-valued, predictable process that is locally bounded.

(iii) $\int_0^t H_s d(aX_s + bY_s) = a\int_0^t H_s dX_s + b\int_0^t H_s dY_s$ for all $t \in [0,\infty)$, a.s., for any constants a and b, where Y is any semimartingale.

(iv) If X is a local martingale, then $\int_0^\cdot H_s dX_s$ is also a local martingale.

(v) If X is a process with finite-variation, then $\int_0^\cdot H_s dX_s$ is also a process with finite-variation and it coincides almost surely with the Stieltjes integral process defined by fixing each $\omega \in \Omega$.

(vi) $\Delta(\int_0^{\cdot} H_s dX_s)_t = H_t \Delta X_t$ for all $t \in [0,\infty)$, a.s.

6.3 Formula for the Integration by Parts

In the usual theory of analysis, a version of the formula for the integration by parts is given as follows: for two real-valued càdlàg functions x and y on $\mathbb{R}$, it holds for any $-\infty < s < t < \infty$ that

$$x(t)y(t) - x(s)y(s) = \int_{(s,t]} x(u-)dy(u) + \int_{(s,t]} y(u-)dx(u) + \sum_{u \in (s,t]} \Delta x(u)\Delta y(u),$$

where all terms on the right-hand side are proved to be finite; in particular, note that a real-valued function on a compact interval with finite-variation has only countably many jumps to have a better understanding of the third term on the right-hand side. See, e.g., Theorem 11 in Section II.6 of Shiryaev (1996) for a proof.

The "quadratic co-variation" of semimartingales that we will study from now may be considered as a "stochastic version" of the last term of the above formula, namely, "$\sum_{u\in(s,t]} \Delta x(u)\Delta y(u)$", which should be compared with the formula (6.3) below.

Definition 6.3.1 (Formula for the integration by parts) The *quadratic co-variation* of two semimartingales X and Y is a semimartingale given by

$$[X,Y]_t := X_t Y_t - X_0 Y_0 - \int_0^t X_{s-} dY_s - \int_0^t Y_{s-} dX_s, \quad \forall t \in [0,\infty),$$

where the two stochastic integrals on the right-hand side are well-defined by Theorem 6.2.6 because the integrands $t \rightsquigarrow X_{t-}$ and $t \rightsquigarrow Y_{t-}$ are locally bounded and predictable. The *quadratic variation* of a semimartingale X is $[X,X]$, which is sometimes denoted simply by $[X]$.

The proofs for many properties of quadratic co-variation are found in Section I.4e of Jacod and Shiryaev (2003). Instead of "copying" them to this monograph, we shall give a résumé that would hopefully help readers who would like to master the usage of the martingale theory in a short period.

Fact X. For any (general) semimartingales X and Y, it is proved that the quadratic co-variation is actually given as

$$[X,Y]_t = \sum_{s\in(0,t]} \Delta X_s \Delta Y_s + \langle X^c, Y^c \rangle_t, \quad \forall t \in [0,\infty), \quad \text{a.s.} \tag{6.3}$$

Here, the first term on the right-hand side is proved to be absolutely convergent;

actually, each path of any given semimartingale X has at most countably many jumps in each bounded interval, and it holds that $\sum_{s\in(0,t]}(\Delta X_s)^2 < \infty$ for every $t \in [0,\infty)$, a.s. Let us state two remarks for some special cases.

(X.1) If a given semimartingale X is a process with finite-variation or a purely discontinuous local martingale, then $X^c = 0$. Hence, if either of X or Y is such a stochastic process, then the second term on the right-hand side of (6.3) vanishes and the formula is reduced to $[X,Y]_t = \sum_{s\in(0,t]} \Delta X_s \Delta Y_s$ for all $t \in [0,\infty)$, a.s.

(X.2) If either of X or Y is continuous or if X and Y have no simultaneous jumps, then the first term on the right-hand side of (6.3) vanishes and the formula is reduced to $[X,Y]_t = \langle X^c, Y^c\rangle_t$ for all $t \in [0,\infty)$, a.s.

Fact M. Let us consider some special cases where X and Y are local martingales; we shall use the notations "M,N" here, instead of "X,Y".

(M.1) If $M,N \in \mathcal{M}_{\text{loc}}$, then $MN - [M,N]$ is a local martingale.

(M.2) In particular, if $M,N \in \mathcal{M}^2_{\text{loc}}$, then the following three processes are local martingales: $MN - [M,N]$, $[M,N] - \langle M,N\rangle$ and $MN - \langle M,N\rangle$.

(M.3) In particular, if $M,N \in \mathcal{M}^2$, then the three processes considered in (M.2) are uniformly integrable martingales.

(M.4) If $M \in \mathcal{M}_{\text{loc}}$, then $M_t = M_0$ for all $t \in [0,\infty)$, a.s., if and only if $[M]_t = 0$ for all $t \in [0,\infty)$, a.s. In particular, if $M \in \mathcal{M}^2_{\text{loc}}$, then it also holds that $M_t = M_0$ for all $t \in [0,\infty)$, a.s., if and only if $\langle M\rangle_t = 0$ for all $t \in [0,\infty)$, a.s.

Fact Q. For (general) semimartingales X and Y, the quadratic co-variation $[X,Y]$ is a process with finite-variation and the quadratic variation $[X]$ is an increasing process; of course, they are adapted processes. It would be good to remember these properties in connection with the ones for $M,N \in \mathcal{M}^2_{\text{loc}}$; the predictable quadratic co-variation $\langle M,N\rangle$ is a predictable process with finite-variation and the predictable quadratic variation $\langle M\rangle$ is a predictable increasing process.

6.4 Itô's Formula

In this section, we try to understand what is Itô's formula. One of the main reasons why semimartingales are so important is that they are given as an "additive form" $X_t = X_0 + A_t + M_t$ which we can treat easily. For example, when we would like to know the expectation of X_T for a given stopping time T, if we have some good integrability conditions for the martingale part M, it suffices to compute the expectations only of X_0 and A_T due to the optional sampling theorem. However, some functionals of semimartingales are not always easy to treat. For example, we often encounter some situations where we would like to compute the expectations of $(X_0 + A_t + M_t)^2$ or $e^{X_0+A_t+M_t}$, etc.; indeed, we need to compute the characteristic function $\mathsf{E}[e^{izX_t}] = \mathsf{E}[\cos(zX_t)] + i\mathsf{E}[\sin(zX_t)]$ to prove the martingale central limit theorem. Itô's formula is a powerful tool to transform $f(X_t)$ for a smooth function f, which is not of an additive form any more, into an additive form that is easy to treat.

Now let us state Itô's formula. We say $X = (X^1, ..., X^d)^{\mathrm{tr}}$ is a d-dimensional semimartingale if each X^i is a (real-valued) semimartingale.

Theorem 6.4.1 (Itô's formula) Let $f : \mathbb{R}^d \to \mathbb{R}$ be a twice continuously differentiable function. For any d-dimensional semimartingale X, it holds that

$$\begin{aligned} f(X_t) - f(X_0) &= \sum_{i=1}^{d} \int_0^t D_i f(X_{s-}) dX_s^i + \frac{1}{2} \sum_{i=1}^{d} \sum_{j=1}^{d} \int_0^t D_{i,j} f(X_{s-}) d\langle X^{i,c}, X^{j,c} \rangle_s \\ &\quad + \sum_{s \in (0,t]} \left\{ f(X_s) - f(X_{s-}) - \sum_{i=1}^{d} D_i f(X_{s-}) \Delta X_s^i \right\}, \end{aligned}$$

for all $t \in [0, \infty)$, a.s., where $D_i f(x) := \frac{\partial}{\partial x_i} f(x)$ and $D_{i,j} f(x) := \frac{\partial^2}{\partial x_i \partial x_j} f(x)$.

The theorem actually says that all terms on the right-hand side exist and the equation between the left- and right-hand sides holds true. To be more precise, since $s \rightsquigarrow D_i f(X_{s-})$'s and $s \rightsquigarrow D_{i,j} f(X_{s-})$'s are locally bounded and predictable, the first two terms on the right-hand side are well-defined as the stochastic integrals with respect to semimartingales. As for the last term on the right-hand side, noting also that any semimartingale has at most countably many jumps on each bounded interval, we have used the notation "$\sum_{s \in (0,t]} \{\bullet\}$" presuming that the (possibly infinite) summation is absolutely convergent, that is, $\sum_{s \in (0,t]} |\bullet| < \infty$. Here, we remark that the last term of the above formula should *not* be written as

$$\sum_{s \in (0,t]} \{f(X_s) - f(X_{s-})\} - \sum_{s \in (0,t]} \left\{ \sum_{i=1}^{d} D_i f(X_{s-}) \Delta X_s^i \right\},$$

because each of the two infinite summations may not be absolutely convergent.

Ikeda and Watanabe (1989) gives an elementary proof starting from the two term Taylor expansion of f. On the other hand, Jacod and Shiryaev (2003) and Kallenberg (2002) give some elegant proofs, where they first prove the case where f is a polynomial on $\mathbb{R}^d$ and generalize the tentative result up to the one for any $f \in C^2(\mathbb{R}^d)$.

Here, let us try to have an intuitive interpretation of Itô's formula. For every $t \in (0, \infty)$, introduce a sequence of finite points $0 = t_0^n < t_1^n < \cdots < t_n^n = t$ such that $\max_k |t_k^n - t_{k-1}^n| \to 0$ as $n \to \infty$. Noting that $f(X_t) - f(X_0) = \sum_{k=1}^n \{f(X_{t_k^n}) - f(X_{t_{k-1}^n})\}$, let us compute each term on the right-hand side by using the Taylor expansion, that is,

$$\begin{aligned} f(X_{t_k^n}) - f(X_{t_{k-1}^n}) &= \sum_{i=1}^{d} D_i f(X_{t_{k-1}^n})(X_{t_k^n}^i - X_{t_{k-1}^n}^i) \\ &\quad + \frac{1}{2} \sum_{i=1}^{d} \sum_{j=1}^{d} D_{i,j} f(\widetilde{X}_{k,n})(X_{t_k^n}^i - X_{t_{k-1}^n}^i)(X_{t_k^n}^j - X_{t_{k-1}^n}^j), \end{aligned} \tag{6.4}$$

where $\widetilde{X}_{k,n}$ is a point between $X_{t_{k-1}^n}$ and $X_{t_k^n}$.

To get an intuitive understanding, suppose that we may use the approximation

$$\begin{aligned}(X_t^i - X_s^i)(X_t^j - X_s^j) &= X_t^i X_t^j - X_s^i X_s^j - X_s^i(X_t^j - X_s^j) - X_s^j(X_t^i - X_s^i) \\ &\approx [X^i, X^j]_t - [X^i, X^j]_s \\ &= \sum_{u\in(s,t]} \Delta X_u^i \Delta X_u^j + \langle X^{i,c}, X^{j,c}\rangle_t - \langle X^{i,c}, X^{j,c}\rangle_s,\end{aligned}$$

which is based on the formula for the integration by parts, to a part of the last term of (6.4). Then, taking $\lim_{n\to\infty}\sum_{k=1}^n$, we might expect a "formula" like

$$\begin{aligned}f(X_t) - f(X_0) &= \sum_{i=1}^d \int_0^t D_i f(X_{s-}) dX_s^i + \frac{1}{2}\sum_{i=1}^d\sum_{j=1}^d \int_0^t D_{i,j} f(\widetilde{X}_s) d\langle X^{i,c}, X^{j,c}\rangle_s \\ &\quad + \frac{1}{2}\sum_{i=1}^d\sum_{j=1}^d \sum_{s\in(0,t]} D_{i,j} f(\widetilde{X}_s)\Delta X_s^i \Delta X_s^j.\end{aligned}$$

However, this informal argument has not reached the correct form of Itô's formula; especially the last term is completely different. This is due to the reason that "$\widetilde{X}_s$" contained in the last term is originally a point between $X_{t_{k-1}^n}$ and $X_{t_k^n}$, which cannot be approximated by "X_{s-}" at time points s where $\Delta X_s^i \Delta X_s^j$ is not zero. On the other hand, the problem concerning "$\widetilde{X}_s$" eventually does not affect the second term because the values of the integrand at countably many points do not make any difference for the computation of the Lebesgue-Stieltjes integral with respect to the continuous function $s \mapsto \langle X^{i,c}, X^{j,c}\rangle_s(\omega)$, hence $\widetilde{X}_s$ in the second term can be replaced by X_{s-}, or even by X_s. To fix the flaw in the last term of the above wrong "formula", we should go back to the Taylor expansion (6.4) and replace the pending term with

$$\sum_{k=1}^n \sum_{s\in(t_{k-1}^n, t_k^n]} \left\{ f(X_{t_k^n}) - f(X_{t_{k-1}^n}) - \sum_{i=1}^d D_i f(X_{t_{k-1}^n})(X_{t_k^n}^i - X_{t_{k-1}^n}^i) \right\},$$

which converges as $n \to \infty$ to the last term of Itô's formula.

Exercise 6.4.1 Verify that the formula for the integration by parts (Definition 6.3.1) can be derived as a special case of Itô's formula.

Exercise 6.4.2 As it is stated in Example 3.5.9, the positive real-valued stochastic process

$$X_t = X_0 \exp(\beta t + \sigma W_t), \quad \forall t \in [0,\infty),$$

where β and σ are given constants, is called an *geometric Brownian motion*. Apply Itô's formula to deduce that this can be rewritten into the form of the stochastic differential equation

$$X_t = X_0 + \int_0^t \left(\beta + \frac{1}{2}\sigma^2\right) X_s ds + \int_0^t \sigma X_s dW_s, \quad \forall t \in [0,\infty).$$

Exercise 6.4.3 A sufficient condition for a continuous local martingale M to be a standard Wiener process in the sense of Definition 5.1.4 is that $\langle M \rangle_t = t$ for all $t \in [0,\infty)$, a.s. When the filtration is the one generated by M and a sub-σ-field $\mathcal{H}$ of $\mathcal{F}$ that is independent of $\sigma(M_s : s \in [0,\infty))$, namely, $\mathbf{F}^{M,\mathcal{H}}$, this condition is necessary. Prove these claims. [**Hint:** The necessity has already been proved in Example 5.9.2 (i), that is, Exercise 5.9.3. To show the sufficiency, compute the characteristic function using Itô's formula.]

Exercise 6.4.4 A sufficient condition for a counting process N to be an inhomogeneous Poisson process with the intensity function $(\lambda(t))_{t\in[0,\infty)}$ is that the predictable compensator for N is given by $A_t = \int_0^t \lambda(s)ds$ for all $t \in [0,\infty)$, a.s. When the filtration is the one generated by N and a sub-σ-field $\mathcal{H}$ of $\mathcal{F}$ that is independent of $\sigma(N_s : s \in [0,\infty))$, this condition is necessary. Prove these claims. [**Hint:** The necessity in the case of a homogeneous Poisson process has already been proved in Example 5.9.2 (ii-a), that is, Exercise 5.9.3. To show the sufficiency, compute the characteristic function using Itô's formula.]

6.5 Likelihood Ratio Processes

6.5.1 Likelihood ratio process and martingale

Let us begin with an illustrative example. Let $(\mathcal{X},\mathcal{A},\mu)$ be a measure space. Let $X_1,X_2,...,X_n$ be a sequence of $\mathcal{X}$-valued random elements defined on a measurable space $(\Omega,\mathcal{F})$, and introduce the filtration $\mathbf{F} = (\mathcal{F}_t)_{t\in[0,1]}$ given by $\mathcal{F}_t = \sigma(X_k : k \leq [nt])$ for every $t \in [0,1]$.

Discussion 6.5.1 (Given P and $\widetilde{\mathsf{P}}$, let us prove that L is an $(\mathbf{F},\mathsf{P})$-martingale)

Let two probability measures P and $\widetilde{\mathsf{P}}$ on $(\Omega,\mathcal{F})$ be given. Suppose that $X_1,...,X_n$ are independent under P. Suppose that the distributions of each X_k under P and $\widetilde{\mathsf{P}}$ have the densities f_k and $\widetilde{f}_k$ with respect to μ, respectively. The likelihood ratio process $L = (L_t)_{t\in[0,1]}$ is then defined by

$$L_t = \begin{cases} 1, & \forall t \in [0,1/n), \\ \displaystyle\prod_{k=1}^{[nt]} \frac{\widetilde{f}_k(X_k)}{f_k(X_k)}, & \forall t \in [1/n,1]. \end{cases} \tag{6.5}$$

This is well-defined in the P-almost sure sense, because each X_k does not take values on the set $\{x : f_k(x) = 0\}$, P-almost surely. We can prove that L is an $(\mathbf{F},\mathsf{P})$-martingale: indeed, for any $1/n \leq s \leq t \leq 1$, it holds that

$$\mathsf{E}[L_t|\mathcal{F}_s] = \mathsf{E}\left[\prod_{k=[ns]+1}^{[nt]} \frac{\widetilde{f}_k(X_k)}{f_k(X_k)}\right] \prod_{k=1}^{[ns]} \frac{\widetilde{f}_k(X_k)}{f_k(X_k)}$$

$$
\begin{aligned}
&= \prod_{k=[ns]+1}^{[nt]} \mathsf{E}\left[\frac{\widetilde{f}_k(X_k)}{f_k(X_k)}\right] \prod_{k=1}^{[ns]} \frac{\widetilde{f}_k(X_k)}{f_k(X_k)} \\
&= \prod_{k=[ns]+1}^{[nt]} \int_{\mathcal{X}} \frac{\widetilde{f}_k(x)}{f_k(x)} f_k(x)\mu(dx) \prod_{k=1}^{[ns]} \frac{\widetilde{f}_k(X_k)}{f_k(X_k)} \\
&= \prod_{k=1}^{[ns]} \frac{\widetilde{f}_k(X_k)}{f_k(X_k)} \\
&= L_s;
\end{aligned}
$$

the computations for the cases where $0 \leq s < 1/n \leq t \leq 1$ and $0 \leq s \leq t < 1/n$ are also easy.

Discussion 6.5.2 (Given P and $(\mathbf{F},\mathsf{P})$-martingale L, let us construct a new $\widetilde{\mathsf{P}}$)

(Step 1.) This time, suppose that only the probability measure P on $(\Omega,\mathcal{F})$ under which $X_1,\ldots,X_n$ are independent is given. Suppose that the distribution of each X_k has a density f_k with respect to μ, and that some probability densities $\widetilde{f}_k$'s on $(\mathcal{X},\mathcal{A},\mu)$ are given. These $\widetilde{f}_k$'s have been given as some candidates for the densities of X_k's under another probability measure, which will be constructed and denoted by $\widetilde{\mathsf{P}}$ later. Let us discuss how to construct such a probability measure. For this purpose, let us define the stochastic process $L = (L_t)_{t\in[0,1]}$ by the same formula as (6.5), which would eventually become the likelihood ratio process in our discussion below. Then, it can be easily proved that L is an $(\mathbf{F},\mathsf{P})$-martingale exactly in the same way as above.

(Step 2.) When we have the $\mathbf{F}$-martingale property for a $[0,\infty)$-valued process $t \rightsquigarrow L_t$ starting from 1 under P, as it is so in the current case, let us define

$$
\widetilde{\mathsf{P}}^t(A) = \mathsf{E}[1_A L_t], \quad \forall A \in \mathcal{F}_t, \quad \forall t \in [0,1].
$$

Then, it is clear that each $\widetilde{\mathsf{P}}^t$ is a probability measure on $(\Omega,\mathcal{F}_t)$ for every $t \in [0,1]$, and moreover, it holds for every $0 \leq s \leq t \leq 1$ that

$$
\widetilde{\mathsf{P}}^t(A) = \mathsf{E}[1_A L_t] = \mathsf{E}[1_A \mathsf{E}[L_t|\mathcal{F}_s]] = \mathsf{E}[1_A L_s] = \widetilde{\mathsf{P}}^s(A), \quad \forall A \in \mathcal{F}_s. \tag{6.6}
$$

Now, let us finally define the probability measure on $(\Omega,\mathcal{F}_1)$ by $\widetilde{\mathsf{P}} := \widetilde{\mathsf{P}}^1$. Then, each $\widetilde{\mathsf{P}}^t$ is a restriction of $\widetilde{\mathsf{P}}$ to $\mathcal{F}_t$. Moreover, denoting the restriction of P to $\mathcal{F}_t$ by P^t, we conclude that $\widetilde{\mathsf{P}}^t$ is absolutely continuous with respect to P with the Radon-Nikodym derivative $\frac{d\widetilde{\mathsf{P}}^t}{d\mathsf{P}^t} = L_t$ for every $t \in [0,1]$.

It is important to note that the method at the Step 2 of Discussion 6.5.2 above works for the problem to construct a new probability measure $\widetilde{\mathsf{P}}$ on any given stochastic basis $(\Omega,\mathcal{F};\mathbf{F},\mathsf{P})$ as far as we succeed in finding a $[0,\infty)$-valued $(\mathbf{F},\mathsf{P})$-martingale L starting from 1. Let us summarise this method in the form of a theorem.

Theorem 6.5.3 (Non-negative martingale L with $L_0 = 1$ is a likelihood ratio) Let a stochastic basis $(\Omega, \mathcal{F}; \mathbf{F} = (\mathcal{F}_t)_{t\in[0,\infty)}, \mathsf{P})$ be given. When a $[0,\infty)$-valued $(\mathbf{F}, \mathsf{P})$-martingale $L = (L_t)_{t\in[0,\infty)}$ starting from 1 is given, the formula

$$\widetilde{\mathsf{P}}^t(A) := \mathsf{E}[1_A L_t], \quad \forall A \in \mathcal{F}_t$$

defines a probability measure on $(\Omega, \mathcal{F}_t)$ for every $t \in [0,\infty)$, and it holds for any $0 \le s < t < \infty$ that $\widetilde{\mathsf{P}}^s$ is the restriction of $\widetilde{\mathsf{P}}^t$ to $\mathcal{F}_s$. Moreover, each $\widetilde{\mathsf{P}}^t$ is absolutely continuous with respect to the restriction of P to $\mathcal{F}_t$, which is denoted by P^t, and its Radon-Nikodym derivative is given by

$$\frac{d\widetilde{\mathsf{P}}^t}{d\mathsf{P}^t} = L_t, \quad \forall t \in [0,\infty).$$

Proof. The first claim is immediate from the assumption that $L_0 = 1$; we just remark that $\widetilde{\mathsf{P}}^t(\Omega) = \mathsf{E}[1_\Omega L_t] = \mathsf{E}[L_0] = 1$, among other properties for being a probability measure. The second claim is clear from the fact that the formula (6.6) holds for every $0 \le s \le t < \infty$. Finally, since L_t is $\mathcal{F}_t$-measurable, the last claim is true because it holds that $\widetilde{\mathsf{P}}^t(A) = \int_A L_t(\omega)\mathsf{P}^t(d\omega)$ holds for every $A \in \mathcal{F}_t$. □

6.5.2 Girsanov's theorem

Let a filtered space $(\Omega, \mathcal{F}; \mathbf{F} = (\mathcal{F}_t)_{t\in[0,\infty)})$ be given. Let X be a d-dimensional $\mathbf{F}$-adapted process defined on this space; recall that when we introduce some adapted processes, no probability measure is necessary. Hereafter, we denote by $\varepsilon_a(dx)$ the Dirac measure at point $a \in \mathcal{X}$ on a measurable space $(\mathcal{X}, \mathcal{A})$; that is, $\varepsilon_a(A)$ is 1 if $a \in A$ and 0 otherwise.

(Step 1.) Now, let a probability measure P on this space be given, and suppose that under this P, the given adapted process X is a d-dimensional special semimartingale $X = (X^1, ..., X^d)^{\mathrm{tr}}$ of the form

$$X_t^i = X_0^i + A_t^i + M_t^{c,i} + M_t^{d,i}, \quad \forall t \in [0,\infty), \quad \mathsf{P}\text{-a.s.},$$

where we assume the following:

(c) The continuous martingale part $X^c = M^c$ of X is a d-dimensional continuous local martingale starting from zero with the predictable quadratic co-variation $C = (C^{i,j})_{(i,j)\in\{1,...,d\}^2}$ that is absolutely continuous with respect to the Lebesgue measure, namely,

$$\langle M^{c,i}, M^{c,j}\rangle_t = C_t^{i,j} = \int_0^t c_s^{i,j} ds,$$

where $c^{i,j}$'s are real-valued predictable processes.

(a) The d-dimensional predictable process $A = (A^1, ..., A^d)^{\mathrm{tr}}$ with finite-variation is of the form

$$A_t^i = \sum_{j=1}^d \int_0^t a_s^j c_s^{i,j} ds,$$

where a^i's are real-valued predictable processes.

(d) The d-dimensional purely discontinuous local martingale $M^d = (M^{d,1},...,M^{d,d})^{\mathrm{tr}}$ is of the form

$$M_t^{d,i} = \int_{[0,t]\times\mathbb{R}^d} x^i(\mu^X(\cdot;ds,dx) - \lambda(\cdot,s,x)ds\eta(dx)),$$

where

$$\mu^X(\omega;dt,dx) = \sum_s 1\{\Delta X_s(\omega) \neq 0\}\varepsilon_{(s,\Delta X_s(\omega))}(dt,dx).$$

Here, λ is a $[0,\infty)$-valued $\mathcal{P}\otimes\mathfrak{B}(\mathbb{R}^d)$-measurable function on $\Omega\times[0,\infty)\times\mathbb{R}^d$, where $\mathcal{P}$ is the predictable σ-field (see Definition 5.4.1), and η is a measure on $(\mathbb{R}^d,\mathfrak{B}(\mathbb{R}^d))$; we refer to Section II.1 of Jacod and Shiryaev (2003) for the theory of (integer-valued) random measures.

(Step 2.) Now, let some predictable objects $(\widetilde{a},\widetilde{\lambda})$ which are some alternatives to (a,λ), be given; recall that when we introduce some predictable processes, it is not necessary to think of probability measures. From now on, let us discuss how to construct a new probability measure $\widetilde{\mathsf{P}}$ under which the adapted process X given above is a d-dimensional special semimartingale of the form

$$X_t^i = X_0^i + \widetilde{A}_t^i + \widetilde{M}_t^{c,i} + \widetilde{M}_t^{d,i}, \quad \forall t\in[0,\infty), \quad \widetilde{\mathsf{P}}\text{-a.s.}, \quad i=1,...,d, \tag{6.7}$$

where $\widetilde{A} = (\widetilde{A}^1,...,\widetilde{A}^d)^{\mathrm{tr}}$ is a d-dimensional predictable process given by

$$\widetilde{A}_t^i = \sum_{j=1}^d \int_0^t \widetilde{a}_s^j c_s^{i,j} ds, \tag{6.8}$$

the continuous $(\mathbf{F},\widetilde{\mathsf{P}})$-martingale part $\widetilde{X}^c = \widetilde{M}^c = (\widetilde{M}^{c,1},...,\widetilde{M}^{c,d})^{\mathrm{tr}}$ is given by

$$\widetilde{M}_t^{c,i} = M_t^{c,i} - \sum_{j=1}^d \int_0^t (\widetilde{a}_s^j - a_s^j)c_s^{i,j} ds, \tag{6.9}$$

and $\widetilde{M}^d = (\widetilde{M}^{d,1},...,\widetilde{M}^{d,d})^{\mathrm{tr}}$ given by

$$\widetilde{M}_t^{d,i} = \int_{[0,t]\times\mathbb{R}^d} x^i(\mu^X(\cdot;ds,dx) - \widetilde{\lambda}(\cdot,s,x)ds\eta(dx)) \tag{6.10}$$

is a d-dimensional purely discontinuous local $(\mathbf{F},\widetilde{\mathsf{P}})$-martingale.

We dare to remark that even if a given càdlàg adapted process is a semimartingale under a certain probability measure, it may not be a semimartingale under another probability measure any more if the new probability measure is introduced in a bad way. What we will do from now is to introduce a "good" probability measure that makes the "canonical decomposition of special semimartingale (6.7) with (6.8), (6.9) and (6.10)" true.

(Step 3(i).) Seeking for an answer, we shall define a stochastic process $L = (L_t)_{t\in[0,\infty)}$ which would be helpful for constructing such a probability measure $\widetilde{\mathsf{P}}^T$

on $(\Omega, \mathcal{F}_T)$, for any fixed constant $T > 0$. Here, we are content with the case where the index set of X is not $[0,\infty)$ but "$[0,T]$ for any fixed $T>0$"; this would not become a real restriction in statistical applications where observation of stochastic processes is usually given only over compact (or bounded) time sets. Based on such an L, we shall introduce the probability measures $\widetilde{\mathsf{P}}^t$'s given by

$$\widetilde{\mathsf{P}}^t(A) = \int_A L_t(\omega)\mathsf{P}(d\omega), \quad \forall A \in \mathcal{F}_t, \quad \forall t \in [0,\infty). \tag{6.11}$$

Once we have proved that this L is an $(\mathbf{F},\mathsf{P})$-martingale, our aim would be accomplished by using Theorem 6.5.3.

(Step 3(ii).) The following definition of L based on the objects $\widetilde{a}$ and $\widetilde{\lambda}$ given at Step 2 will make $(X_t)_{t\in[0,T]}$ a special $(\mathbf{F}^T, \widetilde{\mathsf{P}}^T)$-semimartingale, where $\mathbf{F}^T = (\mathcal{F}_t)_{t\in[0,T]}$ with any fixed time $T>0$:

$$L_t := L_t^c L_t^d, \quad \forall t \in [0,\infty), \tag{6.12}$$

where

$$L_t^c := \exp\left(-\frac{1}{2}\sum_{i=1}^d\sum_{j=1}^d\int_0^t(\widetilde{a}_s^i - a_s^i)c^{i,j}(\widetilde{a}_s^j - a_s^j)ds - \sum_{i=1}^d\int_0^t(\widetilde{a}_s^i - a_s^i)dM_s^{c,i}\right), \tag{6.13}$$

$$L_t^d := \exp\left(\int_{[0,t]\times\mathbb{R}^d}\log\left(\frac{\widetilde{\lambda}(\cdot,s,x)}{\lambda(\cdot,s,x)}\right)\mu^X(\cdot;ds,dx)\right.$$
$$\left. - \int_{[0,t]\times\mathbb{R}^d}(\widetilde{\lambda}(\cdot,s,x) - \lambda(\cdot,s,x))ds\eta(dx)\right). \tag{6.14}$$

Here, we assume that the new candidate $\widetilde{\lambda}$ as an alternative for λ has been introduced in such a way that for every ω from the complement of a P-null set, $\lambda(\omega,s,x) = 0$ implies $\widetilde{\lambda}(\omega,s,x) = 0$, and in this case we read $\frac{\widetilde{\lambda}(\omega,s,x)}{\lambda(\omega,s,x)}$ to be 1.

Notice that the stochastic integrals appearing in the above formula are, of course, defined based on the filtration $\mathbf{F}$ and the original probability measure P. Readers should always pay attention to the choice of the probability measure when they see some notations of stochastic integrals in our discussion.

Theorem 6.5.4 (Girsanov's theorem) (i) The stochastic process $L = (L_t)_{t\in[0,\infty)}$ defined by (6.12) with (6.13) and (6.14) is a local $(\mathbf{F},\mathsf{P})$-martingale such that $L_0 = 1$.

(ii) Suppose that the stronger condition that L is an $(\mathbf{F},\mathsf{P})$-martingale is satisfied. Then, for every $t \in [0,\infty)$ the formula (6.11) defines a probability measure $\widetilde{\mathsf{P}}^t$ on $(\Omega,\mathcal{F}_t)$, and it holds for every $0 \le s < t < \infty$ that $\widetilde{\mathsf{P}}^s$ is the restriction of $\widetilde{\mathsf{P}}^t$ to $\mathcal{F}_s$ and that each $\widetilde{\mathsf{P}}^t$ is absolutely continuous with respect to the restriction of P to $\mathcal{F}_t$, namely, P^t, with the Radon-Nikodym derivative $\frac{d\widetilde{\mathsf{P}}^t}{d\mathsf{P}^t} = L_t$. Moreover, for any fixed time $T>0$, the adapted process $X = (X_t)_{t\in[0,T]}$ on $(\Omega,\mathcal{F})$ is a d-dimensional special $((\mathcal{F}_t)_{t\in[0,T]}, \widetilde{\mathsf{P}}^T)$-semimartingale that has the canonical decomposition given by (6.7) with (6.8), (6.9) and (6.10).

Proof. (i) It follows from Itô's formula that L is a d-dimensional $(\mathbf{F},\mathsf{P})$-semimartingale and that

$$\begin{aligned} L_t &= 1-\sum_{i=1}^{d}\int_0^t L_{s-}(\widetilde{a}_s^i - a_s^i)dM_s^{c,i} \\ &\quad -\int_{[0,t]\times\mathbb{R}^d} L_{s-}\left(\frac{\widetilde{\lambda}(\cdot,s,x)}{\lambda(\cdot,s,x)}-1\right)(\mu^X(\cdot;ds,dx)-\lambda(\cdot,s,x)ds\eta(dx)). \end{aligned}$$

Hence L is a local $(\mathbf{F},\mathsf{P})$-martingale such that $L_0 = 1$.

(ii-1) The term $\widetilde{A}^i$ in (6.7) is a predictable process because it is continuous and adapted, and it is also a process with finite-variation.

(ii-2) It is clear that the term $\widetilde{M}^{c,i}$ in (6.7) is a continuous, adapted process. Let us apply Theorem 5.7.5 to prove that this is a local $(\mathbf{F},\widetilde{\mathsf{P}})$-martingale; so let us take any bounded stopping time T, say $T \leq t_1$ for a constant $t_1 \in (0,\infty)$. Introducing a localizing sequence (T_n) of stopping times to make it possible to apply the optional sampling theorem on the last line of our computation below, and using also the formula for integration by parts, we have

$$\begin{aligned} &\widetilde{\mathsf{E}}[\widetilde{M}^{c,i}_{T\wedge T_n}] \\ &= \mathsf{E}[\widetilde{M}^{c,i}_{T\wedge T_n}L_{t_1}] \\ &= \mathsf{E}[\widetilde{M}^{c,i}_{T\wedge T_n}E[L_{t_1}|\mathcal{F}_{T\wedge T_n}]] \\ &= \mathsf{E}[\widetilde{M}^{c,i}_{T\wedge T_n}L_{T\wedge T_n}] \\ &= \mathsf{E}\left[\int_0^{T\wedge T_n}\widetilde{M}^{c,i}_{s-}dL_s + \int_0^{T\wedge T_n}L_{s-}d\widetilde{M}^{c,i}_s + \sum_{j=1}^{d}\int_0^{T\wedge T_n}L_{s-}(\widetilde{a}_s^j - a_s^j)c_s^{i,j}ds\right] \\ &= \mathsf{E}\left[\int_0^{T\wedge T_n}\widetilde{M}^{c,i}_{s-}dL_s + \int_0^{T\wedge T_n}L_{s-}dM^{c,i}_s\right] \\ &= 0. \end{aligned}$$

Thus $\widetilde{M}^{c,i,T_n}$ is an $(\mathbf{F},\widetilde{\mathsf{P}})$-martingale. Since $\mathcal{M}_{\mathrm{loc}} = \{\text{martingales}\}_{\mathrm{loc}}$ (see Theorem 5.7.2 (iii)), we can conclude that $\widetilde{M}^{c,i}$ is a continuous local $(\mathbf{F},\widetilde{\mathsf{P}})$-martingale starting from zero.

(ii-3) To prove that $\widetilde{M}^{d,i}$ in (6.7) a local $(\mathbf{F},\widetilde{\mathsf{P}})$-martingale starting from zero, let us repeat exactly the same argument as that in (ii-2):

$$\begin{aligned} &\widetilde{\mathsf{E}}[\widetilde{M}^{d,i}_{T\wedge T_n}] \\ &= \mathsf{E}[\widetilde{M}^{d,i}_{T\wedge T_n}L_{t_1}] \\ &= \mathsf{E}[\widetilde{M}^{d,i}_{T\wedge T_n}E[L_{t_1}|\mathcal{F}_{T\wedge T_n}]] \\ &= \mathsf{E}[\widetilde{M}^{d,i}_{T\wedge T_n}L_{T\wedge T_n}] \\ &= \mathsf{E}\left[\int_{[0,T\wedge T_n]\times\mathbb{R}^d} x^i(\mu^X(\cdot;ds,dx)-\widetilde{\lambda}(\cdot,s,x)ds\eta(dx))L_{T\wedge T_n}\right] \end{aligned}$$

$$
\begin{aligned}
&= \mathsf{E}\left[\int_{[0,T\wedge T_n]\times\mathbb{R}^d} x^i(\mu^X(\cdot;ds,dx) - \lambda(\cdot,s,x)ds\eta(dx))L_{T\wedge T_n}\right] \\
&\quad -\mathsf{E}\left[\int_{[0,T\wedge T_n]\times\mathbb{R}^d} x^i(\widetilde{\lambda}(\cdot,s,x) - \lambda(\cdot,s,x))ds\eta(dx)L_{T\wedge T_n}\right] \\
&= 0+0+\mathsf{E}\left[\sum_{t\leq T\wedge T_n} \Delta\left(\int_{[0,t]\times\mathbb{R}^d} x^i(\mu(\cdot;ds,dx) - \lambda(\cdot,s,x)ds\eta(dx))\right)\Delta L_t\right] \\
&\quad -\mathsf{E}\left[\int_{[0,T\wedge T_n]\times\mathbb{R}^d} x^i(\widetilde{\lambda}(\cdot,s,x) - \lambda(\cdot,s,x))ds\eta(dx)L_{T\wedge T_n}\right] \\
&= \mathsf{E}\left[\int_{[0,T\wedge T_n]\times\mathbb{R}^d} x^i L_{s-}\left(\frac{\widetilde{\lambda}(\cdot,s,x)}{\lambda(\cdot,s,x)} - 1\right)\mu^X(\cdot;ds,dx)\right] \\
&\quad -\mathsf{E}\left[\int_{[0,T\wedge T_n]\times\mathbb{R}^d} x^i(\widetilde{\lambda}(\cdot,s,x) - \lambda(\cdot,s,x))ds\eta(dx)L_{T\wedge T_n}\right] \\
&= \mathsf{E}\left[\int_{[0,T\wedge T_n]\times\mathbb{R}^d} x^i L_{s-}(\widetilde{\lambda}(\cdot,s,x) - \lambda(\cdot,s,x))ds\eta(dx)\right] \\
&\quad -\mathsf{E}\left[\int_{[0,T\wedge T_n]\times\mathbb{R}^d} x^i(\widetilde{\lambda}(\cdot,s,x) - \lambda(\cdot,s,x))ds\eta(dx)L_{T\wedge T_n}\right] \\
&= \mathsf{E}\left[\int_0^{T\wedge T_n}\left(\int_{[0,t]\times\mathbb{R}^d} x^i(\widetilde{\lambda}(\cdot,s,x) - \lambda(\cdot,s,x))ds\eta(dx)\right)dL_t\right] \\
&= 0.
\end{aligned}
$$

Since the canonical decomposition of a special semimartingale is unique, all claims have been proved. □

6.5.3 Example: Diffusion processes

Let β be an $\mathbb{R}^d$-valued function on $\mathbb{R}^d$, and let σ be a $(d\times q)$-real matrix valued function on $\mathbb{R}^d$ such that $\sigma\sigma^{\mathrm{tr}}(x) := \sigma(x)\sigma(x)^{\mathrm{tr}}$ is positive definite for all $x\in\mathbb{R}^d$. Let $(\Omega,\mathcal{F},\mathsf{P})$ be a probability space on which a q-dimensional standard Wiener process $W=(W^1,...,W^q)^{\mathrm{tr}}$ is defined. Suppose that W is a q-dimensional local martingale with respect to a complete filtration $\mathbf{F}$; this is true if we set the filtration to be the one generated by W and a σ-field $\mathcal{H}$, namely $\mathbf{F}=\mathbf{F}^{W,\mathcal{H}}$, where $\mathcal{H}$ is independent of W and contains all P-null sets. It is known that if all components of β and σ are Lipschitz continuous, then for any $x\in\mathbb{R}^d$ there exists a unique (up to indistinguishability) d-dimensional $(\mathbf{F},\mathsf{P})$-semimartingale $X=(X^1,...,X^d)^{\mathrm{tr}}$ such that

$$
X_t = x + \int_0^t \beta(X_s)ds + \int_0^t \sigma(X_s)dW_s, \quad \forall t\in[0,\infty), \quad \mathsf{P}\text{-a.s.};
$$

to be more precise, its component-wise expression is

$$
X_t^i = x^i + \int_0^t \beta^i(X_s)ds + \sum_{r=1}^q \int_0^t \sigma^{i,r}(X_s)dW_s^r, \quad \forall t\in[0,\infty), \quad \mathsf{P}\text{-a.s.} \quad i=1,...,d.
$$

See, e.g., Theorem IX.2.1 of Revuz and Yor (1999) for a proof. This is a special case of d-dimensional semimartingales formulated in the first paragraph of Section 6.5.2 with $X_0 = x$, $c_s = \sigma\sigma^{\mathrm{tr}}(X_s)$, $a_s = (\sigma\sigma^{\mathrm{tr}}(X_s))^{-1}\beta(X_s)$, and $M^d = 0$.

When a new candidate $\widetilde{\beta}$ as an alternative for β is given, the log-likelihood ratio formula is calculated as follows; for every $t \in [0,\infty)$,

$$\begin{aligned}
&\log L_t \\
&= -\frac{1}{2}\int_0^t (\widetilde{\beta}(X_s)-\beta(X_s))^{\mathrm{tr}}(\sigma\sigma^{\mathrm{tr}}(X_s))^{-1}(\widetilde{\beta}(X_s)-\beta(X_s))ds \\
&\quad -\int_0^t (\widetilde{\beta}(X_s)-\beta(X_s))^{\mathrm{tr}}(\sigma\sigma^{\mathrm{tr}}(X_s))^{-1}\sigma(X_s)dW_s \\
&= \int_0^t \beta(X_s)^{\mathrm{tr}}(\sigma\sigma^{\mathrm{tr}}(X_s))^{-1}dX_s - \frac{1}{2}\int_0^t \beta(X_s)^{\mathrm{tr}}(\sigma\sigma^{\mathrm{tr}}(X_s))^{-1}\beta(X_s)ds \\
&\quad -\int_0^t \widetilde{\beta}(X_s)^{\mathrm{tr}}(\sigma\sigma^{\mathrm{tr}}(X_s))^{-1}dX_s + \frac{1}{2}\int_0^t \widetilde{\beta}(X_s)^{\mathrm{tr}}(\sigma\sigma^{\mathrm{tr}}(X_s))^{-1}\widetilde{\beta}(X_s)ds,
\end{aligned}$$

where the stochastic integrals are taken with respect to P.

Under the situation where this $L = (L_t)_{t\in[0,\infty)}$ is not only a local P-martingale but also a P-martingale, we can apply Theorem 6.5.4 to deduce that, for any fixed $T > 0$, the given adapted process $X = (X_t)_{t\in[0,T]}$ is a special semimartingale under the probability measure $\widetilde{\mathsf{P}}^T$ defined by (6.11) with the canonical decomposition

$$X_t^i = x^i + \int_0^t \widetilde{\beta}^i(X_s)ds + \sum_{r=1}^q \int_0^t \sigma^{i,r}(X_s)d\widetilde{W}_s^r, \quad \widetilde{\mathsf{P}}\text{-a.s.} \quad i = 1,\dots,d,$$

where $\widetilde{W} = (\widetilde{W}^1,\dots,\widetilde{W}^q)^{\mathrm{tr}}$ given by

$$\widetilde{W}_t^r = W_t^r - \sum_{j=1}^d \int_0^t (\widetilde{\beta}(X_s)-\beta(X_s))ds, \quad r = 1,\dots,q,$$

is proved to be a q-dimensional standard Brownian motion under $\widetilde{\mathsf{P}}^T$. It is known that a sufficient condition for $L = (L_t)_{t\in[0,\infty)}$ being a P-martingale is that

$$\mathsf{E}\left[\exp\left(\frac{1}{2}\int_0^t (\widetilde{\beta}(X_s)-\beta(X_s))^{\mathrm{tr}}(\sigma\sigma^{\mathrm{tr}}(X_s))^{-1}(\widetilde{\beta}(X_s)-\beta(X_s))ds\right)\right] < \infty$$

holds for every $t \in [0,\infty)$. This is called *Novikov's criterion*; see Proposition VIII.1.15 of Revuz and Yor (1999).

6.5.4 Example: Counting processes

Let $(N^1,\dots,N^d)^{\mathrm{tr}}$ be counting processes defined on a stochastic basis $(\Omega,\mathcal{F};\mathbf{F},\mathsf{P})$ such that each N^i admits an intensity process λ^i, and we shall assume that N^i's have no simultaneous jumps; the last assumption does not depend on the choice of a probability measure. We regard $\lambda_s^i(\cdot) = \lambda(\cdot,s,x)$ for $x = a_i$ in the representation

$$N_t^i - \int_0^t \lambda_s^i ds = X_t^i = \int_{[0,t]\times\mathbb{R}^d} x^i(\mu^X(\cdot;ds,dx) - \lambda(\cdot,s,x)ds\eta(dx)),$$

where $\eta(dx) = \sum_{i=1}^{d} \varepsilon_{a_i}(dx)$ with a_i being the vector in $\mathbb{R}^d$ whose i-th component is 1 and the others are zero.

As an alternative for $(\lambda^1, ..., \lambda^d)^{\text{tr}}$, let a new candidate $(\widetilde{\lambda}^1, ..., \widetilde{\lambda}^d)^{\text{tr}}$ satisfying the property stated in the remark after the formula (6.14) be given. Then, the log-likelihood ratio formula is calculated as

$$\begin{aligned}\log L_t &= \sum_{i=1}^{d} \left\{ \int_0^t \log \widetilde{\lambda}_s^i dN_s^i - \int_0^t \widetilde{\lambda}_s^i ds \right\} \\ &\quad - \sum_{i=1}^{d} \left\{ \int_0^t \log \lambda_s^i dN_s^i - \int_0^t \lambda_s^i ds \right\}, \quad \forall t \in [0, \infty).\end{aligned}$$

Under the situation where this $L = (L_t)_{t \in [0,\infty)}$ is a P-martingale, we can apply Theorem 6.5.4 to conclude that, for every $i = 1, ..., d$, the intensity process of $N^i = (N_t^i)_{t \in [0,T]}$, under $\widetilde{\mathsf{P}}^T$ defined by (6.11), is $\widetilde{\lambda}^i$ on each compact $[0, T]$.

6.6 Inequalities for 1-Dimensional Martingales

6.6.1 Lenglart's inequality and its corollaries

Definition 6.6.1 Let X and Y be two càdlàg, adapted processes. We say that X is L-dominated by Y if it holds that $\mathsf{E}[|X_T|] \leq \mathsf{E}[|Y_T|]$ for every bounded stopping time T.

Theorem 6.6.2 (Lenglart's inequality) Let X be a $[0, \infty)$-valued càdlàg adapted process starting from zero, and A a $[0, \infty)$-valued càdlàg adapted process, whose all paths are non-decreasing, such that $A_0 \geq 0$ is deterministic. If X is L-dominated by A, then the following claims hold for any stopping time T.

(i) If A is just adapted, then

$$\mathsf{P}\left(\sup_{t \in [0,T]} X_t \geq \eta \right) \leq \frac{\delta + \mathsf{E}[\sup_{s \in [0,T]} \Delta A_s]}{\eta} + \mathsf{P}(A_T \geq \delta), \quad \forall \eta, \delta > 0.$$

(ii) In particular, if A is predictable, then

$$\mathsf{P}\left(\sup_{t \in [0,T]} X_t \geq \eta \right) \leq \frac{\mathsf{E}[A_T \wedge \delta]}{\eta} + \mathsf{P}(A_T \geq \delta), \quad \forall \eta, \delta > 0$$

and

$$\mathsf{E}\left[\sup_{t \in [0,T]} (X_t)^p \right] \leq \left(\frac{2-p}{1-p} \right) \mathsf{E}[(A_T)^p], \quad \forall p \in (0, 1).$$

Remark. In the standard textbooks on the martingale theory it is usually assumed that $A_0 = 0$. The reason why we extend the inequality to the case of a deterministic $A_0 \geq 0$ is to use the current result as a key tool to apply the "stochastic maximal inequality" given in Section A1.1. The assumption that $A_0 \geq 0$ is deterministic cannot be weakened to that it is $\mathcal{F}_0$-measurable; see Remark after Theorem 4.3.2 for a counter example.

Proof. Let us prove (i) and the first inequality in (ii). By the same reason as that in the proof of Theorem 4.3.2 it suffices to show the inequality with the left-hand side is replaced by $\mathsf{P}\left(\sup_{n \leq T} X_n > \eta\right)$. Also, we may assume that the stopping time T is bounded.

The inequalities are evident for $\delta \leq A_0$ when A_0 is positive, because the second term on the right-hand side is 1. So, we shall consider the case $\delta' := \delta - A_0 > 0$, including the case $A_0 = 0$. Set $R = \inf(t : X_t > \eta)$ and $S = \inf(t : A_t - A_0 \geq \delta') = \inf(t : A_t \geq \delta)$. Then R and S are stopping times (see Theorem 5.5.2 (i_2) and (i_3)), and we have also that $S(\omega) > 0$ for all ω. Observing that $\{\sup_{t \in [0,T]} X_t > \eta\} \subset \{A_T \geq \delta\} \cup \{R \leq T < S\}$, we have

$$\mathsf{P}\left(\sup_{t \in [0,T]} X_t > \eta\right) \leq \mathsf{P}(R \leq T < S) + \mathsf{P}(A_T \geq \delta).$$

To prove (i), the first term on the right-hand side is bounded by

$$\mathsf{P}(X_{R \wedge T \wedge S} \geq \eta) \leq \frac{1}{\eta}\mathsf{E}[A_{R \wedge T \wedge S}],$$

and $A_{R \wedge T \wedge S} \leq \delta + \sup_{t \in [0,T]} \Delta A_t$ yields the result.

To prove the first inequality in (ii), since A is predictable, S is a predictable time (see Theorem 5.5.4 (i_1)), and thus S is announced by a sequence (S_n) of stopping times. Thus we have

$$\mathsf{P}(R \leq T < S) \leq \lim_{(n)} \mathsf{P}(P \leq T < S_n) \leq \lim_{(n)} \mathsf{P}(X_{R \wedge T \wedge S_n} \geq \eta) \leq \frac{1}{\eta}\mathsf{E}[A_{R \wedge T \wedge S_n}].$$

Since $S(\omega) > 0$ for all ω, we have $S_n < S$ a.s., hence $A_{R \wedge T \wedge S_n} \leq A_{T \wedge S_n} \leq A_T \wedge \delta$. The inequality in (ii) has been proved.

The second inequality in (ii) is proved exactly in the same way as that for Theorem 4.3.2. The proof is finished. □

The important point of these inequalities is that the supremum with respect to t is taken in the inside of the probability; while the evaluation of such a probability is usually difficult, the inequalities presented here provides good bounds for those values. Readers should memorize an important usage of the inequalities in the following way: *When we would like to prove that a locally square-integrable martingale starting from zero converges in probability to the degenerate stochastic process "zero", it is sufficient to check that the quadratic variation of the process converges in probability to zero.* We state this fact in the form of a corollary (Corollary 6.6.4 below).

Corollary 6.6.3 (i) If M is a local martingale starting from zero, then it holds that for any stopping time T and any $\varepsilon, \delta > 0$

$$\mathsf{P}\left(\sup_{t\in[0,T]} |M_t| \geq \varepsilon\right) \leq \frac{\delta + \mathsf{E}[\sup_{t\in[0,T]}(\Delta M_t)^2]}{\varepsilon^2} + \mathsf{P}([M]_T \geq \delta).$$

(ii) If M is a locally square-integrable martingale starting from zero, then it holds that for any stopping time T and any $\varepsilon, \delta > 0$

$$\mathsf{P}\left(\sup_{t\in[0,T]} |M_t| \geq \varepsilon\right) \leq \frac{\delta}{\varepsilon^2} + \mathsf{P}(\langle M\rangle_T \geq \delta).$$

Corollary 6.6.4 (Corollary to Lenglart's inequality) For every $n \in \mathbb{N}$, let M^n be a local martingale starting from zero and T_n a stopping time[3], both defined on a stochastic basis $(\Omega^n, \mathcal{F}^n; \mathbf{F}^n, \mathsf{P}^n)$.

(i) It holds as $n \to \infty$ that:

$$[M^n]_{T_n} = o_{\mathsf{P}^n}(1) \text{ and } \mathsf{E}^n\left[\sup_{t\in[0,T_n]} |\Delta[M^n]_t|\right] = o(1) \quad \text{imply} \quad \sup_{t\in[0,T_n]} |M_t^n| = o_{\mathsf{P}^n}(1);$$

$$[M^n]_{T_n} = O_{\mathsf{P}^n}(1) \text{ and } \mathsf{E}^n\left[\sup_{t\in[0,T_n]} |\Delta[M^n]_t|\right] = O(1) \quad \text{imply} \quad \sup_{t\in[0,T_n]} |M_t^n| = O_{\mathsf{P}^n}(1).$$

(ii) When M^n's are locally square-integrable martingales, it holds as $n \to \infty$ that:

$$\langle M^n\rangle_{T_n} = o_{\mathsf{P}^n}(1) \quad \text{implies} \quad \sup_{t\in[0,T_n]} |M_t^n| = o_{\mathsf{P}^n}(1);$$

$$\langle M^n\rangle_{T_n} = O_{\mathsf{P}^n}(1) \quad \text{implies} \quad \sup_{t\in[0,T_n]} |M_t^n| = O_{\mathsf{P}^n}(1).$$

Proofs of Corollaries 6.6.3 and 6.6.4. Let us prove the claim (ii) of Corollary 6.6.3; the claim (i) can be also proved in a similar way, and the claims in Corollary 6.6.4 are immediate from Corollary 6.6.3.

First choose a localizing sequence (T_n), and apply the above theorem to $X := (M^{T_n})^2$ and $A := \langle M^{T_n}\rangle$ to obtain that

$$\begin{aligned}\mathsf{P}\left(\sup_{t\in[0,T]} |M_t^{T_n}| > \varepsilon\right) &= \mathsf{P}\left(\sup_{t\in[0,T]} (M_t^{T_n})^2 > \varepsilon^2\right) \\ &\leq \frac{\delta}{\varepsilon^2} + \mathsf{P}(\langle M^{T_n}\rangle_T \geq \delta).\end{aligned}$$

Next, let $n \to \infty$ to have that

$$\mathsf{P}\left(\sup_{t\in[0,T]} |M_t| > \varepsilon\right) \leq \frac{\delta}{\varepsilon^2} + \mathsf{P}(\langle M\rangle_T \geq \delta)$$

[3]Our setting allows either the case of $T_n \to \infty$ as $n \to \infty$ in some sense or the case of $T_n \equiv T$ where T is a fixed time.

by some monotone arguments.

Finally, replacing ε appearing in the above inequality by $\varepsilon - (1/m)$ and letting $m \to \infty$, we obtain the desired inequality. □

Exercise 6.6.1 For every $n \in \mathbb{N}$, let X^n be a càdlàg adapted process starting from zero which is L-dominated by a predictable, increasing process A^n, and T_n a stopping time, all defined on a stochastic basis $(\Omega^n, \mathcal{F}^n; \mathbf{F}^n, \mathsf{P}^n)$. Prove that $A^n_{T_n} = o_{\mathsf{P}^n}(1)$ implies $\sup_{t\in[0,T_n]} |X^n_t| = o_{\mathsf{P}^n}(1)$.

Exercise 6.6.2 For every $n \in \mathbb{N}$, let X^n be a locally integrable, increasing process, $X^{n,p}$ the predictable compensator for X^n, and T_n a stopping time, all defined on a stochastic basis $(\Omega^n, \mathcal{F}^n; \mathbf{F}^n, \mathsf{P}^n)$.

(i) Prove that $X^{n,p}_{T_n} = o_{\mathsf{P}^n}(1)$ implies $X^n_{T_n} = o_{\mathsf{P}^n}(1)$. The converse is not always true; construct a counter example. Prove that $\mathsf{E}^n[\sup_{t\in[0,T_n]} \Delta X^n_t] = o(1)$ and $X^n_{T_n} = o_{\mathsf{P}^n}(1)$ imply $X^{n,p}_{T_n} = o_{\mathsf{P}^n}(1)$.

(ii) Prove that $X^{n,p}_{T_n} = O_{\mathsf{P}^n}(1)$ implies $X^n_{T_n} = O_{\mathsf{P}^n}(1)$. The converse is not always true; construct a counter example. Prove that $\mathsf{E}^n[\sup_{t\in[0,T_n]} \Delta X^n_t] = O(1)$ and $X^n_{T_n} = O_{\mathsf{P}^n}(1)$ imply $X^{n,p}_{T_n} = O_{\mathsf{P}^n}(1)$.

6.6.2 Bernstein's inequality

Theorem 6.6.5 Let $(M_t)_{t\in[0,\infty)}$ be a locally square-integrable martingale on a stochastic basis such that $|\Delta M| \le a$ for a constant $a \ge 0$. Then, it holds for any stopping time T and any $x, v > 0$ that

$$\mathsf{P}\left(\sup_{t\in[0,T]} |M_t| \ge x,\ \langle M\rangle_T \le v \right) \le 2\exp\left(-\frac{x^2}{2(ax+v)} \right).$$

See Appendix B.6 of Shorack and Wellner (1986) for a proof. See van de Geer (1995) and Dzhaparidze and van Zanten (2001) for some extensions.

6.6.3 Burkholder-Davis-Gundy's inequalities

Theorem 6.6.6 For every $p \ge 1$ there exist some constants $c_p, C_p > 0$, depending only on p, such that for any local martingale $(M_t)_{t\in[0,\infty)}$ and any stopping time T on a stochastic basis $(\Omega, \mathcal{F}; (\mathcal{F}_t)_{t\in[0,\infty)}, \mathsf{P})$, it holds that

$$c_p \mathsf{E}\left[[M]_T^{p/2} \right] \le \mathsf{E}\left[\sup_{t\in[0,T]} |M_t|^p \right] \le C_p \mathsf{E}\left[[M]_T^{p/2} \right].$$

Moreover, it also holds that

$$c_p \mathsf{E}\left[[M]_T^{p/2} \middle| \mathcal{F}_0 \right] \le \mathsf{E}\left[\sup_{t\in[0,T]} |M_t|^p \middle| \mathcal{F}_0 \right] \le C_p \mathsf{E}\left[[M]_T^{p/2} \middle| \mathcal{F}_0 \right] \quad \text{a.s.}$$

Regarding the first displayed inequalities, see Theorems 17.7 and 26.12 of Kallenberg (2002) for elegant proofs, as well as Section IV.4 of Revuz and Yor (1999) for sophisticated treatment of the cases for continuous local martingales.

The second displayed inequalities can be proved in the same way as those in Theorem 4.3.5 by reducing the problem to the first displayed inequalities.

Part III

Asymptotic Statistics with Martingale Methods

7

Tools for Asymptotic Statistics

This chapter is devoted to preparing three tools for asymptotic statistics—the martingale central limit theorems, the uniform convergence of random fields, and some techniques concerning discrete sampling of diffusion processes. Many of the theorems presented in this chapter could be generalized more; for example, the first topic, the limit theory for (semi-)martingales, is explained much in depth by some authoritative books such as Hall and Heyde (1980) and Jacod and Shiryaev (2003). In contrast, we aim to give some elementary, self-contained proofs to which readers can easily access. The tools prepared here will be applied to concrete problems in statistics in the subsequent chapters repeatedly.

Throughout this chapter, the limit operations are taken as $n \to \infty$, unless otherwise stated.

7.1 Martingale Central Limit Theorems

In this section, we present four types of martingale central limit theorems (CLTs). Although the former three CLTs may be viewed as special cases of the last one, learning the proofs for the concrete, special forms of the theorems would be helpful for better understanding of that for the most general one.

7.1.1 Discrete-time martingales

First, we will establish the CLT for discrete-time martingales by two steps in this subsection, where a sequence of discrete-time stochastic bases $\mathbf{B}^n = (\Omega^n, \mathcal{F}^n; \mathbf{F}^n = (\mathcal{F}_k^n)_{k \in \mathbb{N}_0}, \mathsf{P}^n)$ is given. Let us start with proving a special case of the theorem, as a "lemma" for the second step.

Lemma 7.1.1 For every $n \in \mathbb{N}$, let $(\xi_k^n)_{k \in \mathbb{N}}$ be a real-valued martingale difference sequence such that $|\xi_k^n| \le a$ for all k, for a constant $a > 0$ not depending on n, and let T_n be a finite stopping time[1], both defined on a discrete-time stochastic basis $\mathbf{B}^n$.

Suppose that the following conditions (a) and (b) are satisfied:

(a) $\sum_{k=1}^{T_n} \mathsf{E}^n[(\xi_k^n)^2 | \mathcal{F}_{k-1}^n] \xrightarrow{\mathsf{P}^n} C$, where the limit C is a constant;

(b) $\sum_{k=1}^{T_n} \mathsf{E}^n[|\xi_k^n|^3 | \mathcal{F}_{k-1}^n] \xrightarrow{\mathsf{P}^n} 0$.

[1]We allow either the case of $T_n \to \infty$ as $n \to \infty$ in some sense or the case of $T_n \equiv T$ where T is a fixed time. This remark is valid also for Theorems 7.1.2 and 7.1.3 given below.

DOI: 10.1201/9781315117768-7

Then it holds that $\sum_{k=1}^{T_n} \xi_k^n \stackrel{\mathsf{P}^n}{\Longrightarrow} \mathcal{N}(0,C)$ in $\mathbb{R}$.

Remark. The assumption that ξ_k^n's are uniformly bounded is not actually necessary. Also, the condition (b) appearing above is stronger than Lindeberg's condition[2]:

$$\sum_{k=1}^{T_n} \mathsf{E}^n[(\xi_k^n)^2 1\{|\xi_k^n| > \varepsilon\}|\mathcal{F}_{k-1}^n] \stackrel{\mathsf{P}^n}{\longrightarrow} 0, \quad \forall \varepsilon > 0.$$

Although these stronger conditions are temporally assumed here due to some technical reasons, the former will be removed and the latter weakened into Lindeberg's condition in our main theorems below (i.e.,Theorems 7.1.2 and 7.1.3).

Let us prepare a notation for conveniences in the proof given below; we define

$$V_0^n = 0 \quad \text{and} \quad V_m^n = \sum_{k=1}^{m} \mathsf{E}^n[(\xi_k^n)^2|\mathcal{F}_{k-1}^n], \quad \forall m \in \mathbb{N}.$$

Remark. Without loss of generality, we may assume that all paths $m \mapsto V_m^n(\omega)$ are non-decreasing.

Proof of Lemma 7.1.1. Define the stopping time S_n by

$$S_n := \inf\{m \in \mathbb{N}_0 : V_m^n \geq C\} \wedge T_n.$$

Since $(C \wedge V_{T_n}^n) \leq V_{S_n}^n \leq V_{T_n}^n$ and $V_{T_n}^n \stackrel{\mathsf{P}^n}{\longrightarrow} C$, it holds that $V_{S_n}^n \stackrel{\mathsf{P}^n}{\longrightarrow} C$. This implies that

$$\sum_{k=1}^{S_n} \xi_k^n - \sum_{k=1}^{T_n} \xi_k^n \stackrel{\mathsf{P}^n}{\longrightarrow} 0,$$

with the help of the corollary to Lenglart's inequality (Corollary 4.3.3 (i)), because the quadratic covariation of the locally square-integrable martingale $(Y_m^n)_{m\in\mathbb{N}_0}$ defined by

$$Y_m^n := \sum_{k=1}^{m\wedge S_n} \xi_k^n - \sum_{k=1}^{m\wedge T_n} \xi_k^n$$

is computed as

$$((Y^n)^2)_{T_n}^p = V_{S_n}^n + V_{T_n}^n - 2V_{S_n\wedge T_n}^n = V_{T_n}^n - V_{S_n}^n \stackrel{\mathsf{P}^n}{\longrightarrow} 0.$$

It is thus sufficient to prove that

$$\lim_{n\to\infty} \mathsf{E}^n\left[\exp\left(iz\sum_{k=1}^{S_n} \xi_k^n + \frac{z^2}{2}C\right)\right] = 1, \quad \forall z \in \mathbb{R},$$

---

[2]This is because $\sum_{k=1}^{T_n} \mathsf{E}^n[(\xi_k^n)^2 1\{|\xi_k| > \varepsilon\}|\mathcal{F}_{k-1}^n] \leq \sum_{k=1}^{T_n} \mathsf{E}^n[|\xi_k^n|^3|\mathcal{F}_{k-1}^n]/\varepsilon$ for any $\varepsilon > 0$. A generalization of this argument will be given in Exercise 7.1.1 after Theorem 7.1.3 at the end of this subsection. See also Exercise 7.1.3 after Theorem 7.1.6 for Lindeberg's and Lyapunov's conditions for counting processes.

because this implies that

$$\lim_{n\to\infty} \mathsf{E}^n\left[\exp\left(iz\sum_{k=1}^{S_n}\xi_k^n\right)\right] = \exp\left(-\frac{z^2}{2}C\right), \quad \forall z\in\mathbb{R}.$$

For this purpose, it is sufficient to prove that

$$\lim_{n\to\infty} \mathsf{E}^n\left[\exp\left(iz\sum_{k=1}^{S_n}\xi_k^n + \frac{z^2}{2}V_{S_n}^n\right)\right] = 1, \quad \forall z\in\mathbb{R}, \tag{7.1}$$

because the fact $V_{S_n}^n \le C + a^2$ implies that

$$\begin{aligned}
&\left|\mathsf{E}^n\left[\exp\left(iz\sum_{i=1}^{S_n}\xi_i^n + \frac{z^2}{2}C\right)\right] - \mathsf{E}^n\left[\exp\left(iz\sum_{i=1}^{S_n}\xi_i^n + \frac{z^2}{2}V_{S_n}^n\right)\right]\right| \\
\le\ & \mathsf{E}^n\left[\left|\exp\left(\frac{z^2}{2}C\right) - \exp\left(\frac{z^2}{2}V_{S_n}^n\right)\right|\right] \to 0, \quad \text{as } n\to\infty;
\end{aligned}$$

see Exercise 2.3.2 for the convergence on the last line.

To prove (7.1), fix any $z\in\mathbb{R}\setminus\{0\}$ (the case $z=0$ is trivial), and put

$$G_m^n := \exp\left(iz\sum_{k=1}^{m}\xi_k^n + \frac{z^2}{2}V_m^n\right),$$

where $G_0^n = 1$. Observe that

$$\begin{aligned}
G_{S_n}^n - 1 &= \sum_{k=1}^{S_n} G_{k-1}^n\left\{\exp\left(iz\xi_k^n + \frac{z^2}{2}\mathsf{E}^n[(\xi_k^n)^2|\mathcal{F}_{k-1}^n]\right) - 1\right\} \\
&= \sum_{k=1}^{S_n} \widetilde{G}_{k-1}^n\left\{\exp(iz\xi_k^n) - \exp\left(-\frac{z^2}{2}\mathsf{E}^n[(\xi_k^n)^2|\mathcal{F}_{k-1}^n]\right)\right\},
\end{aligned}$$

where

$$\widetilde{G}_{k-1}^n = G_{k-1}^n \exp\left(\frac{z^2}{2}\mathsf{E}^n[(\xi_k^n)^2|\mathcal{F}_{k-1}^n]\right).$$

By using the inequalities $|e^{ix} - (1 + ix - \frac{x^2}{2})| \le |x|^3$ for all $x\in\mathbb{R}$ and $|e^{-x} - (1-x)| \le K_{z,a}x^2$ if $|x| \le \frac{z^2}{2}a^2$, where $K_{z,a}$ is a constant depending only on z and a (see Lemma A1.2.3), we have that

$$G_{S_n}^n - 1 = iM_{S_n}^{n,1} + M_{S_n}^{n,2} + R^{n,1} + R^{n,2},$$

where

$$M_m^{n,1} = \sum_{k=1}^{m} \widetilde{G}_{k-1}^n \xi_k^n,$$

$$
\begin{aligned}
M_m^{n,2} &= -\frac{z^2}{2}\sum_{k=1}^{m}\widetilde{G}_{k-1}^n\{(\xi_k^n)^2 - \mathsf{E}^n[(\xi_k^n)^2|\mathcal{F}_{k-1}^n]\}, \\
|R^{n,1}| &\le \sum_{k=1}^{S_n}|\widetilde{G}_{k-1}^n||\xi_k^n|^3, \\
|R^{n,2}| &\le K_{z,a}\sum_{k=1}^{S_n}|\widetilde{G}_{k-1}^n|\left(\frac{z^2}{2}\mathsf{E}^n[(\xi_k^n)^2|\mathcal{F}_{k-1}^n]\right)^2.
\end{aligned}
$$

Note that $|\widetilde{G}_{k-1}^n| \le \exp\left(\frac{z^2}{2}(C+2a^2)\right) =: \overline{G} < \infty$. Thus, both $(M_{m\wedge S_n}^{n,1})_{m\in\mathbb{N}_0}$ and $(M_{m\wedge S_n}^{n,2})_{m\in\mathbb{N}_0}$ are uniformly integrable martingales starting from zero, and it follows from the optional sampling theorem that $\mathsf{E}^n[M_{S^n}^{n,1}] = \mathsf{E}^n[M_{S_n}^{n,2}] = 0$.

On the other hand, we have that

$$
|R^{n,1}| \le \overline{G}\sum_{k=1}^{S_n}\{|\xi_k^n|^3 - \mathsf{E}^n[|\xi_k^n|^3|\mathcal{F}_{k-1}^n]\} + \overline{G}\sum_{k=1}^{S_n}\mathsf{E}^n[|\xi_k^n|^3|\mathcal{F}_{k-1}^n].
$$

The expectation of the first term on the right-hand side is zero by the optional sampling theorem, while the sequence of the expectations of the second terms converges to zero because the terms are uniformly bounded by $\overline{G}(V+a^2)a$ and converges in probability to zero by the assumption (b) (recall also Exercise 2.3.2). Hence we obtain that $\lim_{n\to\infty}\mathsf{E}^n[|R^{n,1}|] = 0$.

Finally, observe that

$$
\begin{aligned}
|R^{n,2}| &\le K_{z,a}\overline{G}\cdot\frac{z^2}{4}V_{S_n}^n\max_{1\le k\le S_n}\mathsf{E}^n[(\xi_k^n)^2|\mathcal{F}_{k-1}^n] \\
&\le K_{z,a}\overline{G}\cdot\frac{z^4}{4}(C+a^2)\max_{1\le k\le S_n}\mathsf{E}^n[(\xi_k^n)^2|\mathcal{F}_{k-1}^n].
\end{aligned}
$$

Since

$$
\max_{1\le k\le S_n}\mathsf{E}^n[(\xi_k^n)^2|\mathcal{F}_{k-1}^n] \le \varepsilon^2 + \sum_{k=1}^{S_n}\mathsf{E}^n[(\xi_k^n)^2 1\{|\xi_k^n| > \varepsilon\}|\mathcal{F}_{k-1}^n], \quad \forall\varepsilon > 0,
$$

it is proved that $|R^{n,2}|$ converges in probability to zero. Noting that $|R^{n,2}|$ is uniformly bounded, we have that $\lim_{n\to\infty}\mathsf{E}^n[|R^{n,2}|] = 0$ by Exercise 2.3.2.

We have therefore established (7.1), and the proof is finished. □

Now, we shall remove the unnecessary conditions imposed in Lemma 7.1.1, as we previously announced.

Theorem 7.1.2 For every $n \in \mathbb{N}$, let $(\xi_k^n)_{k\in\mathbb{N}}$ be a real-valued martingale difference sequence such that $\mathsf{E}^n[(\xi_k^n)^2] < \infty$ for all k, and let T_n be a finite stopping time, both defined on a discrete-time stochastic basis $\mathbf{B}^n$.

Suppose that the following conditions (a) and (b) are satisfied:

(a) $\sum_{k=1}^{T_n} \mathsf{E}^n[(\xi_k^n)^2|\mathcal{F}_{k-1}^n] \xrightarrow{\mathsf{P}^n} C$, where the limit C is a constant;

(b) $\sum_{k=1}^{T_n} \mathsf{E}^n[(\xi_k^n)^2 1\{|\xi_k^n| > \varepsilon\}|\mathcal{F}_{k-1}^n] \xrightarrow{\mathsf{P}^n} 0$ for every $\varepsilon > 0$.

Then it holds that $\sum_{k=1}^{T_n} \xi_k^n \xRightarrow{\mathsf{P}^n} \mathcal{N}(0,C)$ in $\mathbb{R}$.

Proof. Fix any $a > 0$, and put

$$\widetilde{\xi}_k^n := \xi_k^n 1\{|\xi_k^n| \leq a\} - \mathsf{E}^n[\xi_k^n 1\{|\xi_k^n| \leq a\}|\mathcal{F}_{k-1}^n].$$

It follows again from the corollary to Lenglart's inequality that

$$\sum_{k=1}^{T_n} \xi_k^n - \sum_{k=1}^{T_n} \widetilde{\xi}_k^n \xrightarrow{\mathsf{P}^n} 0,$$

because the predictable quadratic variation of the martingale $(Y_m^n)_{m\in\mathbb{N}_0}$ given by

$$Y_m^n = \sum_{k=1}^{m} (\xi_k^n - \widetilde{\xi}_k^n) = \sum_{k=1}^{m} (\xi_k^n 1\{|\xi_k^n| > a\} - \mathsf{E}^n[\xi_k^n 1\{|\xi_k^n| > a\}|\mathcal{F}_{k-1}^n])$$

stopped at $m = T_n$ is computed as

$$\begin{aligned}
((Y^n)^2)_{T_n}^p &= \sum_{k=1}^{T_n} \mathsf{E}^n[(\xi_k^n 1\{|\xi_k^n| > a\} - \mathsf{E}^n[\xi_k^n 1\{|\xi_k^n| > a\}|\mathcal{F}_{k-1}^n])^2|\mathcal{F}_{k-1}^n] \\
&= \sum_{k=1}^{T_n} \{\mathsf{E}^n[(\xi_k^n 1\{|\xi_k^n| > a\})^2|\mathcal{F}_{k-1}^n] - (\mathsf{E}^n[\xi_k^n 1\{|\xi_k^n| > a\}|\mathcal{F}_{k-1}^n])^2\} \\
&\leq \sum_{k=1}^{T_n} \mathsf{E}^n[(\xi_k^n)^2 1\{|\xi_k^n| > a\}|\mathcal{F}_{k-1}^n] \\
&\xrightarrow{\mathsf{P}^n} 0.
\end{aligned}$$

It is thus sufficient to check the conditions in Lemma 7.1.1 are satisfied for the martingale difference sequences $(\widetilde{\xi}_k^n)_{k\in\mathbb{N}}$.

It is evident that $|\widetilde{\xi}_k^n| \leq 2a$ for all n, k. Next observe that

$$\begin{aligned}
&\sum_{k=1}^{T_n} \mathsf{E}^n[(\widetilde{\xi}_k^n)^2|\mathcal{F}_{k-1}^n] \\
&= \sum_{k=1}^{T_n} \{\mathsf{E}^n[(\xi_k^n 1\{|\xi_k^n| \leq a\})^2|\mathcal{F}_{k-1}^n] - (\mathsf{E}^n[\xi_k^n 1\{|\xi_k^n| \leq a\}|\mathcal{F}_{k-1}^n])^2\} \\
&= \sum_{k=1}^{T_n} \mathsf{E}^n[(\xi_k^n)^2|\mathcal{F}_{k-1}^n] - \sum_{k=1}^{T_n} \mathsf{E}^n[(\xi_k^n)^2 1\{|\xi_k^n| > a\}|\mathcal{F}_{k-1}^n] \\
&\quad - \sum_{k=1}^{T_n} (\mathsf{E}^n[\xi_k^n 1\{|\xi_k^n| \leq a\}|\mathcal{F}_{k-1}^n])^2.
\end{aligned}$$

The last term on the right-hand side is computed as

$$\begin{aligned}
&\sum_{k=1}^{T_n}(\mathsf{E}^n[\xi_k^n 1\{|\xi_k^n| \le a\}|\mathcal{F}_{k-1}^n])^2 \\
&\le a\sum_{k=1}^{T_n}|\mathsf{E}^n[\xi_k^n 1\{|\xi_k^n| \le a\}|\mathcal{F}_{k-1}^n]| \\
&= a\sum_{k=1}^{T_n}|\mathsf{E}^n[\xi_k^n 1\{|\xi_k^n| > a\}|\mathcal{F}_{k-1}^n]| \\
&\le a\sum_{k=1}^{T_n}\mathsf{E}^n[|\xi_k^n| 1\{|\xi_k^n| > a\}|\mathcal{F}_{k-1}^n] \\
&\le a\sum_{k=1}^{T_n}\mathsf{E}^n\left[\frac{(\xi_k^n)^2}{a}1\{|\xi_k^n| > a\}\middle|\mathcal{F}_{k-1}^n\right],
\end{aligned}$$

which converges in probability to zero. It therefore holds that

$$\sum_{k=1}^{T_n}\mathsf{E}^n[(\widetilde{\xi}_k^n)^2|\mathcal{F}_{k-1}^n]-\sum_{k=1}^{T_n}\mathsf{E}^n[(\xi_k^n)^2|\mathcal{F}_{k-1}^n]\xrightarrow{\mathsf{P}^n}0,$$

which implies that the condition (a) in Lemma 7.1.1 is satisfied for $(\widetilde{\xi}_k^n)_{k\in\mathbb{N}}$.

Finally, we have that for any $\varepsilon \in (0,a)$,

$$\begin{aligned}
&\sum_{k=1}^{T_n}\mathsf{E}^n[|\widetilde{\xi}_k^n|^3|\mathcal{F}_{k-1}^n] \\
&\le 4\sum_{k=1}^{T_n}\mathsf{E}^n[(|\xi_k^n 1\{|\xi_k^n| \le a\}|^3+|\mathsf{E}^n[\xi_k^n 1\{|\xi_k^n| \le a\}|\mathcal{F}_{k-1}^n]|^3)|\mathcal{F}_{k-1}^n] \\
&\le 8\sum_{k=1}^{T_n}\mathsf{E}^n[|\xi_k^n|^3 1\{|\xi_k^n| \le a\}|\mathcal{F}_{k-1}^n] \\
&\le 8a\sum_{k=1}^{T_n}\mathsf{E}^n[(\xi_k^n)^2 1\{|\xi_k^n| > \varepsilon\}|\mathcal{F}_{k-1}^n]+8\varepsilon\sum_{k=1}^{T_n}\mathsf{E}^n[(\xi_k^n)^2 1\{|\xi_k^n| \le \varepsilon\}|\mathcal{F}_{k-1}^n] \\
&\le 8a\sum_{k=1}^{T_n}\mathsf{E}^n[(\xi_k^n)^2 1\{|\xi_k^n| > \varepsilon\}|\mathcal{F}_{k-1}^n]+8\varepsilon\sum_{k=1}^{T_n}\mathsf{E}^n[(\xi_k^n)^2|\mathcal{F}_{k-1}^n] \\
&\xrightarrow{\mathsf{P}^n} 8\varepsilon C.
\end{aligned}$$

Since the choice of ε is arbitrary, we may conclude that the condition (b) in Lemma 7.1.1 is satisfied for $(\widetilde{\xi}_k^n)_{k\in\mathbb{N}}$. The proof is finished. □

We finish up this subsection with presenting a multidimensional version of CLT for discrete-time martingales.

Theorem 7.1.3 (The CLT for discrete-time martingales) For every $n \in \mathbb{N}$, let $(\xi_k^n)_{k\in\mathbb{N}} = ((\xi_k^{n,1},...,\xi_k^{n,d})^{\mathrm{tr}})_{k\in\mathbb{N}}$ be a d-dimensional martingale difference sequence such that $\mathsf{E}^n[||\xi_k^n||^2] < \infty$ for all k, and let T_n be a finite stopping time, both defined on a discrete-time stochastic basis $\mathbf{B}^n$.

If

$$\sum_{k=1}^{T_n} \mathsf{E}^n[||\xi_k^n||^2 1\{||\xi_k^n|| > \varepsilon\}|\mathcal{F}_{k-1}^n] \xrightarrow{\mathsf{P}^n} 0, \quad \forall \varepsilon > 0, \tag{7.2}$$

and if

$$\sum_{k=1}^{T_n} \mathsf{E}^n[\xi_k^{n,i}\xi_k^{n,j}|\mathcal{F}_{k-1}^n] \xrightarrow{\mathsf{P}^n} C^{i,j}, \quad \forall (i,j) \in \{1,...,d\}^2, \tag{7.3}$$

where the limits, $C^{i,j}$'s, are some constants, then it holds that

$$\sum_{k=1}^{T_n} \xi_k^n \overset{\mathsf{P}^n}{\Longrightarrow} \mathcal{N}_d(0,C) \quad \text{in } \mathbb{R}^d, \quad \text{where } C = (C^{i,j})_{(i,j)\in\{1,...,d\}^2}.$$

Exercise 7.1.1 The condition (7.2) is called *Lindeberg's condition*. Prove that, for given $\alpha > 0$, a sufficient condition for the *Lindeberg-type condition*

$$\sum_{k=1}^{T_n} \mathsf{E}^n[||\xi_k^n||^\alpha 1\{||\xi_k^n|| > \varepsilon\}|\mathcal{F}_{k-1}^n] \xrightarrow{\mathsf{P}^n} 0, \quad \forall \varepsilon > 0,$$

is that: there exists a $\delta > 0$ such that

$$\sum_{k=1}^{T_n} \mathsf{E}^n[||\xi_k^n||^{\alpha+\delta}|\mathcal{F}_{k-1}^n] \xrightarrow{\mathsf{P}^n} 0,$$

which is called the *Lyapunov-type condition* (the case $\alpha = 2$ is called *Lyapunov's condition*).

Exercise 7.1.2 Prove Theorem 7.1.3 using Cramér-Wold's device (Theorem 2.3.9). [**Hint:** Reduce the problem to the one for the sequence of real-valued martingales $c^{\mathrm{tr}}\sum_{k=1}^{T_n}\xi_k^n$ for any fixed $c \in \mathbb{R}^d$, and check the conditions in Theorem 7.1.2.]

7.1.2 Continuous local martingales

This subsection provides the CLT for continuous local martingales.

Theorem 7.1.4 (The CLT for continuous local martingales) For every $n \in \mathbb{N}$, let $M^n = (M^{n,1},...,M^{n,d})^{\mathrm{tr}}$ be a d-dimensional continuous local martingales starting from zero, and let T_n be a finite stopping time[3], both defined on a stochastic basis $(\Omega^n, \mathcal{F}^n; (\mathcal{F}_t^n)_{t\in[0,\infty)}, \mathsf{P}^n)$.

[3] We allow either the case of $T_n \to \infty$ as $n \to \infty$ in some sense or the case of $T_n \equiv T$ where T is a fixed time.

If

$$\langle M^{n,i}, M^{n,j}\rangle_{T_n} \xrightarrow{\mathsf{P}^n} C^{i,j}, \quad \forall (i,j) \in \{1,...,d\}^2, \tag{7.4}$$

where $C = (C^{i,j})_{(i,j)\in\{1,...,d\}^2}$ is a deterministic matrix, then it holds that

$$M^n_{T_n} \stackrel{\mathsf{P}^n}{\Longrightarrow} \mathcal{N}_d(0,C) \quad \text{in } \mathbb{R}^d.$$

Proof. Let us consider only the 1-dimensional case, where the notations $M^n = M^{n,1}$ and $C = C^{1,1}$ are used; the multidimensional case is reduced to the 1-dimensional case by using Cramér-Wold's device (recall the arguments in Exercise 7.1.2).

Introduce the stopping time

$$S_n = \inf\{t \in [0,\infty) : \langle M^n\rangle_t \geq C\} \wedge T_n.$$

Then, since $C \wedge \langle M^n\rangle_{T_n} = \langle M^n\rangle_{S_n}$ and $\langle M^n\rangle_{T_n} \xrightarrow{\mathsf{P}^n} C$, it holds that $\langle M^n\rangle_{S_n} \xrightarrow{\mathsf{P}^n} C$. Notice also that

$$\langle M^n\rangle_{S_n} \leq C, \quad \forall n \in \mathbb{N}.$$

To compute the characteristic function of $M^n_{S_n}$, fix any $z \in \mathbb{R}\setminus\{0\}$ and put

$$G^n_t = \exp\left(\frac{z^2}{2}\langle M^n\rangle_t + izM^n_t\right).$$

We shall prove that

$$\mathsf{E}^n[G^n_{S_n}] = 1, \quad \forall n \in \mathbb{N}, \tag{7.5}$$

and

$$M^n_{T_n} - M^n_{S_n} \xrightarrow{\mathsf{P}^n} 0, \quad \text{as } n \to \infty. \tag{7.6}$$

Once we have proved this two claims, the former implies that

$$\begin{aligned}
&\left|\mathsf{E}^n\left[\exp\left(\frac{z^2}{2}C + izM^n_{S_n}\right)\right] - 1\right| \\
&= \left|\mathsf{E}^n\left[\exp\left(\frac{z^2}{2}C + izM^n_{S_n}\right)\right] - \mathsf{E}^n\left[\exp\left(\frac{z^2}{2}\langle M^n\rangle_{S_n} + izM^n_{S_n}\right)\right]\right| \\
&\leq \mathsf{E}^n\left[\left|\exp\left(\frac{z^2}{2}C\right) - \exp\left(\frac{z^2}{2}\langle M^n\rangle_{S_n}\right)\right|\right] \to 0, \quad \text{as } n \to \infty,
\end{aligned}$$

which implies that $\lim_{n\to\infty} \mathsf{E}^n[\exp(izM^n_{S_n})] = e^{-\frac{z^2}{2}C}$. Thus we can get $M^n_{T_n} \stackrel{\mathsf{P}^n}{\Longrightarrow} \mathcal{N}(0,C)$ in $\mathbb{R}$ by using the latter and Slutsky's theorem.

Now, let us prove (7.5). Write $M^{n,S_n}_t = M^n_{t\wedge S_n}$, and apply Itô's formula to the 2-dimensional semimartingale $(X^1,X^2) = (\frac{z^2}{2}\langle M^{n,S_n}\rangle, zM^{n,S_n})$ substituted to the function $f(x_1,x_2) = \exp(x_1 + ix_2)$. Since $\frac{\partial}{\partial x_1}f(x_1,x_2) = e^{x_1+ix_2}$, $\frac{\partial}{\partial x_2}f(x_1,x_2) = ie^{x_1+ix_2}$ and $\frac{\partial^2}{\partial x_2^2}f(x_1,x_2) = -e^{x_1+ix_2}$, we have

$$G^n_t - 1 = iz\int_0^t \exp(X^1_s + iX^2_s)dM^{n,S_n}_s.$$

Since the stochastic integral on the right-hand side is, apart from "i", a square-integrable martingale starting from zero (indeed, its predictable quadratic variation is bounded), we thus obtain $\mathsf{E}^n[G^n_{S_n} - 1] = 0$ by the optional sampling theorem.

It remains to prove (7.6). This can be verified by applying the corollary to Lenglart's inequality (Corollary 6.6.4 (ii)) to the square-integrable martingale $Y^n = (Y^n_t)_{t\in[0,\infty)}$ given by

$$Y^n_t = M^n_{t\wedge T_n} - M^n_{t\wedge S_n}.$$

Indeed, by the assumption (7.4) it holds that

$$\begin{aligned} 0 \leq \langle Y^n\rangle_{T_n\vee S_n} &= \langle M^n\rangle_{T_n} + \langle M^n\rangle_{S_n} - 2\langle M^n\rangle_{T_n\wedge S_n} \\ &= \langle M^n\rangle_{T_n} - \langle M^n\rangle_{S_n} \\ &\xrightarrow{\mathsf{P}^n} 0. \end{aligned}$$

The proof is finished. □

7.1.3 Stochastic integrals w.r.t. counting processes

Throughout this subsection, let a sequence of stochastic bases $\mathbf{B}^n$ be given.

Theorem 7.1.5 (The CLT for stochastic integrals, I) For every $n \in \mathbb{N}$, let N^n be a counting process with the intensity λ^n, and let $H^n = (H^{n,1},...,H^{n,d})^{\mathrm{tr}}$ be an $\mathbb{R}^d$-valued predictable processes such that $t \rightsquigarrow \int_0^t ||H^n_s||^2\lambda^n_s ds$ is locally integrable, both defined on a stochastic basis $\mathbf{B}^n$. Introduce the sequence of the d-dimensional locally square-integrable martingales $M^n = (M^{n,1},...,M^{n,d})^{\mathrm{tr}}$ given by

$$M^{n,i}_t = \int_0^t H^{n,i}_s(dN^n_s - \lambda^n_s ds), \quad i = 1,...,d.$$

Let T_n be a finite stopping time[4] on $\mathbf{B}^n$, and suppose that

$$\int_0^{T_n} ||H^n_s||^2 1\{||H^n_s|| > \varepsilon\}\lambda^n_s ds \xrightarrow{\mathsf{P}^n} 0, \quad \forall\varepsilon > 0,$$

and that

$$\int_0^{T_n} H^{n,i}_s H^{n,j}_s \lambda^n_s ds \xrightarrow{\mathsf{P}^n} C^{i,j}, \quad \forall(i,j) \in \{1,...,d\}^2, \tag{7.7}$$

where $C = (C^{i,j})_{(i,j)\in\{1,...,d\}^2}$ is a deterministic matrix. Then it holds that

$$M^n_{T_n} \xrightarrow{\mathsf{P}^n} \mathcal{N}_d(0,C) \quad \text{in } \mathbb{R}^d.$$

[4]We allow either of the case of $T_n \to \infty$ as $n \to \infty$ in some sense or the case of $T_n \equiv T$ where T is a fixed time. This remark is valid also for Theorem 7.1.6 given below

Proof. We shall prove the case $d = 1$ only, using the notations

$$M_t^n = \int_0^t H_s^n (dN_s^n - \lambda_s^n ds), \quad C = C^{1,1};$$

the multidimensional case is reduced to the case $d = 1$ by Cramér-Wold's device (recall the arguments in Exercise 7.1.2).

Introduce the stopping time

$$S_n = \inf\{t \in [0,\infty) : \langle M^n \rangle_t \geq C\} \wedge T_n.$$

Then, since $C \wedge \langle M^n \rangle_{T_n} = \langle M^n \rangle_{S_n}$, $\langle M^n \rangle_{T_n} \xrightarrow{\mathsf{P}^n} C$ implies that $\langle M^n \rangle_{S_n} \xrightarrow{\mathsf{P}^n} C$. Notice also that

$$\langle M^n \rangle_{S_n} \leq C.$$

To compute the characteristic function, fix any $z \in \mathbb{R} \setminus \{0\}$ and define

$$G_t^n = \exp\left(\frac{z^2}{2}\langle M^{n,S_n}\rangle_t + izM_t^{n,S_n}\right),$$

where $M_t^{n,S_n} = M_{t\wedge S_n}$. We shall prove that

$$\lim_{n\to\infty} \mathsf{E}^n[G_{S_n}^n] = 1 \tag{7.8}$$

and that

$$M_{T_n}^n - M_{S_n}^n \xrightarrow{\mathsf{P}^n} 0, \quad \text{as } n \to \infty. \tag{7.9}$$

Once we have proved these two claims, the former yields that

$$\lim_{n\to\infty} \mathsf{E}^n[\exp(izM_{S_n}^n)] = \exp\left(-\frac{z^2}{2}C\right)$$

by the same argument as that in the proofs of Lemma 7.1.1 and Theorem 7.1.4, which implies that $M_{S_n}^n \xrightarrow{\mathsf{P}^n} \mathcal{N}(0,C)$ in $\mathbb{R}$. Thus it follows from the latter and Slutsky's theorem that $M_{T_n}^n \xrightarrow{\mathsf{P}^n} \mathcal{N}(0,C)$ in $\mathbb{R}$.

Now, let us prove (7.8). It follows from Itô's formula that

$$\begin{aligned} G_t^n - 1 &= \int_0^{t\wedge S_n} G_{s-}^n (\exp(izH_s^n) - 1) dM_s^n \\ &\quad + \int_0^{t\wedge S_n} G_{s-}^n \left\{\exp(izH_s^n) - 1 - izH_s^n + \frac{(zH_s^n)^2}{2}\right\} \lambda_s^n ds. \end{aligned}$$

Notice that the first term on the right-hand side is a square-integrable martingale; indeed, the fact that

$$\sup_{s\in[0,S_n]} |G_{s-}^n| \leq \exp\left(\frac{z^2}{2}C\right) \quad \text{and} \quad |e^{ix} - 1| \leq 2, \ \forall x \in \mathbb{R},$$

yields that the predictable quadratic variation is bounded. Thus it follows from the optional sampling theorem that the expectation of the first term on the right-hand side with $t = S_n$ is zero. We therefore have that $\mathsf{E}[G^n_{S_n} - 1]$ is equal to the expectation of the second term on the right-hand side with $t = S_n$. Its value is, by using the well-known inequality $|e^{ix} - 1 - ix + \frac{x^2}{2}| \leq |x|^3 \wedge |x|^2$ (see Lemma A1.2.3), bounded by (for any $\varepsilon > 0$)

$$
\begin{aligned}
&\mathsf{E}^n\left[\int_0^{S_n} |G^n_{s-}|\left|\exp(izH^n_s) - 1 - izH^n_s + \frac{(zH^n_s)^2}{2}\right|\lambda^n_s ds\right] \\
&\leq \exp\left(\frac{z^2}{2}C\right)\left\{\mathsf{E}^n\left[\int_0^{S_n} |zH^n_s|^3 1\{|zH^n_s| \leq \varepsilon\}\lambda^n_s ds\right]\right. \\
&\qquad \left. +\mathsf{E}^n\left[\int_0^{S_n} (zH^n_s)^2 1\{|zH^n_s| > \varepsilon\}\lambda^n_s ds\right]\right\} \\
&\leq \exp\left(\frac{z^2}{2}C\right)\left\{\varepsilon\frac{z^2}{2}C + \mathsf{E}^n\left[\int_0^{S_n} (zH^n_s)^2 1\{|zH^n_s| > \varepsilon\}\lambda^n_s ds\right]\right\}.
\end{aligned}
$$

Here, noting that the sequence of random variables $\int_0^{S_n}(zH^n_s)^2 1\{|zH^n_s| > \varepsilon\}\lambda^n_s ds$ is bounded by z^2C, the assumption that the sequence converges in probability to zero implies that the sequence of the expectations also converges to zero by Exercise 2.3.2. In conclusion, the last expectation on the right-hand side converges to zero, and since the choice of $\varepsilon > 0$ is arbitrary we obtain that $\lim_{n\to\infty} \mathsf{E}^n[G^n_{S_n} - 1] = 0$. The claim (7.8) has been proved.

The claim (7.9) can be proved in a similar way to (7.6). The proof is finished. □

For the purpose of some applications (see, e.g., Section 8.5.4 for Cox's regression models), let us state a slight generalization of the above theorem here. The proof is similar to that of the preceding one, so it is omitted.

Theorem 7.1.6 (The CLT for stochastic integrals, II) For every $n \in \mathbb{N}$ and every $k = 1, ..., m_n$, let $N^{n,k}$ be a counting process with the intensity $\lambda^{n,k}$, and let $H^{n,k} = (H^{n,k,1}, ..., H^{n,k,d})^{\mathrm{tr}}$ be $\mathbb{R}^d$-valued predictable processes such that $t \leadsto \int_0^t ||H^{n,k}_s||^2\lambda^{n,k}_s ds$ is locally integrable, all defined on a stochastic basis $\mathbf{B}^n$. For every $n \in \mathbb{N}$, assume that $N^{n,k}$'s have no simultaneous jumps, and introduce the d-dimensional locally square-integrable martingale $M^n = (M^{n,1}, ..., M^{n,d})^{\mathrm{tr}}$ given by

$$M^{n,i}_t = \sum_{k=1}^{m_n}\int_0^t H^{n,k,i}_s(dN^{n,k}_s - \lambda^{n,k}_s ds), \quad i = 1, ..., d.$$

Let T_n be a finite stopping time on $\mathbf{B}^n$, and suppose that

$$\sum_{k=1}^{m_n}\int_0^{T_n} ||H^{n,k}_s||^2 1\{||H^{n,k}_s|| > \varepsilon\}\lambda^{n,k}_s ds \xrightarrow{\mathsf{P}^n} 0, \quad \forall\varepsilon > 0, \tag{7.10}$$

and that

$$\sum_{k=1}^{m_n}\int_0^{T_n} H^{n,k,i}_s H^{n,k,j}_s \lambda^{n,k}_s ds \xrightarrow{\mathsf{P}^n} C^{i,j}, \quad \forall(i,j) \in \{1, ..., d\}^2,$$

where $C = (C^{i,j})_{(i,j)\in\{1,\dots,d\}^2}$ is a deterministic matrix. Then it holds that

$$M^n_{T_n} \stackrel{\mathsf{P}^n}{\Longrightarrow} \mathcal{N}_d(0,C) \quad \text{in } \mathbb{R}^d.$$

Note that Theorem 7.1.5 is a special case of Theorem 7.1.6 with $m_n = 1$.

Exercise 7.1.3 The condition (7.10) is called *Lindeberg's condition*. Prove that, for given $\alpha > 0$, a sufficient condition for the *Lindeberg-type condition*

$$\sum_{k=1}^{m_n} \int_0^{T_n} ||H^{n,k}_t||^{\alpha} 1\{||H^{n,k}_t|| > \varepsilon\} \lambda^{n,k}_t dt \stackrel{\mathsf{P}^n}{\longrightarrow} 0, \quad \forall \varepsilon > 0,$$

is that there exists a $\delta > 0$ such that

$$\sum_{k=1}^{m_n} \int_0^{T_n} ||H^{n,k}_t||^{\alpha+\delta} \lambda^{n,k}_t dt \stackrel{\mathsf{P}^n}{\longrightarrow} 0,$$

which is called the *Lyapunov-type condition*. (The case $\alpha = 2$ is called *Lyapunov's condition*.)

7.1.4 Local martingales

For every $n \in \mathbb{N}$, let M^n be a local martingale starting from zero defined on a stochastic basis $\mathbf{B}^n = (\Omega^n, \mathcal{F}^n; \mathbf{F}^n = (\mathcal{F}^n_t)_{t\in[0,\infty)}, \mathsf{P}^n)$. For every $n \in \mathbb{N}$ and $a > 0$, recalling Lemma 5.10.5, introduce the decomposition $M^n = M^{\prime,n,a} + M^{\prime\prime,n,a}$, where $M^{\prime,n,a} = A^{n,a} - A^{n,a,p}$ with

$$A^{n,a} = \sum_{t\le\cdot} \Delta M^n_t 1\{|\Delta M^n_t| > a/2\}$$

and $A^{n,a,p}$ is the predictable compensator for $A^{n,a}$. Notice that $M^{\prime,n,a}$ is a local martingale with finite-variation starting from zero and that $M^{\prime\prime,n,a}$ is a locally square-integrable martingale starting from zero such that $|\Delta M^{\prime\prime,n,a}| \le a$.

Furthermore, for every $n \in \mathbb{N}$ and $\varepsilon > 0$, put

$$V^{1,n,\varepsilon} = \sum_{t\le\cdot} |\Delta M^n_t| 1\{|\Delta M^n_t| > \varepsilon\} \quad \text{and} \quad V^{2,n,\varepsilon} = \sum_{t\le\cdot} (\Delta M^n_t)^2 1\{|\Delta M^n_t| > \varepsilon\}.$$

Denote by $V^{1,n,\varepsilon,p}$ the predictable compensator for $V^{1,n,\varepsilon}$; denote by $V^{2,n,\varepsilon,p}$ the predictable compensator for $V^{2,n,\varepsilon}$ when M^n is a locally square-integrable martingale.

Lemma 7.1.7 For every $n \in \mathbb{N}$, let M^n be a local martingale starting from zero, and let T_n be a finite stopping time, both defined on $\mathbf{B}^n$. Then, the following (i) and (ii) hold true.

(i) If $V^{1,n,\varepsilon,p}_{T_n} \stackrel{\mathsf{P}^n}{\longrightarrow} 0$ for any $\varepsilon > 0$, then

$$V^{1,n,\varepsilon}_{T_n} \stackrel{\mathsf{P}^n}{\longrightarrow} 0, \quad \forall \varepsilon > 0, \tag{7.11}$$

$$\sup_{t\in[0,T_n]} |M_t'^{,n,a}| \xrightarrow{\mathsf{P}^n} 0, \quad \forall a>0, \tag{7.12}$$

$$\sup_{t\in[0,T_n]} |\Delta M_t''^{,n,a}| \xrightarrow{\mathsf{P}^n} 0, \quad \forall a>0, \tag{7.13}$$

and

$$\sup_{t\in[0,T_n]} |[M^n]_t - [M''^{,n,a}]_t| \xrightarrow{\mathsf{P}^n} 0, \quad \forall a>0. \tag{7.14}$$

(ii) If all M^n's are locally square-integrable martingales and if $V_{T_n}^{2,n,\varepsilon,p} \xrightarrow{\mathsf{P}^n} 0$ for any $\varepsilon>0$, then

$$V_{T_n}^{1,n,\varepsilon,p} \xrightarrow{\mathsf{P}^n} 0, \quad \forall \varepsilon>0, \tag{7.15}$$

and

$$\sup_{t\in[0,T_n]} |[M^n]_t - \langle M^n\rangle_t| \xrightarrow{\mathsf{P}^n} 0. \tag{7.16}$$

Proof of (i). The claim (7.11) is a special case of Exercise 6.6.2 (i).

As for the claim (7.12), observe that for any $a>0$,

$$\sup_{t\in[0,T_n]} |M_t'^{,n,a}| = \sup_{t\in[0,T_n]} |A^{n,a} - A^{n,a,p}| \le V_{T_n}^{1,n,a/2} + V_{T_n}^{1,n,a/2,p} \xrightarrow{\mathsf{P}^n} 0.$$

As for the claim (7.13), observe that (7.11) yields that

$$\mathsf{P}^n\left(\sup_{t\in[0,T_n]} |\Delta M_t^n| > \varepsilon\right) \le \mathsf{P}^n\left(V_{T_n}^{1,n,\varepsilon} > \varepsilon\right) \to 0, \quad \forall \varepsilon>0,$$

and that (7.12) implies that $\sup_{t\in[0,T_n]} |\Delta M_t'^{,n,a}| \xrightarrow{\mathsf{P}^n} 0$ for any $a>0$. The claim (7.13) follows from these two facts.

As for the claim (7.14), observe that

$$\begin{aligned}
[M^n] - [M''^{,n,a}] &= [M^n] - [M^n - M'^{,n,a}] \\
&= -[M'^{,n,a}] + 2[M^n, M'^{,n,a}] \\
&= -[M'^{,n,a}] + 2[M'^{,n,a} + M''^{,n,a}, M'^{,n,a}] \\
&= [M'^{,n,a}] + 2[M'^{,n,a}, M''^{,n,a}] \\
&= \sum_{t\le\cdot} (\Delta M_t'^{,n,a})^2 + 2\sum_{t\le\cdot} \Delta M_t'^{,n,a} \Delta M_t''^{,n,a},
\end{aligned}$$

thus

$$\begin{aligned}
\sup_{t\in[0,T_n]} |[M^n]_t - [M''^{,n,a}]_t| &\le \left\{\sup_{t\in[0,T_n]} |\Delta M_t'^{,n,a}| + 2a\right\} \sum_{t\le T_n} |\Delta M_t'^{,n,a}| \\
&= \left\{\sup_{t\in[0,T_n]} |\Delta M_t'^{,n,a}| + 2a\right\} \sum_{t\le T_n} |\Delta A_t^{n,a} - \Delta A_t^{n,a,p}|
\end{aligned}$$

$$\leq \left\{ \sup_{t\in[0,T_n]} |\Delta M_t'^{,n,a}| + 2a \right\} \left\{ V_{T_n}^{1,n,a/2} + V_{T_n}^{1,n,a/2,p} \right\}$$
$$\stackrel{\mathsf{P}^n}{\longrightarrow} 0.$$

Proof of (ii). The claim (7.15) is immediate from $V_{T_n}^{1,n,\varepsilon,p} \leq V_{T_n}^{2,n,\varepsilon}/\varepsilon$.

As for the claim (7.16), observe first that $V^{2,n,\varepsilon,p} \stackrel{\mathsf{P}^n}{\longrightarrow} 0$ yields that $V_{T_n}^{2,n,\varepsilon} \stackrel{\mathsf{P}^n}{\longrightarrow} 0$ by using Lenglart's inequality; see Exercise 6.6.2 (i). Put

$$\widetilde{V}^{n,\varepsilon} = \sum_{t\leq \cdot} (\Delta M_t^n)^2 1\{|M_t^n| \leq \varepsilon\}, \quad \forall \varepsilon \in (0,\infty],$$

and denote by $\widetilde{V}^{n,\varepsilon,p}$ the predictable compensator for $\widetilde{V}^{n,\varepsilon}$. Observe that $\Delta\widetilde{V}^{n,\varepsilon} \leq \varepsilon^2$; moreover, it can be proved[5] that $\Delta\widetilde{V}^{n,\varepsilon,p} \leq \varepsilon^2$. Now, we have that

$$\begin{aligned}
\sup_{t\in[0,T_n]} |[M^n]_t - \langle M^n\rangle_t| &= \sup_{t\in[0,T_n]} |\widetilde{V}_t^{n,\infty} - \widetilde{V}_t^{n,\infty,p}| \\
&= \sup_{t\in[0,T_n]} |\widetilde{V}_t^{n,\varepsilon} - \widetilde{V}_t^{n,\varepsilon,p}| + o_{\mathsf{P}^n}(1), \quad \forall \varepsilon > 0.
\end{aligned}$$

Thus, it is sufficient to prove that for any $\eta > 0$ there exits an $\varepsilon > 0$ such that

$$\limsup_{n\to\infty} \mathsf{P}^n \left(\sup_{t\in[0,T_n]} |\widetilde{V}_t^{n,\varepsilon} - \widetilde{V}_t^{n,\varepsilon,p}| \geq \eta \right) \leq \eta. \tag{7.17}$$

Now fix any $\eta > 0$. By Lenglart's inequality, we have that for any $\delta > 0$ and any $\varepsilon \in (0,1]$,

$$\begin{aligned}
&\mathsf{P}^n \left(\sup_{t\in[0,T_n]} |\widetilde{V}_t^{n,\varepsilon} - \widetilde{V}_t^{n,\varepsilon,p}| \geq \eta \right) \\
&\leq \frac{\delta + \mathsf{E}^n[\sup_{t\in[0,T_n]} (\Delta(\widetilde{V}^{n,\varepsilon} - \widetilde{V}^{n,\varepsilon,p})_t)^2]}{\eta^2} + \mathsf{P}^n \left([\widetilde{V}^{n,\varepsilon} - \widetilde{V}^{n,\varepsilon,p}]_{T_n} \geq \delta \right) \\
&\leq \frac{\delta + \varepsilon^4}{\eta^2} + \mathsf{P}^n \left([\widetilde{V}^{n,\varepsilon} - \widetilde{V}^{n,\varepsilon,p}]_{T_n} \geq \delta \right).
\end{aligned}$$

Noting that

$$\begin{aligned}
[\widetilde{V}^{n,\varepsilon} - \widetilde{V}^{n,\varepsilon,p}]_{T_n} &\leq \sup_{t\leq T_n} \Delta(|\widetilde{V}_t^{n,\varepsilon} - \widetilde{V}_t^{n,\varepsilon,p}|) \cdot \left\{ \widetilde{V}_{T_n}^{n,\varepsilon} + \widetilde{V}_{T_n}^{n,\varepsilon,p} \right\} \\
&\leq \varepsilon^2 \left\{ \widetilde{V}_{T_n}^{n,1} + \widetilde{V}_{T_n}^{n,1,p} \right\},
\end{aligned}$$

[5] In general, if A is a process with locally integrable-variation such that $|\Delta A| \leq a$ for a constant $a > 0$, then $|\Delta(A^p)| \leq a$. This fact may look trivial at first sight, but it needs a proof based on the operation called "predictable projection" which is not treated in this monograph. See I.3.21 and related parts of Jacod and Shiryaev (2003).

we have that

$$\begin{aligned}
& \mathsf{P}^n\left(\sup_{t\in[0,T_n]}|\widetilde{V}_t^{n,\varepsilon}-\widetilde{V}_t^{n,\varepsilon,p}|\geq\eta\right) \\
\leq\ & \frac{\delta+\varepsilon^4}{\eta^2}+\mathsf{P}^n\left(\widetilde{V}_{T_n}^{n,1}+\widetilde{V}_{T_n}^{n,1,p}\geq\frac{\delta}{\varepsilon^2}\right).
\end{aligned}$$

Since $|\Delta\widetilde{V}^{n,1}|\leq 1$, $\widetilde{V}^{n,1}\leq[M^n]$ and $\widetilde{V}^{n,1,p}\leq\langle M^n\rangle$, either of "$[M^n]_{T_n}\xrightarrow{\mathsf{P}^n}C$" or "$\langle M^n\rangle_{T_n}\xrightarrow{\mathsf{P}^n}C$" implies that $\widetilde{V}_{T_n}^{n,1}+\widetilde{V}_{T_n}^{n,1,p}=O_{\mathsf{P}^n}(1)$ by Exercise 6.6.2 (ii). To establish the claim (7.17), for the given $\eta>0$, choose sufficiently small $\delta>0$ and very small $\varepsilon>0$, so that $(\delta+\varepsilon^4)/\eta^2\leq\eta/2$ and that $\limsup_{n\to\infty}\mathsf{P}^n(\widetilde{V}_{T_n}^{n,1}+\widetilde{V}_{T_n}^{n,1,p}\geq\delta/\varepsilon^2)\leq\eta/2$. The proof is finished. □

Theorem 7.1.8 (Rebolledo's CLT for local martingales) For every $n\in\mathbb{N}$, let M^n be a d-dimensional local martingale starting from zero defined on $\mathbf{B}^n$. For every $n\in\mathbb{N}$ and $\varepsilon>0$, put

$$\mathbf{V}^{1,n,\varepsilon}=\sum_{t\leq\cdot}||\Delta M_t^n||1\{||\Delta M_t||>\varepsilon\}\quad\text{and}\quad\mathbf{V}^{2,n,\varepsilon}=\sum_{t\leq\cdot}||\Delta M_t^n||^2 1\{||\Delta M_t||>\varepsilon\}.$$

Denote by $\mathbf{V}^{1,n,\varepsilon,p}$ the predictable compensator of $\mathbf{V}^{1,n,\varepsilon}$; denote by $\mathbf{V}^{2,n,\varepsilon,p}$ the predictable compensator of $\mathbf{V}^{2,n,\varepsilon}$ when M^n is a d-dimensional locally square-integrable martingale.

For every $n\in\mathbb{N}$, let T_n be a finite stopping time[6] on $\mathbf{B}^n$, and suppose that either of the following (a) or (b) is satisfied.

(a) $\mathbf{V}_{T_n}^{1,n,\varepsilon,p}\xrightarrow{\mathsf{P}^n}0$ for any $\varepsilon>0$. Moreover, it holds that

$$[M^{n,i},M^{n,j}]_{T_n}\xrightarrow{\mathsf{P}^n}C^{i,j},\quad\forall(i,j)\in\{1,...,d\}^2,$$

where the limit $C^{i,j}$ is a constant.

(b) All M^n's are locally square-integrable martingales, and $\mathbf{V}_{T_n}^{2,n,\varepsilon,p}\xrightarrow{\mathsf{P}^n}0$ for any $\varepsilon>0$. Moreover, it holds that

$$\langle M^{n,i},M^{n,j}\rangle_{T_n}\xrightarrow{\mathsf{P}^n}C^{i,j},\quad\forall(i,j)\in\{1,...,d\}^2,$$

where the limit $C^{i,j}$ is a constant.

Then, it holds that

$$M_{T_n}^n\overset{\mathsf{P}^n}{\Longrightarrow}\mathcal{N}_d(0,C)\quad\text{in }\mathbb{R}^d,\quad\text{where }C=(C^{i,j})_{(i,j)\in\{1,...,d\}^2}.$$

[6]We allow either of the case of $T_n\to\infty$ as $n\to\infty$ in some sense or the case of $T_n\equiv T$ where T is a fixed time.

Proof. We shall give a proof for the 1-dimensional case only; use Cramér-Wold device to extend the proof here to that for the d-dimensional case. We use the notation $V^{1,n,\varepsilon}$, $V^{2,n,\varepsilon}$, $V^{1,n,\varepsilon,p}$, $V^{2,n,\varepsilon,p}$, M^n and C instead of the d-dimensional notations $\mathbf{V}^{1,n,\varepsilon}$, $\mathbf{V}^{2,n,\varepsilon}$, $\mathbf{V}^{1,n,\varepsilon,p}$, $\mathbf{V}^{2,n,\varepsilon,p}$, $M^{n,1}$ and $C^{1,1}$, respectively.

Fix any $a>0$. Introduce the stopping time S_n by

$$S_n = \inf\{t\in[0,\infty):\ [M'',^{n,a}]_t \geq C\}\wedge T_n.$$

Since $(C\wedge[M'',^{n,a}]_{T_n}) \leq [M'',^{n,a}]_{S_n} \leq [M'',^{n,a}]_{T_n}$, it holds that $[M'',^{n,a}]_{S_n} \xrightarrow{\mathsf{P}^n} C$. This implies that

$$M''^{,n,a}_{S_n} - M''^{,n,a}_{T_n} \xrightarrow{\mathsf{P}^n} 0,$$

with the help of Corollary 6.6.4 (i), because

$$\begin{aligned}
&[M'',^{n,a,S_n} - M'',^{n,a,T_n}]_{S_n\vee T_n}\\
&\quad= [M'',^{n,a,S_n}]_{S_n\vee T_n} - 2[M'',^{n,a,S_n}, M'',^{n,a,T_n}]_{S_n\vee T_n} + [M'',^{n,a,T_n}]_{S_n\vee T_n}\\
&\quad= [M'',^{n,a}]_{S_n} - 2[M'',^{n,a}]_{S_n} + [M'',^{n,a}]_{T_n}\\
&\quad\xrightarrow{\mathsf{P}^n} 0
\end{aligned}$$

and $\mathsf{E}^n[\sup_{t\in[0,T_n]}|\Delta[M'',^{n,a,S_n} - M'',^{n,a,T_n}]_t|] \to 0$ (this follows from the fact $|\Delta M'',^{n,a}| \leq a$ and (7.13) by Exercise 2.3.2). Hence, with the help of (7.12), it is sufficient to prove that for some $a>0$,

$$\lim_{n\to\infty}\mathsf{E}^n\left[\exp\left(izM''^{,n,a}_{S_n} + \frac{z^2}{2}C\right)\right] = 1, \quad \forall z\in\mathbb{R}.$$

For this purpose, noting $[M'',^{n,a}]_{S_n} \leq C + a^2$, by the same argument as that in the proof of Lemma 7.1.1 it is sufficient to prove that for some $a>0$,

$$\lim_{n\to\infty}\mathsf{E}^n\left[\exp\left(izM''^{,n,a}_{S_n} + \frac{z^2}{2}[M'',^{n,a}]_{S_n}\right)\right] = 1, \quad \forall z\in\mathbb{R}. \tag{7.18}$$

To prove (7.18), fix any $z\in\mathbb{R}$ and $a\in(0,1]$, and put

$$G^{n,a} = \exp\left(izM'',^{n,a} + \frac{z^2}{2}[M'',^{n,a}]\right);$$

note that $G^{n,a}_0 = 1$. It follows from Itô's formula that

$$G^{n,a}_{S_n} - 1 = iz\int_0^{S_n} G^{n,a}_{s-}dM''^{,n,a}_s + R^a_n, \tag{7.19}$$

where

$$R^a_n = \sum_{s\leq S_n} G^{n,a}_{s-}\left\{\exp\left(iz\Delta M''^{,n,a}_s + \frac{z^2}{2}(\Delta M''^{,n,a}_s)^2\right) - 1 - iz\Delta M''^{,n,a}_s\right\}.$$

Since $|e^{ix+x^2/2}-1-ix|\le K_z|x|^3$ whenever $|x|\le|z|$, where $K_z>0$ is a constant depending only on z (see Lemma A1.2.3), and $\sup_{s\in[0,S_n]}|G^{n,a}_{s-}|\le e^{\frac{z^2}{2}C}$, we have that

$$
\begin{aligned}
|R_n^a| &\le e^{\frac{z^2}{2}C}K_z\sum_{s\in[0,S_n]}|z\Delta M''^{,n,a}_s|^3\\
&\le e^{\frac{z^2}{2}C}K_z|z|^3[M''^{,n,a}]_{S_n}\sup_{s\in[0,S_n]}|\Delta M''^{,n,a}_s|\\
&\le e^{\frac{z^2}{2}C}K_z|z|^3(C+1)\sup_{s\in[0,S_n]}|\Delta M''^{,n,a}_s|.
\end{aligned}
$$

Since $\lim_{n\to\infty}\mathsf{E}^n[\sup_{s\in[0,S_n]}|\Delta M''^{,n,a}_s|]=0$ due to $|\Delta M''^{,n,a}|\le a$ and (7.13), we obtain that $\lim_{n\to\infty}\mathsf{E}^n[|R_n^a|]=0$. On the other hand, the expectation of the first term on the right-hand side of (7.19) is zero by the optional sampling theorem, because $s\rightsquigarrow M''^{,n,a}_{s\wedge S_n}$ is a square-integrable martingale.

We therefore obtained that $\lim_{n\to\infty}\mathsf{E}^n[G^{n,a}_{S_n}-1]=0$ for any $a\in(0,1]$, and the proof under the assumption (a) is finished. The claim with the assumption (b) follows from the result of (a) and Lemma 7.1.7 (ii). □

7.2 Functional Martingale Central Limit Theorems

This section is devoted to developing the weak convergence theory for martingales in a "functional sense".

Fix a finite time $T>0$, and denote by $D[0,T]$ the set of càdlàg functions on $[0,T]$. Any random element X taking values in $D[0,T]$ satisfies that $\sup_{t\in[0,T]}|X_t(\omega)|<\infty$ for all ω. Thus, we may regard X as a random element taking values in the space $\ell^\infty([0,T])$ as well. Here, in general, we denote by $\ell^\infty(\Theta)$ the set of real-valued, bounded functions defined on the set Θ, and equip it with the uniform metric, that is, $d(x,y)=\sup_{\theta\in\Theta}|x(\theta)-y(\theta)|$ for $x,y\in\ell^\infty(\Theta)$. In the case of $\mathbb{R}^d$-valued càdlàg process $X=\{(X_t^i)_{t\in[0,T]}\}_{i\in\{1,\dots,d\}}$, we may regard $(t,i)\rightsquigarrow X_t^i$ as a random element taking values in the space $\ell^\infty([0,T]\times\{1,\dots,d\})$.

Below, we first give a brief review for the modern version of Prohorov's (1956) theory developed by J. Hoffmann-Jørgensen and R.M. Dudley. The original version of Prohorov's (1956) theory deals with a sequence of random variables X_n taking values in a Polish space (a complete separable metric space), where each X_n is assumed to be Borel measurable. The space $D[0,T]$ with the *Skorokhod topology* is a Polish space, so the random element $X^n=(X_t^n)_{t\in[0,T]}$ taking values in $D[0,T]$ with this topology is Borel measurable if and only if every X_t^n is Borel measurable in $\mathbb{R}$. So is not the case where we equip the space $D[0,T]$ with the uniform topology. Hoffmann-Jørgensen and Dudley's theory is an attempt to overcome this problem. We remark that both the Skorokhod and uniform topologies for the space $D[0,T]$

have some merits and demerits, respectively[7]. The argument in this monograph will be based on the weak convergence theory under the uniform topology (see, e.g., van der Vaart and Wellner (1996) for the details); readers who are interested in that under the Skorokhod topology should consult with, e.g., Jacod and Shiryaev (2003). See also Billingsley (1968, 1999) and Pollard (1984).

Next, we present a functional CLT for $\mathbb{R}^d$-valued local martingales $M^n = \{(M_t^{n,i})_{t\in[0,T]}\}_{i\in\{1,...,d\}}$ as a sequence of random elements taking values in $\ell^\infty([0,T]\times\{1,...,d\})$. We then apply the general theorem to some important special cases.

It is well-known that the functional CLTs have rich applications; indeed, in this monograph we will apply them to derive the weak convergence of "Z-processes" for the change point problems in Chapter 10.

7.2.1 Preliminaries

In this subsection, we present two types of sufficient conditions for weak convergence in $\ell^\infty(\Theta)$ under the situation where the convergence in law of every finite-dimensional marginal has been established.

Let us first consider the situation where a pseudo-metric ρ on Θ is given in such a way that Θ is totally bounded with respect to ρ. A sequence $(X_n)_{n=1,2,...}$ of $\ell^\infty(\Theta)$-valued (possibly, non-measurable) maps $X_n = \{X_n(\theta);\theta\in\Theta\}$ defined on probability spaces $(\Omega_n,\mathcal{F}_n,\mathsf{P}_n)$ is *asymptotically ρ-equicontinuous in probability* if for any $\varepsilon,\eta>0$ there exists a $\delta>0$ such that

$$\limsup_{n\to\infty}\mathsf{P}_n^*\left(\sup_{\rho(\theta,\theta')<\delta}|X_n(\theta)-X_n(\theta')|>\varepsilon\right)<\eta,$$

where P_n^* denotes the outer probability measure of P_n. Since we will always consider some separable[8] random fields $X_n = \{X_n(\theta);\theta\in\Theta\}$ indexed by a totally bounded pseudometric space (Θ,ρ), the above condition is equivalent to that for any $\varepsilon,\eta>0$ there exists a $\delta>0$ such that

$$\limsup_{n\to\infty}\mathsf{P}_n\left(\sup_{\substack{\theta,\theta'\in\Theta^*\\ \rho(\theta,\theta')<\delta}}|X_n(\theta)-X_n(\theta')|>\varepsilon\right)<\eta,$$

for a countable, ρ-dense subset Θ^* of Θ.

[7] While the elaborate definition of the Skorokhod topology is not easy to understand, adopting it enables us to treat the weak convergence to a stochastic process with jumps. On the other hand, we may have $\alpha_n\to\alpha$, $\beta_n\to\beta$ under the Skorokhod topology and yet $\alpha_n+\beta_n$ not converging to $\alpha+\beta$; this does not happen under the uniform topology.

[8] A random field $X=\{X(\theta);\theta\in\Theta\}$ is said to be ρ-*separable* if there exist a negligible set $N\subset\Omega$ and a countable subset Θ^* of Θ such that

$$X(\theta)(\omega)\in\overline{\{X(\theta')(\omega);\theta'\in\Theta^*,\rho(\theta,\theta')<\varepsilon\}},\quad \forall\theta\in\Theta,\ \forall\varepsilon>0,\quad \forall\omega\in N^c,$$

where the closure is taken in $\mathbb{R}\cup\{\infty\}$. Any random field $\{X(t);t\in[0,T]\}$ that takes values in $D[0,T]$ is separable (with respect to the Euclidean metric on $[0,T]$).

Next let us consider the situation where no pseudo-metric ρ is equipped with Θ in advance. The second type of condition is the following: for any $\varepsilon, \eta > 0$ there exists a finite partition $\Theta = \bigcup_{i=1}^{N} \Theta_i$ such that

$$\limsup_{n\to\infty} \mathsf{P}_n^* \left(\max_{1\leq i\leq N} \sup_{\theta,\theta'\in\Theta_i} |X_n(\theta) - X_n(\theta')| > \varepsilon \right) < \eta. \tag{7.20}$$

Now we are ready to state the main theorem in this subsection.

Theorem 7.2.1 A sequence $(X_n)_{n=1,2,...}$ of $\ell^\infty(\Theta)$-valued random elements converges weakly to a tight[9] Borel law if and only if every finite-dimensional marginal

$$(X_n(\theta_1), ..., X_n(\theta_d))^{\mathrm{tr}}, \quad n = 1, 2, ...$$

converges weakly to a (tight) Borel law, and either of the following (a) or (b) is satisfied:

(a) there exists a pseudometric ρ with respect to which Θ is totally bounded, such that $(X_n)_{n=1,2,...}$ is asymptotically ρ-equicontinuous in probability;

(b) for any $\varepsilon, \eta > 0$ there exists a finite partition $\Theta = \bigcup_{i=1}^{N} \Theta_i$ such that the condition (7.20) is satisfied.

If, moreover, the tight Borel law on $\ell^\infty(\Theta)$ appearing as the limit is that of a random field $X = \{X(\theta); \theta \in \Theta\}$, then almost all paths $\theta \rightsquigarrow X(\theta)$ are uniformly ρ-continuous, where ρ is any pseudo-metric for which the conditions in (a) are met[10].

See Theorems 1.5.4, 1.5.6 and 1.5.7 of van der Vaart and Wellner (1996) for the proofs.

A tight Borel law in $\ell^\infty(\Theta)$ is characterized by all of the (tight) Borel laws of finite-dimensional marginals (see Lemma 1.5.3 of van der Vaart and Wellner (1996)). If the limit random field X is Gaussian, then the pseudo-metric ρ_X defined by $\rho_X(\theta, \theta') = \sqrt{\mathsf{E}[(X(\theta) - X(\theta'))^2]}$ satisfies the conditions (a) in Theorem 7.2.1, and therefore almost all paths $\theta \rightsquigarrow X(\theta)$ are uniformly ρ_X-continuous (see Example 1.5.10 of van der Vaart and Wellner (1996)).

7.2.2 The functional CLT for local martingales

Let us extend Theorem 7.1.8 up to a theorem in the functional sense.

Theorem 7.2.2 (Rebolledo's functional CLT for local martingales) Consider the same setting as that in the first paragraph of Theorem 7.1.8.

---

[9] A Borel law $\mathcal{L}$ on a metric space $(\mathcal{X}, d)$ is said to be *tight* if for any $\varepsilon > 0$ there exists a compact set $K \subset \mathcal{X}$ such that $\mathcal{L}(X \in K^c) < \varepsilon$. The law of any Borel random variable on a complete separable metric space is tight. Remark however that the space $\ell^\infty(\Theta)$ with the uniform metric is complete, but is *not* separable; a key point of Hoffmann-Jørgensen and Dudley's weak convergence theory is that, although any measurability is not required for X_n's, the law of the limit X is assumed to be (tight) Borel measurable.

[10] Once the weak convergence in $\ell^\infty(\Theta)$ is established, every finite-dimensional marginal converges weakly, and *both* of (a) and (b) in Theorem 7.2.1 are satisfied.

Let $T > 0$ be a fixed time, and suppose that either of the following (a) or (b) is satisfied.

(a) $\mathbf{V}_T^{1,n,\varepsilon,p} \xrightarrow{\mathsf{P}^n} 0$ for any $\varepsilon > 0$. Moreover, it holds that

$$[M^{n,i}, M^{n,j}]_t \xrightarrow{\mathsf{P}^n} C^{i,j}(t), \quad \forall t \in [0,T], \ \forall (i,j) \in \{1,...,d\}^2,$$

where the limit $t \mapsto C^{i,j}(t)$ is a deterministic, continuous function.

(b) All M^n's are locally square-integrable martingales, and $\mathbf{V}_{T_n}^{2,n,\varepsilon,p} \xrightarrow{\mathsf{P}^n} 0$ for any $\varepsilon > 0$. Moreover, it holds that

$$\langle M^{n,i}, M^{n,j} \rangle_t \xrightarrow{\mathsf{P}^n} C^{i,j}(t), \quad \forall t \in [0,T], \ \forall (i,j) \in \{1,...,d\}^2,$$

where the limit $t \mapsto C^{i,j}(t)$ is a deterministic, continuous function.

Then, it holds that

$$M^n \xRightarrow{\mathsf{P}^n} G = \{(G_t^i)_{t\in[0,T]}\}_{i\in\{1,...,d\}} \quad \text{in } \ell^\infty([0,T] \times \{1,...,d\}),$$

where each $t \rightsquigarrow G_t^i$ is a Gaussian process, whose almost all paths are continuous on $[0,T]$, such that $\mathsf{E}[G_t^i] = 0$ and that $\mathsf{E}[G_s^i G_t^j] = C^{i,j}(s \wedge t)$.

Proof. In view of (7.16) in Lemma 7.1.7, it suffices to show the claim under the assumption (a).

The convergence in law of every finite-dimensional marginal is deduced from Theorem 7.1.8 (see Exercise 7.2.1 below). Let us check the criterion (7.20).

It is sufficient to consider the 1-dimensional case (see Exercise 7.2.2 below). Consider the decomposition $M^n = M'^{,n,a} + M''^{,n,a}$ with $a = m^{-1}$, for any given $m \in \mathbb{N}$, introduced at the beginning of Subsection 7.1.4; recall that $\sup_{t\in[0,T]} |M_t'^{,n,a}| \xrightarrow{\mathsf{P}^n} 0$ for any $a > 0$ (see (7.12) in Lemma 7.1.7).

Put $Y^{n,a} = [M''^{,n,a}] - \langle M''^{,n,a} \rangle$. Since $|\Delta Y^{n,a}| \leq \max\{\Delta[M''^{,n,a}], \Delta\langle M''^{,n,a}\rangle\} \leq a^2$ and $|\Delta[Y^{n,a}]| \leq a^4$, the same argument as the proof of Lemma 7.1.7 (ii) yields that, for any $\delta, \eta > 0$,

$$\begin{aligned}
&\mathsf{P}^n\left(\sup_{t\in[0,T]} |Y_t^{n,a}| \geq \eta \right) \\
&\leq \frac{\delta + a^4}{\eta^2} + \mathsf{P}^n([Y^{n,a}]_T \geq \delta) \\
&\leq \frac{\delta + a^4}{\eta^2} + \mathsf{P}^n\left(\sup_{t\in[0,T]} |\Delta Y_t^{n,a}| \{[M''^{,n,a}]_T + \langle M''^{,n,a}\rangle_T\} \geq \delta \right) \\
&\leq \frac{\delta + a^4}{\eta^2} + \mathsf{P}^n\left([M''^{,n,a}]_T + \langle M''^{,n,a}\rangle_T \geq \frac{\delta}{a^2} \right) \\
&\leq \frac{\delta + a^4}{\eta^2} + \frac{3C(T) + 2a^2}{\delta/a^2} + \mathsf{P}^n(2[M''^{,n,a}]_T \geq 3C(T)).
\end{aligned}$$

Due to the assumption that $t \mapsto C(t)$ is continuous, we can choose some finite points $0 = t_0 < t_1 < \cdots < t_{N_m} = T$ such that $C(t_j) - C(t_{j-1}) \le m^{-1}$ for all j; this can be done with $N_m \le Km$ for a constant K depending only on $C(T)$. Then we have that for any $\delta > 0$, $m \in \mathbb{N}$ and $q \in (0,1)$,

$$\begin{aligned} p_{n,m,q} &:= \mathsf{P}^n\left(\max_{1\le j\le N_m}(\langle M''^{,n,a}\rangle_{t_j} - \langle M''^{,n,a}\rangle_{t_{j-1}}) > 3m^{-q}\right) \\ &\le \mathsf{P}^n\left(\max_{1\le j\le N_m}([M''^{,n,a}]_{t_j} - [M''^{,n,a}]_{t_{j-1}}) > m^{-q}\right) \\ &\quad + \mathsf{P}^n\left(\sup_{t\in[0,T]}|Y_t^{n,a}| \ge m^{-q}\right) \\ &\le \mathsf{P}^n\left(\max_{1\le j\le N_m}([M''^{,n,a}]_{t_j} - [M''^{,n,a}]_{t_{j-1}}) > m^{-q}\right) \\ &\quad + \frac{\delta + m^{-4}}{(m^{-q})^2} + \frac{3C(T) + 2m^{-2}}{\delta m^2} + \mathsf{P}^n(2[M''^{,n,a}]_T \ge 3C(T)). \end{aligned}$$

Since (7.14) in Lemma 7.1.7 implies that $[M''^{,n,a}]_t \xrightarrow{\mathsf{P}^n} C(t)$ for every $t \in [0,T]$ and any $a > 0$, it follows from Bernsten's inequality (Theorem 6.6.5) that, for any $\varepsilon \in (0,1]$,

$$\begin{aligned} &\mathsf{P}^n\left(\max_{1\le j\le N_m}\sup_{s,t\in[t_{j-1},t_j]}|M_s''^{,n,a} - M_t''^{,n,a}| > 2\varepsilon\right) \\ &\le \mathsf{P}^n\left(\max_{1\le j\le N_m}\sup_{t\in[t_{j-1},t_j]}|M_t''^{,n,a} - M_{t_{j-1}}''^{,n,a}| > \varepsilon\right) \\ &\le N_m \cdot 2\exp\left(-\frac{\varepsilon^2}{2(m^{-1}\varepsilon + 3m^{-q})}\right) + p_{n,m,q}, \end{aligned}$$

thus we obtain that

$$\begin{aligned} &\limsup_{n\to\infty}\mathsf{P}^n\left(\max_{1\le j\le N_m}\sup_{s,t\in[t_{j-1},t_j]}|M_s''^{,n,a} - M_t''^{,n,a}| > 2\varepsilon\right) \\ \le\ & 2Km\exp\left(-\frac{\varepsilon^2 m^q}{8}\right) \\ & + \limsup_{n\to\infty}\mathsf{P}^n\left(\max_{1\le j\le N_m}([M''^{,n,a}]_{t_j} - [M''^{,n,a}]_{t_{j-1}}) > m^{-q}\right) \\ & + m^{2q}\{\delta + m^{-4}\} + \frac{3C(T) + 2m^{-2}}{\delta m^2} + \limsup_{n\to\infty}\mathsf{P}^n(2[M''^{,n,a}]_T \ge 3C(T)) \\ =\ & 2Km\exp\left(-\frac{\varepsilon^2 m^q}{8}\right) + m^{2q}\{\delta + m^{-4}\} + \frac{3C(T) + 2m^{-2}}{\delta m^2}. \end{aligned}$$

Putting $\delta = m^{-1}$ and $q = 1/4$, choose a large $m \in \mathbb{N}$ so that the right-hand side is smaller than any given constant $\eta' > 0$. The proof is finished. □

Exercise 7.2.1 In the first paragraph of the proof of Theorem 7.2.2 (a), it is stated that "The convergence in law of every finite-dimensional marginal is deduced from Theorem 7.1.8". Explain the details.

Exercise 7.2.2 At the beginning of the second paragraph of Theorem 7.2.2 (a), it is stated that "It is sufficient to consider the 1-dimensional case". Explain the details.

7.2.3 Special cases

In this subsection, we apply Rebolledo's functional CLT for locally square-integrable martingales (Theorem 7.2.2(b)) to deduce the functional versions of the CLTs for various, concrete forms of martingales which we have discussed in the previous section.

Let us first extend Theorem 7.1.3 to the functional version.

Theorem 7.2.3 (The functional CLT for discrete-time martingales) For every $n \in \mathbb{N}$, let $(\xi_k^n)_{k\in\mathbb{N}} = ((\xi_k^{n,1}, \dots, \xi_k^{n,d})^{\rm tr})_{k\in\mathbb{N}}$ be a d-dimensional martingale difference sequence on a discrete-time stochastic basis $(\Omega^n, \mathcal{F}^n; (\mathcal{F}_k^n)_{k\in\mathbb{N}_0}, \mathsf{P}^n)$, and let m_n be a positive integer. Define

$$M_u^n = \sum_{k=1}^{[um_n]} \xi_k, \quad \forall u \in [0,1].$$

If

$$\sum_{k=1}^{m_n} \mathsf{E}^n[||\xi_k^n||^2 1\{||\xi_k^n|| > \varepsilon\} | \mathcal{F}_{k-1}^n] \xrightarrow{\mathsf{P}^n} 0, \quad \forall \varepsilon > 0, \tag{7.21}$$

and if

$$\sum_{k=1}^{[um_n]} \mathsf{E}^n[\xi_k^{n,i}\xi_k^{n,j} | \mathcal{F}_{k-1}^n] \xrightarrow{\mathsf{P}^n} C^{i,j}(u), \quad \forall u \in [0,1], \ \forall (i,j) \in \{1,\dots,d\}^2,$$

where the limits, $u \mapsto C^{i,j}(u)$'s, are some continuous functions, then it holds that

$$M^n \overset{\mathsf{P}^n}{\Longrightarrow} G = \{(G_u^i)_{u\in[0,1]}\}_{i\in\{1,\dots,d\}} \quad \text{in } \ell^\infty([0,1] \times \{1,\dots,d\}),$$

where each $u \rightsquigarrow G_u^i$ is a Gaussian process, whose almost all paths are continuous on $[0,1]$, such that $\mathsf{E}[G_u^i] = 0$ and that $\mathsf{E}[G_u^i G_v^j] = C^{i,j}(u \wedge v)$.

Proof. Introduce the filtration $\mathbf{F}^{c,n} = (\mathcal{F}_u^{c,n})_{u\in[0,1]}$ by the 2nd method in Section 5.2. Then, the càdlàg process $(M_u^n)_{u\in[0,1]}$ is a square-integrable martingale with respect to the filtration $\mathbf{F}^{c,n}$. When we apply Theorem 7.2.2, it remains only to check the condition "$\mathbf{V}_1^{2,n,\varepsilon,p} \xrightarrow{\mathsf{P}} 0$ for any $\varepsilon > 0$".

To check it, observe that

$$\begin{aligned}\mathbf{V}_u^{2,n,\varepsilon} &= \sum_{s\le u} ||\Delta M_s^n||^2 1\{||\Delta M_s|| > \varepsilon\} \\ &= \sum_{k=1}^{[um_n]} ||\xi_k^n||^2 1\{||\xi_k^n|| > \varepsilon\},\end{aligned}$$

and thus

$$\begin{aligned}
\mathbf{V}_1^{2,n,\varepsilon,p} &= \sum_{u\le 1} ||\Delta M_u^n||^2 1\{||\Delta M_u|| > \varepsilon\} \\
&= \sum_{k=1}^{m_n} \mathsf{E}^n[||\xi_k^n||^2 1\{||\xi_k^n|| > \varepsilon\}|\mathcal{F}^{c,n}_{(k-1)/m_n}] \\
&= \sum_{k=1}^{m_n} \mathsf{E}^n[||\xi_k^n||^2 1\{||\xi_k^n|| > \varepsilon\}|\mathcal{F}^{n}_{k-1}].
\end{aligned}$$

Hence, the condition (7.21) is nothing else than the condition "$\mathbf{V}_1^{2,n,\varepsilon,p} \xrightarrow{\mathsf{P}} 0$ for any $\varepsilon > 0$". The proof is finished. □

Corollary 7.2.4 (Donsker's theorem) Let $X, X_1, X_2, ...$ be a sequence of $\mathbb{R}^d$-valued, independently, identically distributed random variables such that $\mathsf{E}[X] = 0$ and that $\mathsf{E}[||X||^2] < \infty$. Define the rescaled partial sum process $u \rightsquigarrow M_u^n$ by

$$M_u^n = \frac{1}{\sqrt{n}} \sum_{k=1}^{[un]} X_k, \quad \forall u \in [0,1].$$

Then, it holds that

$$M^n \stackrel{\mathsf{P}}{\Longrightarrow} G \quad \text{in } \ell^\infty([0,1]\times\{1,...,d\}),$$

where $G = C^{1/2}B$ with $C = \mathsf{E}[XX^{\mathrm{tr}}]$ and $u \rightsquigarrow B(u)$ being the vector of independent, standard Brownian motions.

Remark. Since the matrix C is proved to be non-negative definite, the matrix $C^{1/2}$ is well-defined.

Proof. We shall apply Theorem 7.2.3 to $\xi_k^n = X_k/\sqrt{n}$ and $m_n = n$. Notice that for any $\varepsilon > 0$,

$$\begin{aligned}
\mathbf{V}_1^{2,n,\varepsilon,p} &= \frac{1}{n}\sum_{k=1}^{n} \mathsf{E}[||X_k||^2 1\{||X_k/\sqrt{n}|| > \varepsilon\}] \\
&= \mathsf{E}[||X||^2 1\{||X/\sqrt{n}|| > \varepsilon\}] \\
&\to 0.
\end{aligned}$$

The other condition in Theorem 7.2.3 is satisfied with $C^{i,j}(u) = uC^{i,j}$. Thus we obtain the weak convergence result with the limit $G = \{(G_u^i)_{u\in[0,1]}\}_{i\in\{1,...,d\}}$ satisfying that $u \rightsquigarrow G_u^i$ is a Gaussian process, whose almost all paths are continuous on $[0,1]$, such that $\mathsf{E}[G_u^i] = 0$ and that $\mathsf{E}[G_u^i G_v^j] = (u\wedge v)C^{i,j}$. Since the process $u \rightsquigarrow C^{1/2}B(u)$ meets these properties, and since the tight, Borel law in $\ell^\infty([0,1]\times\{1,...,d\})$ is characterized by the laws of its marginals, we obtain the desired conclusion. □

Next, let us extend the CLT for continuous local martingales (Theorem 7.1.4) to the functional version.

Theorem 7.2.5 (The functional CLT for continuous local martingales) For every $n \in \mathbb{N}$, let $M^n = (M^{n,1}, ..., M^{n,d})^{\rm tr}$ be a d-dimensional continuous local martingale starting from zero defined on a stochastic basis $(\Omega^n, \mathcal{F}^n, (\mathcal{F}_t^n)_{t\in[0,\infty)}, \mathsf{P}^n)$. Let $T > 0$ be a fixed time. If

$$\langle M^{n,i}, M^{n,j} \rangle_t \xrightarrow{\mathsf{P}^n} C^{i,j}(t), \quad \forall t \in [0,T], \ \forall (i,j) \in \{1,...,d\}^2,$$

where the limit $t \mapsto C^{i,j}(t)$ is a deterministic, continuous function, then it holds that

$$M^n \overset{\mathsf{P}^n}{\Longrightarrow} G \quad \text{in } \ell^\infty([0,T] \times \{1,...,d\}),$$

where each $t \rightsquigarrow G_t^i$ is a Gaussian process, whose almost all paths are continuous on $[0,T]$, such that $\mathsf{E}[G_t^i] = 0$ and that $\mathsf{E}[G_s^i G_t^j] = C^{i,j}(s \wedge t)$.

Proof. Since $t \rightsquigarrow M_t^n$ has no jump, the result is immediate from Theorem 7.2.2 (b). □

To close this subsection, we shall extend Theorem 7.1.6 to the functional version.

Theorem 7.2.6 (The functional CLT for stochastic integrals) For every $n = 1, 2, ...$ and every $k = 1, ..., m_n$, let $N^{n,k}$ be a counting process with the intensity $\lambda^{n,k}$ and $H^{n,k,i}$, $i = 1, ..., d$, be predictable processes such that $t \rightsquigarrow \int_0^t (H_s^{n,k,i})^2 \lambda_s^{n,k} ds$ are locally integrable, all defined on a stochastic basis $\mathbf{B}^n$. For every $n \in \mathbb{N}$, suppose that $N^{n,k}$'s have no simultaneous jumps, and introduce the d-dimensional locally square-integrable martingale $M^n = (M^{n,1}, ..., M^{n,d})^{\rm tr}$, where

$$M_t^{n,i} = \sum_{k=1}^{m_n} \int_0^t H_s^{n,k,i} (dN_s^{n,k} - \lambda_s^{n,k} ds), \quad i = 1, ..., d.$$

Let T be a fixed time. If

$$\sum_{k=1}^{m_n} \int_0^T ||H_s^{n,k}||^2 1\{||H_s^{n,k}|| > \varepsilon\} \lambda_s^{n,k} ds \xrightarrow{\mathsf{P}^n} 0, \quad \forall \varepsilon > 0, \tag{7.22}$$

and if

$$\sum_{k=1}^{m_n} \int_0^t H_s^{n,k,i} H_s^{n,k,j} \lambda_s^{n,k} ds \xrightarrow{\mathsf{P}^n} C^{i,j}(t), \quad \forall t \in [0,T], \ \forall (i,j) \in \{1,...,d\}^2,$$

where the limit $t \mapsto C^{i,j}(t)$ is a deterministic, continuous function on $[0,T]$, then it holds that

$$M^n \overset{\mathsf{P}^n}{\Longrightarrow} G \quad \text{in } \ell^\infty([0,T] \times \{1,...,d\}),$$

where each $t \rightsquigarrow G_t^i$ is a Gaussian process, whose almost all paths are continuous on $[0,T]$, such that $\mathsf{E}[G_t^i] = 0$ and that $\mathsf{E}[G_s^i G_t^j] = C^{i,j}(s \wedge t)$.

Proof. Since

$$\mathbf{V}_t^{2,n,\varepsilon} = \sum_{k=1}^{m_n} \int_0^t ||H_s^{n,k}||^2 1\{||H_s^{n,k}|| > \varepsilon\} dN_s^{n,k},$$

we have that

$$\mathbf{V}_T^{2,n,\varepsilon,p} = \sum_{k=1}^{m_n} \int_0^T ||H_s^{n,k}||^2 1\{||H_s^{n,k}|| > \varepsilon\} \lambda_s^{n,k} ds.$$

Thus, the condition "$\mathbf{V}_T^{2,n,\varepsilon,p} \xrightarrow{\mathsf{P}^n} 0$ for any $\varepsilon > 0$" in Theorem 7.2.2 is nothing else than the assumption (7.22). The other conditions are straightforward. □

7.3 Uniform Convergence of Random Fields

Let Θ be an infinite set. Given sequence of random fields $X_n = \{X_n(\theta); \theta \in \Theta\}$, the pointwise convergence

$$X_n(\theta) \xrightarrow{\mathsf{P}} x(\theta), \quad \text{as } n \to \infty, \quad \forall \theta \in \Theta,$$

does *not* always imply the uniform convergence

$$\sup_{\theta \in \Theta} |X_n(\theta) - x(\theta)| \xrightarrow{\mathsf{P}} 0,$$

where the limit $x = \{x(\theta); \theta \in \Theta\}$ is deterministic.

The goal of this section is to provide some sufficient conditions under which the former implies the latter. Although the idea to prove the theorems in Subsection 7.3.1 will be applied to more general cases in Subsection 7.3.2, let us first start with a rather special case in order to grasp the outline to prove the "uniform convergence" of random fields.

7.3.1 Uniform law of large numbers for ergodic random fields

Let $(\mathcal{X}, \mathcal{A})$ be a measurable space. If an $\mathcal{X}$-valued stochastic process $(X_t)_{t \in [0,\infty)}$ is *ergodic*, then there exists a probability measure P° on $(\mathcal{X}, \mathcal{A})$ called the *invariant measure* such that it holds for any P°-integrable function f that, as $T \to \infty$,

$$\frac{1}{T} \int_0^T f(X_t) dt \xrightarrow{\mathsf{P}} \int_{\mathcal{X}} f(x) P^\circ(dx).$$

In this case, it holds for any finite P°-integrable functions $f_1, ..., f_p$

$$\max_{1 \le i \le p} \left| \frac{1}{T} \int_0^T f_i(X_t) dt - \int_{\mathcal{X}} f_i(x) P^\circ(dx) \right| \xrightarrow{\mathsf{P}} 0.$$

However, in the case of a class $\{f_\theta;\theta\in\Theta\}$ of infinitely many P°-integrable functions, some additional conditions are needed to ensure that

$$\sup_{\theta\in\Theta}\left|\frac{1}{T}\int_0^T f_\theta(X_t)dt-\int_{\mathcal{X}} f_\theta(x)P^\circ(dx)\right|\xrightarrow{\mathsf{P}}0.$$

The next theorem gives a sufficient condition under which this convergence holds true.

Theorem 7.3.1 Let $(\mathcal{X},\mathcal{A})$ be a measurable space. Let Θ be a bounded subset of $\mathbb{R}^p$. For a given family $\{f(\cdot;\theta);\theta\in\Theta\}$ of measurable functions on $\mathcal{X}$, suppose that there exist a measurable function K and a constant $\alpha\in(0,1]$ such that

$$|f(x;\theta)-f(x;\theta')|\le K(x)||\theta-\theta'||^\alpha,\quad \forall\theta,\theta'\in\Theta. \tag{7.23}$$

(i) For a given $\mathcal{X}$-valued ergodic stochastic process $(X_t)_{t\in[0,\infty)}$ with the invariant measure P°, if all $f(\cdot;\theta)$'s and K are P°-integrable, then it holds that, as $T\to\infty$,

$$\sup_{\theta\in\Theta}\left|\frac{1}{T}\int_0^T f(X_t;\theta)dt-\int_{\mathcal{X}} f(x;\theta)P^\circ(dx)\right|\xrightarrow{\mathsf{P}}0.$$

(ii) For a given $\mathcal{X}$-valued ergodic stochastic process $(X_k)_{k\in\mathbb{N}}$ with the invariant measure P°, if all $f(\cdot;\theta)$'s and K are P°-integrable, then it holds that, as $n\to\infty$,

$$\sup_{\theta\in\Theta}\left|\frac{1}{n}\sum_{k=1}^n f(X_k;\theta)-\int_{\mathcal{X}} f(x;\theta)P^\circ(dx)\right|\xrightarrow{\mathsf{P}}0.$$

Proof. We shall prove the claim (i) only; the claim (ii) is also proved in a similar way.

For any given constant $\varepsilon>0$, let $B_1,\dots,B_{N(\varepsilon)}$ be closed balls with radius ε that cover the bounded subset Θ of $\mathbb{R}^p$; such a covering is possible with a finite integer $N(\varepsilon)$ depending on ε. Choosing an arbitrary point θ_i from each B_i, put

$$\begin{aligned} u_i(x) &= f(x;\theta_i)+K(x)\varepsilon^\alpha,\\ l_i(x) &= f(x;\theta_i)-K(x)\varepsilon^\alpha. \end{aligned}$$

Then it holds that

$$l_i(x)\le\inf_{\theta\in B_i} f(x;\theta)\le\sup_{\theta\in B_i} f(x;\theta)\le u_i(x).$$

Since we have for any $\theta\in B_i$ that

$$\begin{aligned} &\frac{1}{T}\int_0^T f(X_t;\theta)dt-\int_{\mathcal{X}} f(x;\theta)P^\circ(dx)\\ &\le \frac{1}{T}\int_0^T u_i(X_t)dt-\int_{\mathcal{X}} u_i(x)P^\circ(dx)+\int_{\mathcal{X}} u_i(x)P^\circ(dx)-\int_{\mathcal{X}} f(x;\theta)P^\circ(dx)\\ &\le \frac{1}{T}\int_0^T u_i(X_t)dt-\int_{\mathcal{X}} u_i(x)P^\circ(dx)+\int_{\mathcal{X}}\{u_i(x)-l_i(x)\}P^\circ(dx)\\ &= \frac{1}{T}\int_0^T u_i(X_t)dt-\int_{\mathcal{X}} u_i(x)P^\circ(dx)+2\int_{\mathcal{X}} K(x)P^\circ(dx)\varepsilon^\alpha, \end{aligned}$$

it holds that

$$\sup_{\theta\in\Theta}\left\{\frac{1}{T}\int_0^T f(X_t;\theta)dt-\int_{\mathcal{X}} f(x;\theta)P^\circ(dx)\right\}$$
$$\leq \max_{1\leq i\leq N(\varepsilon)}\left\{\frac{1}{T}\int_0^T u_i(X_t)dt-\int_{\mathcal{X}} u_i(x)P^\circ(dx)\right\}+2\int_{\mathcal{X}} K(x)P^\circ(dx)\varepsilon^\alpha.$$

By preforming an evaluation from below in a similar way, we finally obtain that

$$\sup_{\theta\in\Theta}\left|\frac{1}{T}\int_0^T f(X_t;\theta)dt-\int_{\mathcal{X}} f(x;\theta)P^\circ(dx)\right|$$
$$\leq \max_{1\leq i\leq N(\varepsilon)}\left|\frac{1}{T}\int_0^T u_i(X_t)dt-\int_{\mathcal{X}} u_i(x)P^\circ(dx)\right|$$
$$+\max_{1\leq i\leq N(\varepsilon)}\left|\frac{1}{T}\int_0^T l_i(X_t)dt-\int_{\mathcal{X}} l_i(x)P^\circ(dx)\right|+2\int_{\mathcal{X}} K(x)P^\circ(dx)\varepsilon^\alpha.$$

We can now prove the desired claim by choosing a sufficiently small $\varepsilon>0$ and then letting $T\to\infty$. □

It would be clear from the above proof that Θ need not be an Euclidean space and that the smoothness condition (7.23) may be replaced by the so-called "bracketing condition", as it is actually seen in the next theorem.

Theorem 7.3.2 Let $(\mathcal{X},\mathcal{A})$ be a measurable space. For a given $\mathcal{X}$-valued ergodic stochastic process $(X_t)_{t\in[0,\infty)}$ with the invariant measure P° and a given family $\mathcal{F}$ of P°-integrable functions on $\mathcal{X}$, suppose that the "bracketing condition", described as follows, is satisfied: for any constant $\varepsilon>0$ there exist finitely many pairs $[l_i,u_i]$, $i=1,...,N(\varepsilon)$, of P°-integrable functions on $\mathcal{X}$ such that for any $f\in\mathcal{F}$ it holds that $l_i\leq f\leq u_i$ for some i and that $\int_{\mathcal{X}}\{u_i(x)-l_i(x)\}P^\circ(dx)<\varepsilon$ holds for every i. Then it holds that, as $T\to\infty$,

$$\sup_{f\in\mathcal{F}}\left|\frac{1}{T}\int_0^T f(X_t;\theta)dt-\int_{\mathcal{X}} f(x;\theta)P^\circ(dx)\right|\xrightarrow{\mathsf{P}}0.$$

A similar claim holds also in the case of discrete-time ergodic stochastic process $(X_k)_{k\in\mathbb{N}}$.

Remark. The bracketing condition is satisfied, e.g., for the case where $\mathcal{X}=\mathbb{R}$ and $\mathcal{F}=\{1_{(-\infty,x]};x\in\mathbb{R}\}$. It is well-known that many other classes of functions satisfy the bracketing condition; see van der Vaart and Wellner (1996) for the details.

Proof. For every f such that $l_i\leq f\leq u_i$, it holds that

$$\frac{1}{T}\int_0^T f(X_t)dt-\int_{\mathcal{X}} f(x)P^\circ(dx)$$
$$\leq \frac{1}{T}\int_0^T u_i(X_t)dt-\int_{\mathcal{X}} u_i(x)P^\circ(dx)+\int_{\mathcal{X}} u_i(x)P^\circ(dx)-\int_{\mathcal{X}} f(x)P^\circ(dx)$$
$$\leq \frac{1}{T}\int_0^T u_i(X_t)dt-\int_{\mathcal{X}} u_i(x)P^\circ(dx)+\int_{\mathcal{X}}\{u_i(x)-l_i(x)\}P^\circ(dx).$$

Thus we have

$$\sup_{f\in\mathcal{F}}\left\{\frac{1}{T}\int_0^T f(X_t)dt-\int_{\mathcal{X}} f(x)P^\circ(dx)\right\}$$
$$\leq \max_{1\leq i\leq N(\varepsilon)}\left\{\frac{1}{T}\int_0^T u_i(X_t)dt-\int_{\mathcal{X}} u_i(x)P^\circ(dx)\right\}+\varepsilon.$$

By performing an evaluation from below in a similar way, we finally obtain that

$$\sup_{f\in\mathcal{F}}\left|\frac{1}{T}\int_0^T f(X_t)dt-\int_{\mathcal{X}} f(x)P^\circ(dx)\right|$$
$$\leq \max_{1\leq i\leq N(\varepsilon)}\left|\frac{1}{T}\int_0^T u_i(X_t)dt-\int_{\mathcal{X}} u_i(x)P^\circ(dx)\right|$$
$$+\max_{1\leq i\leq N(\varepsilon)}\left|\frac{1}{T}\int_0^T l_i(X_t)dt-\int_{\mathcal{X}} l_i(x)P^\circ(dx)\right|+\varepsilon.$$

We can now prove the desired claim by choosing a sufficiently small $\varepsilon>0$ and then letting $T\to\infty$. □

Next, let us turn to the problem of uniform convergence of partial sum processes.

Lemma 7.3.3 Let $X_1,X_2,...$ be a sequence of $[0,\infty)$-valued random variables, and let $(Y_t)_{t\in[0,T]}$ be a $[0,\infty)$-valued càdlàg process, where $T>0$ is a fixed time. If

$$\frac{1}{n}\sum_{k=1}^{[un]}X_k \xrightarrow{\mathsf{P}} \int_0^{uT} Y_t dt, \quad \forall u\in[0,1],$$

then it holds that

$$\sup_{u\in[0,1]}\left|\frac{1}{n}\sum_{k=1}^{[un]}X_k-\int_0^{uT}Y_t dt\right| \xrightarrow{\mathsf{P}} 0.$$

Proof. Fix any $\varepsilon>0$. Since the random variable $\bar{Y}=\sup_{t\in[0,T]}Y_t$ is tight, we can find a (large) constant $K>0$ such that $\mathsf{P}(\bar{Y}>K)<\varepsilon$. Choose some girds $0=u_0<u_1<\cdots<u_N=1$ such that $(u_j-u_{j-1})TK\leq\varepsilon$ for every $j=1,...,N$.

Since it holds for any $u\in(u_{j-1},u_j]$ that

$$\frac{1}{n}\sum_{k=1}^{[un]}X_k-\int_0^{uT}Y_t dt\leq\frac{1}{n}\sum_{k=1}^{[u_jn]}X_k-\int_0^{u_jT}Y_t dt+\int_{u_{j-1}T}^{u_jT}Y_t dt,$$

we have that

$$\sup_{u\in[0,1]}\left\{\frac{1}{n}\sum_{k=1}^{[un]}X_k-\int_0^{uT}Y_t dt\right\}\leq\max_{1\leq j\leq N}\left\{\frac{1}{n}\sum_{k=1}^{[u_jn]}X_k-\int_0^{u_jT}Y_t dt\right\}+\max_{1\leq j\leq N}\int_{u_{j-1}T}^{u_jT}Y_t dt.$$

By performing an evaluation from below in a similar way, we finally have that

$$\begin{aligned}
&\sup_{u\in[0,1]}\left|\frac{1}{n}\sum_{k=1}^{[un]}X_k-\int_0^{uT}Y_t dt\right| \\
&\leq \max_{1\leq j\leq N}\left|\frac{1}{n}\sum_{k=1}^{[u_jn]}X_k-\int_0^{u_jT}Y_t dt\right| \\
&\quad+\max_{1\leq j\leq N}\left|\frac{1}{n}\sum_{k=1}^{[u_{j-1}n]}X_k-\int_0^{u_{j-1}T}Y_t dt\right| \\
&\quad+\max_{1\leq j\leq N}\int_{u_{j-1}T}^{u_jT}Y_t dt \\
&\xrightarrow{\mathsf{P}} 0+0+\max_{1\leq j\leq N}\int_{u_{j-1}T}^{u_jT}Y_t dt \\
&\leq \varepsilon, \quad \text{on the set } \{\bar{Y}\leq K\}.
\end{aligned}$$

Thus we obtain that

$$\begin{aligned}
&\limsup_{n\to\infty}\mathsf{P}\left(\sup_{u\in[0,1]}\left|\frac{1}{n}\sum_{k=1}^{[un]}X_k-\int_0^{uT}Y_t dt\right|>2\varepsilon\right) \\
&\leq \mathsf{P}\left(\max_{1\leq j\leq N}\int_{u_{j-1}T}^{u_jT}Y_t dt>\varepsilon \text{ and } \bar{Y}\leq K\right)+\mathsf{P}(\bar{Y}>K) \\
&\leq 0+\varepsilon.
\end{aligned}$$

The proof is finished. □

Theorem 7.3.4 Let $(\mathcal{X},\mathcal{A})$ be a measurable space. Let $X_1,X_2,\ldots$ be a sequence of $\mathcal{X}$-valued random variables which is ergodic with the invariant measure P°. For any P°-integrable function f, it holds that

$$\sup_{u\in[0,1]}\left|\frac{1}{n}\sum_{k=1}^{[un]}f(X_k)-u\int_{\mathcal{X}}f(x)P^\circ(dx)\right|\xrightarrow{\mathsf{P}}0.$$

Proof. Introducing the decomposition $f=f^+-f^-$, where $f^+=f\vee 0$ and $f^-=-(f\wedge 0)$, we apply Lemma 7.3.3 to each of f^+ and f^-.

Choose any $u\in(0,1]$ (the case $u=0$ is trivial). Observe that

$$\begin{aligned}
\frac{1}{n}\sum_{k=1}^{[un]}f^+(X_k) &= \frac{[un]}{n}\frac{1}{[un]}\sum_{k=1}^{[un]}f^+(X_k) \\
&\xrightarrow{\mathsf{P}} u\int_{\mathcal{X}}f^+(x)P^\circ(dx),
\end{aligned}$$

thus Lemma 7.3.3 yields that

$$\sup_{u\in[0,1]}\left|\frac{1}{n}\sum_{k=1}^{[un]} f^+(X_k) - u\int_{\mathcal{X}} f^+(x)P^\circ(dx)\right| \xrightarrow{\mathsf{P}} 0.$$

Since we can deduce the same conclusion also for f^-, we obtain the desired result. □

7.3.2 Uniform convergence of smooth random fields

Let us generalize the idea appearing in the previous subsection.

A pseudo-metric space (Θ,ρ) is said to be *totally bounded* if for any constant $\varepsilon > 0$ there exist finitely many closed balls with radius ε such that the union of them covers Θ. For example, a bounded subset of an Euclidean space is totally bounded.

Theorem 7.3.5 (Device for uniform convergence) Let (Θ,ρ) be a totally bounded pseudo-metric space. For a given sequence of random fields $X_n = \{X_n(\theta);\theta\in\Theta\}$, suppose that

$$X_n(\theta) \xrightarrow{\mathsf{P}} x(\theta), \quad \forall\theta\in\Theta,$$

where the limits $x(\theta)$'s are assumed to be deterministic.

(i) If there exist a sequence of real-valued random variables K_n and some constants $K>0$ and $\alpha\in(0,1]$ such that

$$\begin{cases} |X_n(\theta)-X_n(\theta')| \le K_n\rho(\theta,\theta')^\alpha, & \forall\theta,\theta'\in\Theta, \\ \text{and} \quad K_n = O_{\mathsf{P}}(1), \end{cases} \tag{7.24}$$

$$|x(\theta)-x(\theta')| \le K\rho(\theta,\theta')^\alpha, \quad \forall\theta,\theta'\in\Theta,$$

then it holds that

$$\sup_{\theta\in\Theta}|X_n(\theta)-x(\theta)| \xrightarrow{\mathsf{P}} 0.$$

(ii) In particular, if Θ is a bounded, open, convex subset of p-dimensional Euclidean space, and if all paths of the random random filed $\theta \rightsquigarrow X_n(\theta)$ are continuously differentiable with the derivatives satisfying

$$\sup_{\theta\in\Theta}\left|\frac{\partial}{\partial\theta_i}X_n(\theta)\right| = O_{\mathsf{P}}(1), \quad i=1,\ldots,p, \tag{7.25}$$

then the condition (7.24) holds true. It is sufficient for (7.25) that the condition

$$\limsup_{n\to\infty}\mathsf{E}\left[\sup_{\theta\in\Theta}\left|\frac{\partial}{\partial\theta_i}X_n(\theta)\right|\right] < \infty, \quad i=1,\ldots,p, \tag{7.26}$$

holds true[11].

[11] See Exercise 2.3.4.

Proof. For any constant $\delta > 0$, choose finitely many sets $B_1, ..., B_{N(\delta)}$ with the diameter being at most δ that cover Θ, and then choose an arbitrary pointy $\theta_i \in B_i$ for every $i = 1, ..., N(\delta)$. Then it holds that

$$\sup_{\theta\in\Theta} |X_n(\theta) - x(\theta)| \leq \max_{1\leq i\leq N(\delta)} |X_n(\theta_i) - x(\theta_i)| + (K_n + K)\delta^{\alpha}.$$

Now, for any $\varepsilon, \eta > 0$, first choose a constant $H_\eta > 0$ such that

$$\limsup_{n\to\infty} \mathsf{P}(K_n > H_\eta) \leq \eta,$$

and then choose a constant $\delta > 0$ such that $(H_\eta + K)\delta^{\gamma} \leq \varepsilon/2$. It then holds that

$$\begin{aligned}
&\limsup_{n\to\infty} \mathsf{P}\left(\sup_{\theta\in\Theta} |X_n(\theta) - x(\theta)| > \varepsilon\right) \\
\leq\ & \limsup_{n\to\infty} \mathsf{P}(K_n > H_\eta) + \sum_{i=1}^{N(\delta)} \limsup_{n\to\infty} \mathsf{P}(|X_n(\theta_i) - x(\theta_i)| > \varepsilon/(2N(\delta))) \\
\leq\ & \eta,
\end{aligned}$$

which completes the proof of (i). The claims in (ii) are immediate from (i). □

Exercise 7.3.1 Let a sequence of random fields $X_n = \{X_n(\theta); \theta \in \Theta\}$ indexed by a set Θ be given.

(i) Under which condition does

$$X_n(\theta) \xrightarrow{\mathsf{P}} 0, \quad \forall \theta \in \Theta,$$

imply

$$\sup_{\theta\in\Theta} |X_n(\theta)| = o_{\mathsf{P}^*}(1)?$$

(ii) Under which condition does

$$X_n(\theta) \xrightarrow{\mathsf{P}} X(\theta), \quad \forall \theta \in \Theta,$$

imply

$$\sup_{\theta\in\Theta} |X_n(\theta) - X(\theta)| = o_{\mathsf{P}^*}(1),$$

in the case where the limit $X = \{X(\theta); \theta \in \Theta\}$ is also a random field indexed by Θ?

[**Comment:** There are several ways to describe a sufficient condition for (i) or (ii). Find your own sufficient conditions by using also the ideas presented in this section.]

7.4 Tools for Discrete Sampling of Diffusion Processes

This section provides some preliminary tools for our discussions in Subsections 8.5.2, 8.5.3, 8.7.1 and 10.3.3.

Let us consider the 1-dimensional stochastic differential equation given by

$$X_t = X_0 + \int_0^t \beta(X_s)ds + \int_0^t \sigma(X_s)dW_s, \tag{7.27}$$

where β and σ are some measurable functions satisfying

$$|\beta(x)| \leq K(1+|x|) \quad \text{and} \quad |\sigma(x)| \leq K(1+|x|), \quad \forall x \in \mathbb{R}, \tag{7.28}$$

for a constant $K > 0$, and $s \rightsquigarrow W_s$ is a standard Wiener process.

Remark. The condition (7.28) is satisfied if the functions $x \mapsto \beta(x)$ and $x \mapsto \sigma(x)$ are Lipschitz continuous. The constant K in (7.28) will repeatedly appear in the description of the following theorems during this section.

In the sequel, we assume that the stochastic differential equation (7.27) has a strong solution X. The proofs of the following theorems will be given later.

Theorem 7.4.1 Let $p \geq 1$ be given, and assume that $\sup_{t\in[0,\infty)} \mathsf{E}[|X_t|^{p\vee 2}] < \infty$[12].

There exists a constant $C = C_{p,K} > 0$, depending only on p and K, such that it holds for any $0 \leq t < t'$ satisfying $|t'-t| \leq 1$ that

$$\mathsf{E}\left[\sup_{u\in[t,t']} |X_u - X_t|^p \middle| \mathcal{F}_t \right] \leq C|t'-t|^{p/2}(1+|X_t|)^p$$

and that

$$\mathsf{E}\left[\sup_{u\in[t,t']} |X_u|^p \middle| \mathcal{F}_t \right] \leq C(1+|X_t|)^p.$$

Definition 7.4.2 (Function of polynomial growth) A function $f : \mathbb{R} \to \mathbb{R}$ is said to be of *polynomial growth* if there exist some constants $C, p \geq 1$ such that

$$|f(x)| \leq C(1+|x|)^p, \quad \forall x \in \mathbb{R}.$$

Theorem 7.4.3 Suppose that $\sup_{t\in[0,\infty)} \mathsf{E}[|X_t|^q] < \infty$ holds for any constant $q \geq 1$. Let $f : \mathbb{R} \to \mathbb{R}$ be a twice continuously differentiable function whose derivatives f' and f'' are functions of polynomial growth. Then, there exist some constants $C = C_{f,K} > 0$ and $p = p_{f,K} \geq 1$, depending only on f and K, such that it holds for any $0 \leq t < t'$ satisfying $|t'-t| \leq 1$ that

$$|\mathsf{E}[f(X_{t'}) - f(X_t)|\mathcal{F}_t]| \leq C|t'-t|(1+|X_t|)^p.$$

[12] This assumption is not so strong. For example, when the stochastic process $t \rightsquigarrow X_t$ is stationary, the assumption is reduced to the simple condition that $\mathsf{E}[|X_0|^{p\vee 2}] < \infty$. Compare this assumption with the very strong condition "$\mathsf{E}[\sup_{t\in[0,\infty)} |X_t|^{p\vee 2}] < \infty$", which is not usually satisfied.

Theorem 7.4.4 Suppose that $\mathcal{K}_q := \sup_{t\in[0,\infty)} \mathsf{E}[|X_t|^q] < \infty$ holds for any constant $q \geq 1$. Let $f : \mathbb{R} \to \mathbb{R}$ be a function such that

$$|f(x) - f(y)| \leq C|x-y|(1+|x|+|y|)^p, \quad \forall x,y \in \mathbb{R},$$

for some constants $C, p \geq 1$; this condition is satisfied if the function $f : \mathbb{R} \to \mathbb{R}$ is differentiable and its derivative f' is a function of polynomial growth.

(i) There exists a constant $C' = C'_{f,K,\mathcal{K}_q} > 0$, depending only on f, K and $\mathcal{K}_q$ for some q, such that it holds for any time grids $0 = t_0^n < t_1^n < \cdots < t_n^n$ satisfying $\Delta_n = \max_{1\leq k\leq n} |t_k^n - t_{k-1}^n| \leq 1$ that

$$\mathsf{E}\left[\left|\frac{1}{t_n^n}\sum_{k=1}^n f(X_{t_{k-1}^n})|t_k^n - t_{k-1}^n| - \frac{1}{t_n^n}\int_0^{t_n^n} f(X_t)dt\right|\right] \leq C'\sqrt{\Delta_n}.$$

(ii) In particular, if the stochastic process $t \rightsquigarrow X_t$ is ergodic with the invariant measure P° and if $t_n^n \to \infty$ and $\Delta_n \to 0$ as $n \to \infty$, then it holds that

$$\frac{1}{t_n^n}\sum_{k=1}^n f(X_{t_{k-1}^n})|t_k^n - t_{k-1}^n| \xrightarrow{\mathsf{P}} \int_{\mathbb{R}} f(x)P^\circ(dx).$$

Further if

$$\sum_{k=1}^n \left|\frac{|t_k^n - t_{k-1}^n|}{t_n^n} - \frac{1}{n}\right| \to 0 \quad \text{as } n \to \infty, \tag{7.29}$$

then it also holds that

$$\frac{1}{n}\sum_{k=1}^n f(X_{t_{k-1}^n}) \xrightarrow{\mathsf{P}} \int_{\mathbb{R}} f(x)P^\circ(dx).$$

Remark. In the typical case of the *equidistant sampling*, the time grids are given by $t_k^n = (k/n)t_n^n$ for every $k = 0,1,...,n$. In this case, it holds that $\Delta_n = t_n^n/n$, and the condition (7.29) is clearly satisfied.

Now, let us proceed with proving the above theorems.

Proof of Theorem 7.4.1. As for the first inequality, it is sufficient to prove the case where $p \geq 2$, because the case of $p \in [1,2)$ can be reduced to the case of $2p$ by Jensen's inequality. We shall prove that there exists a constant $C = C_{p,K}$ depending only on p and K such that

$$\begin{aligned}
&\mathsf{E}\left[\sup_{u\in[t,t']} |X_u - X_t|^p \,\middle|\, \mathcal{F}_t\right] \\
&\leq C\left\{|t'-t|^{p/2}(1+|X_t|)^p + \int_t^{t'} \mathsf{E}\left[\sup_{u\in[t,s]} |X_u - X_t|^p \,\middle|\, \mathcal{F}_t\right] ds\right\};
\end{aligned} \tag{7.30}$$

then the first inequality would follow from Gronwall's inequality (Lemma A1.2.2). In the sequel, we will often use the inequalities

$$|x+y|^p \leq \left|2\cdot\frac{|x|+|y|}{2}\right|^p \leq 2^p\frac{|x|^p+|y|^p}{2} = 2^{p-1}(|x|^p+|y|^p).$$

First, noting that

$$X_u - X_t = \int_t^u \beta(X_s)ds + \int_t^u \sigma(X_s)dW_s,$$

we have

$$\mathsf{E}\left[\sup_{u\in[t,t']}|X_u - X_t|^p \middle| \mathcal{F}_t\right] \le 2^{p-1}\{(I)+(II)\},$$

where

$$\begin{aligned} (I) &= \mathsf{E}\left[\sup_{u\in[t,t']}\left|\int_t^u \beta(X_s)ds\right|^p \middle| \mathcal{F}_t\right], \\ (II) &= \mathsf{E}\left[\sup_{u\in[t,t']}\left|\int_t^u \sigma(X_s)dW_s\right|^p \middle| \mathcal{F}_t\right]. \end{aligned}$$

Let us evaluate the terms (I) and (II).

As for the term (I), it follows from Hölder's inequality that

$$\begin{aligned} \left|\int_t^u \beta(X_s)ds\right|^p &\le |u-t|^{p-1}\int_t^u |\beta(X_s)|^p ds \\ &\le |u-t|^{p-1}K^p\int_t^u (1+|X_s|)^p ds. \\ &\le |u-t|^{p-1}K^p\int_t^u (1+|X_t|+|X_s-X_t|)^p ds \\ &\le |u-t|^{p-1}K^p\int_t^u 2^{p-1}\{(1+|X_t|)^p + |X_s-X_t|^p\}ds. \end{aligned}$$

Thus, noting also that $|t'-t| \le 1$ we have

$$\begin{aligned} &\mathsf{E}\left[\sup_{u\in[t,t']}\left|\int_t^u \beta(X_s)ds\right|^p \middle| \mathcal{F}_t\right] \\ &\le 2^{p-1}|t'-t|^p K^p(1+|X_t|)^p + 2^{p-1}K^p\int_t^{t'} \mathsf{E}\left[\sup_{v\in[t,s]}|X_v - X_t|^p \middle| \mathcal{F}_t\right] ds. \end{aligned}$$

On the other hand, it follows from the extended Burkholder-Davis-Gundy inequality (Theorem 6.6.6) that there exists a constant c_p depending only on p such that

$$\mathsf{E}\left[\sup_{u\in[t,t']}\left|\int_t^u \sigma(X_s)dW_s\right|^p \middle| \mathcal{F}_t\right] \le c_p\mathsf{E}\left[\left|\int_t^{t'} \sigma(X_s)^2 ds\right|^{p/2} \middle| \mathcal{F}_t\right].$$

Now, if $p > 2$ then by Hölder's inequality, we have

$$\left|\int_t^{t'} \sigma(X_s)^2 ds\right|^{p/2}$$

$$
\begin{aligned}
&\leq |t'-t|^{(p/2)-1}\int_t^{t'}\sup_{v\in[t,s]}|\sigma(X_v)|^p ds \\
&\leq |t'-t|^{(p/2)-1}\int_t^{t'}K^p(1+|X_t|+\sup_{v\in[t,s]}|X_v-X_t|)^p ds \\
&\leq 2^{p-1}K^p\left\{|t'-t|^{p/2}(1+|X_t|)^p+\int_t^{t'}\sup_{v\in[t,s]}|X_v-X_t|^p ds\right\}.
\end{aligned}
$$

Since this inequality clearly holds true also in the case of $p=2$, the evaluation of the form (7.30) can be proved also for the term (II). This completes the proof of the first inequality.

The second inequality is proved by observing that

$$|X_u|^p \leq 2^{p-1}(|X_t|^p+|X_u-X_t|^p)$$

with the help of the first inequality. □

Proof of Theorem 7.4.3. Apply Itô's formula and the optional sampling theorem to have that

$$
\begin{aligned}
&|\mathsf{E}[f(X_{t'})-f(X_t)|\mathcal{F}_t]| \\
&= \left|\mathsf{E}\left[\int_t^{t'}\{f'(X_s)\beta(X_s)+\frac{1}{2}f''(X_s)\sigma(X_s)\}ds\middle|\mathcal{F}_t\right]\right| \\
&\leq \mathsf{E}\left[\int_t^{t'}C(1+|X_s|)^p ds\middle|\mathcal{F}_t\right] \\
&\leq |t'-t|\mathsf{E}\left[\sup_{s\in[t,t']}C(1+|X_s|)^p\middle|\mathcal{F}_t\right],
\end{aligned}
$$

for some constants $C>0$ and $p\geq 1$ depending only on f and K, and use the second inequality in Theorem 7.4.1 (ii) to obtain the desired bound. □

Proof of Theorem 7.4.4. The claim (i) is proved by using Theorem 7.4.1 and Hölder's inequality. The claim (ii) is immediate from (i). □

8

Parametric Z-Estimators

This chapter is the core part of this monograph. Using the martingale methods and the tools for asymptotic statistics which we have studied so far, we will prove the consistency and the asymptotic normality of Z-estimators in various parametric models in statistics.

We first give an intuitive explanation of our approach with an example of the i.i.d. model in Section 8.1, and then develop the approach up to a general theory for Z-estimators in Section 8.2. Logically, readers may start their study from Section 8.2, where the rigorous description actually begins. However, it is highly recommended that readers first read the intuitive explanation in Section 8.1 quickly to get a clear overview of the approach. The rigorous arguments for the i.i.d. model will be completed in Subsection 8.3.1.

Next, we deal with Markov chain models (Subsection 8.3.2), method of moment estimators (8.5.1), diffusion process models (Subsections 8.5.2 and 8.5.3), and Cox's regression models (Subsection 8.5.4). Just after the explanation for our treatment of Markov chain models, we give an interim summary of our approach in Section 8.4.

Finally, we give a remark to treat the cases where the components of Z-estimators may have different rates of convergence in Section 8.6; the corresponding example of diffusion process models will be given in Subsection 8.7.1.

Throughout this chapter, the limit operations are taken as $n \to \infty$, unless otherwise stated.

8.1 Illustrations with MLEs in I.I.D. Models

Let $(\mathcal{X}, \mathcal{A}, \mu)$ be a σ-finite[1] measure space. Let $X, X_1, X_2, \ldots$ be independent, $\mathcal{X}$-valued random variables identically distributed to a common distribution with the density $p(\cdot;\theta)$ with respect to μ, defined on $(\Omega, \mathcal{F}; \{\mathsf{P}_\theta; \theta \in \Theta\})$; that is, for every $\theta \in \Theta$,

$$\mathsf{P}_\theta(X \in A) = \int_A p(x;\theta)\mu(dx), \quad \forall A \in \mathcal{A}.$$

Here, θ is a parameter of interest from a subset Θ of $\mathbb{R}^p$ for an integer $p \geq 1$.

In this section, let us learn the outlines to prove some asymptotic properties of *maximum likelihood estimators* (MLEs) for θ in this model. Throughout this section, for simplicity we do not dare to state all the conditions for the densities $\{p(\cdot;\theta); \theta \in$

[1]We assume that the measure μ is σ-finite in order to guarantee the existence of the density p of the probability distribution $A \mapsto \mathsf{P}(X \in A)$ with respect to μ, namely, $\mathsf{P}(X \in A) = \int_A p(x)\mu(dx)$, $A \in \mathcal{A}$, for any given random variable X on $(\mathcal{X}, \mathcal{A})$; recall the Radon-Nikodym theorem.

DOI: 10.1201/9781315117768-8

$\Theta\}$ explicitly. The rigorous arguments for this model will be presented in Subsection 8.3.1.

The log-likelihood function based on the data $X_1,...,X_n$ is given by

$$\ell_n(\theta) = \sum_{k=1}^{n} \log p(X_k;\theta), \quad \theta \in \Theta.$$

There are at least two ways to define the "maximum likelihood estimator (MLE)". The origin of the name for the estimator comes from the (first) way to define it as the *maximizer* of the log-likelihood function $\theta \rightsquigarrow \ell_n(\theta)$ as a special case of *contrast function*:

$$\text{MLE is an argmax } \widehat{\theta}_n \text{ of the log-likelihood function}$$
$$\theta \rightsquigarrow \ell_n(\theta) = \sum_{k=1}^{n} \log p(X_k;\theta).$$

In this section, however, we shall regard the first derivatives of the log-likelihood function with respect to θ (that is, the so-called *gradient vector*[2] of $\ell_n(\theta)$) as *estimating functions*, and call a zero point (if it exists) the *maximum likelihood estimator*:

$$\text{MLE is a solution } \widehat{\theta}_n \text{ to the estimating equation}$$
$$\partial_i \ell_n(\theta) = \sum_{k=1}^{n} \partial_i \log p(X_k;\theta) = 0, \quad i = 1,...,p.$$

Here and in the sequel, let us use the notations $\partial_i = \frac{\partial}{\partial\theta_i}$, $\partial_{i,j} = \frac{\partial^2}{\partial\theta_i\partial\theta_j}$.

The both definitions are equivalent in many concrete models, including the current case, under some regularity conditions. However, distinguishing and generalizing the two different ideas for the definitions, we shall call:

- an estimator $\widehat{\theta}_n$ defined as the *maximizer* of a certain real-valued random function $\theta \rightsquigarrow \mathbb{M}_n(\theta)$ an *M-estimator*;

- an estimator $\widehat{\theta}_n$ defined as the *zero point* of a certain vector-valued random function $\theta \rightsquigarrow \mathbb{Z}_n(\theta)$ a *Z-estimator*.

The two definitions in the current model may be regarded as special cases of

$$\mathbb{M}_n(\theta) = \sum_{k=1}^{n} \log p(X_k;\theta)$$

and

$$\mathbb{Z}_n(\theta) = \dot{\mathbb{M}}_n(\theta) := (\partial_1 \mathbb{M}_n(\theta),...,\partial_p \mathbb{M}_n(\theta))^{\mathrm{tr}},$$

respectively. In general, however, we do *not* always assume that a given vector-valued

[2]Given a (sufficiently smooth) function $f:\mathbb{R}^p \to \mathbb{R}$, the $\mathbb{R}^p$-valued function $(\frac{\partial}{\partial\theta_1}f(\theta),...,\frac{\partial}{\partial\theta_p}f(\theta))^{\mathrm{tr}}$ is called the *gradient vector* of $f(\theta)$, and the $(p\times p)$-matrix valued function $(\frac{\partial^2}{\partial\theta_i\partial\theta_j}f(\theta))_{(i,j)\in\{1,...,p\}^2}$ is called the *Hessian matrix* of $f(\theta)$.

estimation function $\mathbb{Z}_n(\theta)$ is the one obtained as the first derivatives of a certain real-valued contrast function $\mathbb{M}_n(\theta)$; this set-up allows us to treat, e.g., the *method of moment estimators* (see Subsection 8.5.1), which cannot be treated in the framework of M-estimators.

8.1.1 Intuitive arguments for consistency of MLEs

Let us now turn back to our discussion on MLEs (as a special case of Z-estimators) in the i.i.d. model. The following two points are important to prove the consistency of the MLEs in the current model (and also in the general cases as we will find out in the next section).

[c1] Under the probability measure P_{θ_*} for the true point $\theta_* \in \Theta$, the estimating functions multiplied by $1/n$ converge in probability to

$$\mathsf{E}_{\theta_*}[\partial_i \log p(X;\theta)] = \int_{\mathcal{X}} \partial_i \log p(x;\theta) p(x;\theta_*)\mu(dx), \quad i = 1,...,p,$$

by the law of large numbers. Actually, it is possible even to prove a stronger claim that the convergence is "uniform in $\theta \in \Theta$" under some additional conditions; that is,

$$\sup_{\theta\in\Theta} \left| \frac{1}{n}\sum_{k=1}^{n} \partial_i \log p(X_k;\theta) - \mathsf{E}_{\theta_*}[\partial_i \log p(X;\theta)] \right| \xrightarrow{\mathsf{P}_{\theta_*}} 0, \quad i = 1,...,p.$$

[c2] Under some regularity conditions, it holds for the functions appearing in the limit that

$$\theta \mapsto \mathsf{E}_{\theta_*}[\partial_i \log p(X;\theta)], \quad i = 1,...,p, \quad \text{are zero} \quad \text{if and only if} \quad \theta = \theta_*.$$

In fact, if $\theta = \theta_*$ then it holds that

$$\begin{aligned}
& \mathsf{E}_{\theta_*}[\partial_i \log p(X;\theta)]|_{\theta=\theta_*} \\
= & \left. \int_{\mathcal{X}} \partial_i \log p(x;\theta) p(x;\theta_*)\mu(dx) \right|_{\theta=\theta_*} \\
= & \left. \int_{\mathcal{X}} \frac{\partial_i p(x;\theta)}{p(x;\theta)} p(x;\theta_*)\mu(dx) \right|_{\theta=\theta_*} \\
= & \left. \int_{\mathcal{X}} \partial_i p(x;\theta)\mu(dx) \right|_{\theta=\theta_*} \\
= & \left. \frac{\partial}{\partial\theta_i} \int_{\mathcal{X}} p(x;\theta)\mu(dx) \right|_{\theta=\theta_*} \qquad \text{(under some regularity conditions)} \quad (8.1) \\
= & \left. \frac{\partial}{\partial\theta_i} 1 \right|_{\theta=\theta_*} \\
= & \ 0.
\end{aligned}$$

Conversely, the condition that the limit of the estimating functions is not zero for any $\theta \in \Theta \setminus \{\theta_*\}$ is often assumed as a regularity condition called an *identifiability condition*.

These two facts would imply that

$$\theta \rightsquigarrow \textstyle\sum_{k=1}^n \partial_i \log p(X_k;\theta), \quad i=1,\dots,p, \quad \text{are close to zero} \\ \text{if and only if} \qquad \theta \quad \text{is close to} \quad \theta_*.$$

This approximately tells us that the MLEs $\widehat{\theta}_n$ (i.e., the sequence of zero points of the estimating functions) would approach to the true value θ_* (i.e., the zero point of the limit functions) when the sample size n is large.

This idea will be extended up to a theorem in a more general framework in Subsection 8.2.1, and we will prove the consistency of Z-estimators in various statistical models based on the theorem in the subsequent parts of this chapter.

8.1.2 Intuitive arguments for asymptotic normality of MLEs

Next let us grasp the outline of a proof of the fact that MLEs are asymptotically normal. Here, we assume for simplicity that Θ is an open interval of 1-dimensional Euclidean space (i.e., $p=1$), and use the notations $\partial_1 = \frac{\partial}{\partial\theta}$ and $\partial_{1,1} = \frac{\partial^2}{\partial^2\theta}$. Observe the formula obtained by the Taylor expansion:

$$\begin{aligned} \frac{1}{\sqrt{n}}\sum_{k=1}^n \partial \log p(X_k;\theta)\Big|_{\theta=\widehat{\theta}_n} &= \frac{1}{\sqrt{n}}\sum_{k=1}^n \partial_1 \log p(X_k;\theta_*) \\ &\quad + \frac{1}{n}\sum_{k=1}^n \partial_{1,1}\log p(X_k;\theta)\Big|_{\theta=\widetilde{\theta}_n} \cdot \sqrt{n}(\widehat{\theta}_n - \theta_*), \end{aligned}$$

where $\widetilde{\theta}_n$ is a random point on the segment connecting θ_* and $\widehat{\theta}_n$.

The essence of the proof of asymptotic normality consists of the following two points, where we denote by $I(\theta_*)$ the *Fisher information* defined later.

[an1] The first term on the right-hand side is proved to converge weakly to the Gaussian distribution $G(\theta_*) \sim \mathcal{N}(0, I(\theta_*))$.

[an2] The "coefficient part" of the second term on the right-hand side is proved to converge in probability to $-I(\theta_*)$.

By the definition of MLE $\widehat{\theta}_n$, the left-hand side of the Taylor expansion given above is zero. We may thus expect the relationship

$$0 \;\approx\; G(\theta_*) + (-I(\theta_*)) \cdot \sqrt{n}(\widehat{\theta}_n - \theta_*)$$

would be true when the sample size n is large. By "solving" this "asymptotic equation", we have that

$$\begin{aligned} \sqrt{n}(\widehat{\theta}_n - \theta_*) &\approx I(\theta_*)^{-1} G(\theta_*) \\ &\sim I(\theta_*)^{-1}\mathcal{N}(0, I(\theta_*)) \stackrel{d}{=} \mathcal{N}(0, I(\theta_*)^{-1}). \end{aligned}$$

We will sublimate this idea up to an *asymptotic representation theorem* for general Z-estimators in Subsection 8.2.2, and the theorem will be repeatedly applied to a lot of statistical models throughout this chapter.

8.2 General Theory for Z-estimators

In this section two general theorems for estimators defined as solutions to estimating functions will be established. The common set-up for Sections 8.2 and 8.6 is the following.

(Common set-up.) Let Θ be a non-empty subset of $\mathbb{R}^p$ with an integer $p \geq 1$; it may be an arbitrary subset of $\mathbb{R}^p$ in Subsections 8.2.1 and 8.6.1. Let a sequence $(\mathbb{Z}_n)_{n=1,2,...}$ of $\mathbb{R}^p$-valued random fields $\mathbb{Z}_n = \{\mathbb{Z}_n(\theta); \theta \in \Theta\}$, where $\mathbb{Z}_n(\theta) = (\mathbb{Z}_n^1(\theta), ..., \mathbb{Z}_n^p(\theta))^{\mathrm{tr}}$, be given.

In Subsections 8.2.2 and 8.6.2, the set Θ is assumed to be an open, convex subset of $\mathbb{R}^p$, and it is also assumed that there exists a sequence of $(p \times p)$-matrix valued random fields $\dot{\mathbb{Z}}_n = \{\dot{\mathbb{Z}}_n(\theta); \theta \in \Theta\}$, where $\dot{\mathbb{Z}}_n(\theta) = (\dot{\mathbb{Z}}_n^{i,j}(\theta))_{(i,j)\in\{1,...,p\}^2}$, such that it holds for the true value θ_* of the parameter $\theta \in \Theta$ that

$$\mathbb{Z}_n^i(\theta) = \mathbb{Z}_n^i(\theta_*) + \sum_{j=1}^{p} \dot{\mathbb{Z}}_n^{i,j}(\widetilde{\theta}_n)(\theta^j - \theta_*^j), \quad i = 1, ..., p, \quad \forall \theta \in \Theta,$$

or, equivalently,

$$\mathbb{Z}_n(\theta) = \mathbb{Z}_n(\theta_*) + \dot{\mathbb{Z}}_n(\widetilde{\theta}_n)(\theta - \theta_*), \quad \forall \theta \in \Theta, \tag{8.2}$$

where $\widetilde{\theta}_n = \widetilde{\theta}_n(\theta, \theta_*, n)$ is a random point on the segment connecting θ and θ_*.

Remark. To perform the Taylor expansion (8.2), we assume that the set Θ is open (otherwise, some pathological discussion is required for the differentiability with respect to θ at the boundary of the set Θ) and convex (this allows us to involve "$\widetilde{\theta}_n$" in the second term of the expansion).

Remark. (Typical case.) Since the notations in this section are abstract, it may not be easy to understand the materials only through the explanations written here. It is thus recommended for readers to follow the contexts of this section recalling the case of the i.i.d. model discussed in the previous section with the following interpretation of the abstract notations:

$$\begin{aligned}
\mathbb{Z}_n^i(\theta) &= \sum_{k=1}^{n} \partial_i \log p(X_k; \theta), \quad i = 1, ..., p; \\
Z_n^i(\theta) &= \sum_{k=1}^{n} \mathsf{E}_{\theta_*}[\partial_i \log p(X_k; \theta)], \quad i = 1, ..., p; \\
s_n &= n; \\
z^i(\theta) &= \lim_{n\to\infty} s_n^{-1} Z_n^i(\theta) = \mathsf{E}_{\theta_*}[\partial_i \log p(X; \theta)], \quad i = 1, ..., p;
\end{aligned}$$

$$\begin{aligned}
\dot{\mathbb{Z}}_n^{i,j}(\theta) &= \sum_{k=1}^{n} \partial_{i,j} \log p(X_k;\theta), \quad (i,j) \in \{1,\dots,p\}^2; \\
\dot{Z}_n^{i,j}(\theta) &= \sum_{k=1}^{n} \mathsf{E}_{\theta_*}[\partial_{i,j} \log p(X_k;\theta)], \quad (i,j) \in \{1,\dots,p\}^2; \\
q_n, r_n &= \sqrt{n}; \\
-J^{i,j}(\theta_*) &= \lim_{n\to\infty} (q_n r_n)^{-1} \dot{Z}_n^{i,j}(\widetilde{\theta}_n) = \mathsf{E}_{\theta_*}[\partial_{i,j} \log p(X;\theta_*)], \ (i,j) \in \{1,\dots,p\}^2,
\end{aligned}$$

where $(\widetilde{\theta}_n)_{n=1,2,\dots}$ is any sequence of Θ-valued random elements such that $\widetilde{\theta}_n \xrightarrow{\mathsf{P}^*} \theta_*$ (since we do not require any measurability of $\widetilde{\theta}_n$, this convergence is taken in *outer-probability* P^*).

Readers would soon find that the abstract-looking description given here makes the lines of the proofs clear, and moreover that this description would help us to imagine how far the methods explained here could be applied beyond the i.i.d. model. In fact, all examples studied in the subsequent sections will be analyzed by using the general theories developed in the current section.

Let us continue the description of the common set-up.

(Common set-up, continued.) We define the *Z-estimator* $\widehat{\theta}_n$ corresponding to the estimating function $\theta \rightsquigarrow \mathbb{Z}_n(\theta)$ as any (approximate) solution to the equation

$$\mathbb{Z}_n^i(\theta) = 0, \quad i = 1,\dots,p,$$

which is Borel measurable in Θ; in other words, the *Z-estimator* $\widehat{\theta}_n$ is defined as any Θ-valued random variable such that $\mathbb{Z}_n(\widehat{\theta}_n)$ is "zero" or "close to zero" in an appropriate sense, which will be clearly specified in each theorem below.

8.2.1 Consistency of Z-estimators, I

Here, we intend to extend our discussion in Subsection 8.1.1 in such a way that: the fact **[c1]** is replaced by the requirement that the rescaled estimating functions $s_n^{-1}\mathbb{Z}_n(\theta)$, where the sequence $(s_n)_{n=1,2,\dots}$ of positive numbers is typically $s_n = n$, should converge in probability to a (deterministic) $\mathbb{R}^p$-valued function, say $z(\theta)$, uniformly in $\theta \in \Theta$; the fact **[c2]** is replaced by the requirement that $z(\theta)$ should be zero if and only if $\theta = \theta_*$. These requirements are formulated as **[C1]** and **[C2]** below, respectively.

Theorem 8.2.1 (Consistency of Z-estimators) Consider the common set-up described at the beginning of this section. Suppose that there exist a sequence $(s_n)_{n=1,2,\dots}$ of real numbers, an $\mathbb{R}^p$-valued function $\theta \mapsto z(\theta)$ on Θ, and a (deterministic) point $\theta_* \in \Theta$ satisfying the following conditions **[C1]** and **[C2]**.

[C1][3] $\sup_{\theta\in\Theta}||s_n^{-1}\mathbb{Z}_n(\theta)-z(\theta)||\xrightarrow{\mathsf{P}^*}0$.
[C2] It holds that

$$\inf_{\theta:||\theta-\theta_*||>\varepsilon}||z(\theta)||>0=||z(\theta_*)||,\quad \forall\varepsilon>0.$$

Then, for any sequence of Θ-valued random variables $\widehat{\theta}_n$ satisfying

$$s_n^{-1}\mathbb{Z}_n(\widehat{\theta}_n)=o_{\mathsf{P}}(1),$$

it holds that $\widehat{\theta}_n\xrightarrow{\mathsf{P}}\theta_*$.

Proof. First observe that

$$\begin{aligned}||z(\widehat{\theta}_n)|| &\le ||z(\widehat{\theta}_n)-s_n^{-1}\mathbb{Z}_n(\widehat{\theta}_n)||+||s_n^{-1}\mathbb{Z}_n(\widehat{\theta}_n)||\\ &\le \sup_{\theta\in\Theta}||z(\theta)-s_n^{-1}\mathbb{Z}_n(\theta)||+||s_n^{-1}\mathbb{Z}_n(\widehat{\theta}_n)||,\end{aligned}$$

which converges in outer-probability to zero due to the condition **[C1]** and the requirement for $\widehat{\theta}_n$. Next observe that the condition **[C2]** implies that for any $\varepsilon>0$ there exists a $\delta>0$ such that $||z(\theta)||>\delta$ whenever $||\theta-\theta_*||>\varepsilon$. Thus, the event $\{||\widehat{\theta}_n-\theta_*||>\varepsilon\}$ is included in the event $\{||z(\widehat{\theta}_n)||>\delta\}$. Note that both events are measurable sets. As a consequence from the above argument, we have that

$$\mathsf{P}(||\widehat{\theta}_n-\theta_*||>\varepsilon)\le\mathsf{P}(||z(\widehat{\theta}_n)||>\delta)\to 0\quad\text{as } n\to\infty.$$

The proof is finished. □

8.2.2 Asymptotic representation of Z-estimators, I

Noting that (8.2) is equivalent to

$$\frac{1}{q_n}\mathbb{Z}_n(\theta)=\frac{1}{q_n}\mathbb{Z}_n(\theta_*)+\frac{1}{q_nr_n}\dot{\mathbb{Z}}_n(\widetilde{\theta}_n)(r_n(\theta-\theta_*)),\quad i=1,\dots,p, \tag{8.3}$$

we shall extend our discussion in Subsection 8.1.2 in such a way that: the fact **[an1]** is replaced by the requirement that the rescaled random vectors $q_n^{-1}\mathbb{Z}_n(\theta_*)$, which is the first term on the right-hand side of (8.3), should converge weakly to a Borel law $L(\theta_*)$; the fact **[an2]** is replaced by the requirement that the $(p\times p)$-matrix valued random sequence $(q_nr_n)^{-1}\dot{\mathbb{Z}}_n(\widetilde{\theta}_n)$ appearing in the second term on the right-hand side of (8.3) converges in outer-probability to a (deterministic) regular matrix $-J(\theta_*)$, as far as $\widetilde{\theta}_n$ converges in outer-probability to θ_*. The matrix $J(\theta_*)$ in the limit of the second requirement should coincide typically with the Fisher information matrix $I(\theta_*)$. These requirements are formulated as **[AR1]** and **[AR2]** below, respectively.

[3] This convergence is formally considered in outer-probability P^* because the supremum over $\theta\in\Theta$ may hurt the measurability. However, since the random fields $\theta\rightsquigarrow\mathbb{Z}_n(\theta)$ are separable in most applications, this problem of measurability would hardly take place.

Theorem 8.2.2 (Asymptotic representation of Z-estimators) Consider the common set-up described at the beginning of this section. Suppose that there exist two sequences $(q_n)_{n=1,2,\ldots}$, $(r_n)_{n=1,2,\ldots}$ of positive numbers satisfying the following **[AR1]** and **[AR2]**.

[AR1] The sequence of $\mathbb{R}^p$-valued random variables $q_n^{-1}\mathbb{Z}_n(\theta_*)$ converges weakly in $\mathbb{R}^p$ to a random variable $L(\theta_*)$.

[AR2] For any sequence of Θ-valued random elements $\widetilde{\theta}_n$ converging in outer-probability to θ_*, the sequence of $(p \times p)$-matrix valued random elements $(q_n r_n)^{-1}\dot{\mathbb{Z}}_n(\widetilde{\theta}_n)$ converges in outer-probability to a (deterministic) $(p \times p)$-matrix $-J(\theta_*)$ that is regular.

Then, for any sequence of Θ-valued random variables $\widehat{\theta}_n$ satisfying

$$q_n^{-1}\mathbb{Z}_n(\widehat{\theta}_n) = o_{\mathsf{P}}(1) \quad \text{and} \quad \widehat{\theta}_n \xrightarrow{\mathsf{P}} \theta_*,$$

it holds that

$$r_n(\widehat{\theta}_n - \theta_*) = J(\theta_*)^{-1}(q_n^{-1}\mathbb{Z}_n(\theta_*)) + o_{\mathsf{P}}(1).$$

In particular, it holds also that

$$r_n(\widehat{\theta}_n - \theta_*) \stackrel{\mathsf{P}}{\Longrightarrow} J(\theta_*)^{-1}L(\theta_*) \quad \text{in } \mathbb{R}^p.$$

Proof. Let $\widetilde{\theta}_n$ be the one when $\widehat{\theta}_n$ is substituted to θ in (8.2); that is, $\widetilde{\theta}_n$ is a (random) point on the segment connecting θ_* and $\widehat{\theta}_n$ such that

$$q_n^{-1}\mathbb{Z}_n(\widehat{\theta}_n) = q_n^{-1}\mathbb{Z}_n(\theta_*) + (q_n r_n)^{-1}\dot{\mathbb{Z}}_n(\widetilde{\theta}_n)(r_n(\widehat{\theta}_n - \theta_*)). \tag{8.4}$$

By the assumption **[AR2]**, this equation can be further rewritten as

$$q_n^{-1}\mathbb{Z}_n(\widehat{\theta}_n) = q_n^{-1}\mathbb{Z}_n(\theta_*) - J(\theta_*)(r_n(\widehat{\theta}_n - \theta_*)) + o_{\mathsf{P}^*}(||r_n(\widehat{\theta}_n - \theta_*)||).$$

Noting that the left-hand side is $o_{\mathsf{P}}(1)$ due to the requirement for $\widehat{\theta}_n$, move the second term on the right-hand side above to the left-hand side and then multiply both sides by $J(\theta_*)^{-1}$ to obtain that

$$r_n(\widehat{\theta}_n - \theta_*) = J(\theta_*)^{-1}(q_n^{-1}\mathbb{Z}_n(\theta_*)) + o_{\mathsf{P}^*}(1 + ||r_n(\widehat{\theta}_n - \theta_*)||).$$

Since the assumption **[AR1]** implies that the first term on the right-hand side is $O_{\mathsf{P}}(1)$, we have that $r_n(\widehat{\theta}_n - \theta_*) = O_{\mathsf{P}^*}(1)$. This implies further that the second term on the right-hand side is $o_{\mathsf{P}^*}(1)$. Here, noting that this term is given as the left-hand side minus the first term on the right-hand side which is measurable, we actually have that this reminder term is measurable and it is $o_{\mathsf{P}}(1)$. The first claim has now been proved.

The second claim is immediate from the first one. □

8.3 Examples, I-1 (Fundamental Models)

8.3.1 Rigorous arguments for MLEs in i.i.d. models

In this subsection, we shall present some rigorous arguments to derive the consistency and the asymptotic normality of the maximum likelihood estimators (MLEs) discussed intuitively in Section 8.1.

Recall the set-up given in the first paragraph of Section 8.1, and assume that Θ is an open, convex subset of $\mathbb{R}^p$. Assuming also that the densities $p(x;\theta)$ are differentiable with respect to θ on Θ, we introduce the estimating function, which is the first derivatives of the log-likelihood function based on the data $X_1,...,X_n$, given by

$$\mathbb{Z}_n(\theta) = \sum_{k=1}^{n}(\partial_1 \log p(X_k;\theta),...,\partial_p \log p(X_k;\theta))^{\rm tr}, \quad \theta \in \Theta.$$

Let us first derive the consistency of the MLEs.

Discussion 8.3.1 (Consistency) Assume further that the set Θ is bounded (this is against our will; see the last paragraph of the current discussion). Suppose that there exist a measurable function K on $(\mathcal{X},\mathcal{A})$ and a constant $\alpha \in (0,1]$ such that for every $i = 1,...,p$,

$$|\partial_i \log p(x;\theta) - \partial_i \log p(x;\theta')| \leq K(x)||\theta-\theta'||^{\alpha}, \quad \forall \theta,\theta' \in \Theta, \tag{8.5}$$

and that

$$\int_{\mathcal{X}} K(x)p(x;\theta_*)\mu(dx) < \infty, \tag{8.6}$$

where θ_* denotes the true value of the parameter $\theta \in \Theta$. Suppose also that for any $\varepsilon > 0$ there exist some $i = 1,...,,p$ such that

$$\inf_{\theta:||\theta-\theta_*||>\varepsilon}\left|\int_{\mathcal{X}} \partial_i \log p(x;\theta)p(x;\theta_*)\mu(dx)\right| > 0, \tag{8.7}$$

and that some regularity conditions under which the operation (8.1) are guaranteed.

Then, for any sequence of Θ-valued random variables $\widehat{\theta}_n$ satisfying

$$n^{-1}\mathbb{Z}_n(\widehat{\theta}_n) = o_{\mathsf{P}_{\theta*}}(1),$$

it holds that $\widehat{\theta}_n \xrightarrow{\mathsf{P}_{\theta*}} \theta_*$.

To prove this claim, we shall apply Theorem 8.2.1. Under the assumption that Θ is bounded, (8.5) and (8.6), Theorem 7.3.1 yields that

$$\sup_{\theta\in\Theta}||n^{-1}\mathbb{Z}_n(\theta) - z(\theta)|| \xrightarrow{\mathsf{P}_{\theta*}} 0,$$

where

$$z^i(\theta) = \int_{\mathcal{X}} \partial_i \log p(x;\theta)p(x;\theta_*)\mu(dx), \quad i = 1,...,p,$$

and thus the condition **[C1]** is satisfied. The condition **[C2]** is nothing else than (8.7). The claim has been proved.

To close the current discussion, it should be noticed that the assumptions appearing in the above approach are merely a set of "sufficient conditions", and that they are not always necessary. For an illustration, let us consider a toy example of the parametric model "$\{\mathcal{N}(\theta,1);\theta\in\Theta\}$", where $\Theta\subset\mathbb{R}$. In this model, the MLE based on the data $X_1,...,X_n$ is explicitly computed as $\widehat{\theta}_n=\frac{1}{n}\sum_{k=1}^n X_k$, and its consistency is obtained directly by the law of large numbers; thus the assumption that "Θ is bounded" is unnecessary. This assumption was imposed only for the use of Theorem 7.3.1, which is not always the best or necessary tool to derive the consistency of Z-estimators.

Next let us derive the asymptotic normality through the asymptotic representation theorem.

Discussion 8.3.2 (Asymptotic representation) Assuming that $\theta\mapsto p(x;\theta)$ is twice continuously differentiable, introduce the following: for every $(i,j)\in\{1,...,p\}^2$,

$$\begin{aligned}\dot{\mathbb{Z}}_n^{i,j}(\theta) &= \sum_{k=1}^n \partial_{i,j}\log p(X_k;\theta);\\ \dot{z}^{i,j}(\theta) &= \int_{\mathcal{X}}\partial_{i,j}\log p(x;\theta)p(x;\theta_*)\mu(dx)\\ &= \int_{\mathcal{X}}\frac{p(x;\theta)\partial_{i,j}p(x;\theta)-\partial_i p(x;\theta)\partial_j p(x;\theta)}{p(x;\theta)^2}p(x;\theta_*)\mu(dx).\end{aligned}$$

Under some regularity conditions, the latter with $\theta=\theta_*$ being substituted coincides with the (i,j) entry of the Fisher information matrix $I(\theta_*)=(I^{i,j}(\theta_*))_{(i,j)\in\{1,...,p\}^2}$ multiplied by -1; that is,

$$I^{i,j}(\theta_*)=-\dot{z}^{i,j}(\theta_*)=\int_{\mathcal{X}}\frac{\partial_i p(x;\theta_*)\partial_j p(x;\theta_*)}{p(x;\theta_*)}\mu(dx).$$

Suppose that there exists a neighborhood N of the true value θ_*, a measurable function K on $(\mathcal{X},\mathcal{A})$ and a constant $\alpha\in(0,1]$ such that for every $(i,j)\in\{1,...,p\}^2$,

$$|\partial_{i,j}\log p(x;\theta_1)-\partial_{i,j}\log p(x;\theta')|\le K(x)||\theta-\theta'||^\alpha,\quad \forall\theta,\theta'\in N, \tag{8.8}$$

and that

$$\int_{\mathcal{X}}K(x)p(x;\theta_*)\mu(dx)<\infty. \tag{8.9}$$

Suppose also that the Fisher information matrix $I(\theta_*)$ is positive definite.

Then, for any sequence of Θ-valued random variables $\widehat{\theta}_n$ satisfying

$$n^{-1/2}\mathbb{Z}_n(\widehat{\theta}_n)=o_{P_{\theta_*}}(1)\quad\text{and}\quad \widehat{\theta}_n\xrightarrow{P_{\theta_*}}\theta_*,$$

it holds that

$$\begin{aligned}\sqrt{n}(\widehat{\theta}_n - \theta_*) &= I(\theta_*)^{-1}(n^{-1/2}\mathbb{Z}_n(\theta_*)) + o_{\mathsf{P}_{\theta_*}}(1) \\ &\stackrel{\mathsf{P}_{\theta_*}}{\Longrightarrow} \mathcal{N}_p(0, I(\theta_*)^{-1}) \quad \text{in } \mathbb{R}^p.\end{aligned}$$

To prove this claim, we shall apply Theorem 8.2.2. Notice that

$$\mathbb{Z}_n(\theta_*) = \sum_{k=1}^{n} \left(\frac{\partial_1 p(X_k;\theta_*)}{p(X_k;\theta_*)}, \dots, \frac{\partial_p p(X_k;\theta_*)}{p(X_k;\theta_*)} \right)^{\mathrm{tr}},$$

that

$$\mathsf{E}_{\theta_*}\left[\frac{\partial_i p(X;\theta_*)}{p(X;\theta_*)} \right] = 0, \quad i = 1, \dots, p,$$

and that

$$\mathsf{E}_{\theta_*}\left[\frac{\partial_i p(X;\theta_*)}{p(X;\theta_*)} \frac{\partial_j p(X;\theta_*)}{p(X;\theta_*)} \right] = I^{i,j}(\theta_*), \quad (i,j) \in \{1, \dots, p\}^2.$$

Hence, the classical central limit theorem yields that

$$n^{-1/2}\mathbb{Z}_n(\theta_*) \stackrel{\mathsf{P}_{\theta_*}}{\Longrightarrow} \mathcal{N}_p(0, I(\theta_*)) \quad \text{in } \mathbb{R}^p, \tag{8.10}$$

and thus the condition **[AR1]** is satisfied. On the other hand, since we may assume without loss of generality that N is bounded, it follows from the assumption (8.8), (8.9) and Theorem 7.3.1 that

$$\sup_{\theta \in N} ||n^{-1}\dot{\mathbb{Z}}_n(\theta) - \dot{z}(\theta)|| \xrightarrow{\mathsf{P}_{\theta_*}} 0,$$

while it is clear that $\theta \mapsto \dot{z}(\theta)$ is continuous on N. Consequently, we obtain that

$$n^{-1}\dot{\mathbb{Z}}_n(\widetilde{\theta}_n) \xrightarrow{\mathsf{P}^*_{\theta_*}} \dot{z}(\theta_*) = -I(\theta_*) \tag{8.11}$$

for any sequence of Θ-valued random elements $\widetilde{\theta}_n$ converging in outer-probability to θ_*, and thus the condition **[AR2]** is satisfied. The claim has been proved.

8.3.2 MLEs in Markov chain models

Let $(\mathcal{X}, \mathcal{A}, \mu)$ be a σ-finite measure space. Let $X_0, X_1, X_2, \dots$ be a Markov chain with the state space $\mathcal{X}$, the initial density q, and the parametric family of transition densities $\{p(\cdot,\cdot;\theta); \theta \in \Theta\}$, where Θ is an open, convex subset of $\mathbb{R}^p$, defined on a parametric family of probability space $(\Omega, \mathcal{F}, \{\mathsf{P}_\theta; \theta \in \Theta\})$; that is, for any $\theta \in \Theta$,

$$\mathsf{P}_\theta(X_0 \in A) = \int_A q(x)\mu(dx), \quad \forall A \in \mathcal{A},$$

$$\mathsf{P}_\theta(X_k \in A | X_{k-1} = x) = \int_A p(y,x;\theta)\mu(dy), \quad \forall A \in \mathcal{A}, \quad \forall x \in \mathcal{X}, \quad k = 1, 2, \dots.$$

Suppose that, under the probability measure P_θ, the Markov chain is ergodic with the invariant measure P_θ°; then, for any P_θ°-integrable function f on $\mathcal{X}$, it holds that

$$\frac{1}{n}\sum_{k=1}^{n} f(X_k) \xrightarrow{\mathsf{P}_\theta} \int_{\mathcal{X}} f(x) P_\theta^\circ(dx).$$

The likelihood function $L(\theta)$ based on the data $X_0, X_1, X_2, ..., X_n$ is given by

$$L_n(\theta) = \prod_{k=1}^{n} p(X_k, X_{k-1}; \theta),$$

and thus the log-likelihood function $\ell_n(\theta)$ is given by

$$\ell_n(\theta) = \sum_{k=1}^{n} \log p(X_k, X_{k-1}; \theta).$$

From now on, let us simultaneously discuss how to derive the consistency and the asymptotic representation of the maximum likelihood estimators (MLEs).

Discussion 8.3.3 ($\mathbb{Z}_n(\theta)$, $\dot{\mathbb{Z}}_n(\theta)$, their "compensators", and the limits) Assuming that the function $\theta \mapsto p(y,x;\theta)$ is twice continuously differentiable and that $p(y,x;\theta) > 0$ for all x, y, θ, we introduce the estimating functions $\mathbb{Z}_n(\theta)$ and its derivative matrix $\dot{\mathbb{Z}}_n(\theta)$ by

$$\mathbb{Z}_n^i(\theta) = \partial_i \ell_n(\theta) = \sum_{k=1}^{n} G^i(X_k, X_{k-1}; \theta), \quad i = 1, ..., p,$$

and

$$\dot{\mathbb{Z}}_n^{i,j}(\theta) = \partial_{i,j} \ell_n(\theta) = \sum_{k=1}^{n} H^{i,j}(X_k, X_{k-1}; \theta), \quad (i,j) \in \{1, ..., p\}^2,$$

where

$$G^i(y,x;\theta) = \frac{\partial_i p(y,x;\theta)}{p(y,x;\theta)}$$

and

$$H^{i,j}(y,x;\theta) = \frac{p(y,x;\theta)\partial_{i,j} p(y,x;\theta) - \partial_i p(y,x;\theta)\partial_j p(y,x;\theta)}{p(y,x;\theta)^2},$$

with the notations $\partial_i = \frac{\partial}{\partial\theta_i}$ and $\partial_{i,j} = \frac{\partial^2}{\partial\theta_i \partial\theta_j}$.

Here, introduce the "predictable compensators" Z_n and $\dot{Z}_n$ for $\mathbb{Z}_n$ and $\dot{\mathbb{Z}}_n$ under the probability measure P_{θ_*}, respectively, as follows.

$$Z_n^i(\theta) = \sum_{k=1}^{n} \int_{\mathcal{X}} G^i(y, X_{k-1}; \theta) p(y, X_{k-1}; \theta_*) \mu(dy), \quad i = 1, ..., p,$$

$$\dot{Z}_n^{i,j}(\theta) = \sum_{k=1}^{n} \int_{\mathcal{X}} H^{i,j}(y, X_{k-1}; \theta) p(y, X_{k-1}; \theta_*) \mu(dy), \quad (i,j) \in \{1, ..., p\}.$$

Under some mild conditions, the following (8.12), (8.13), (8.14) and (8.15) hold true:

$$n^{-1}(\mathbb{Z}_n^i(\theta) - Z_n^i(\theta)) \xrightarrow{\mathsf{P}_{\theta_*}} 0, \quad \forall\theta \in \Theta,\ i = 1, ..., p, \tag{8.12}$$

$$n^{-1}Z_n^i(\theta) - z^i(\theta) \xrightarrow{\mathsf{P}_{\theta_*}} 0, \quad \forall\theta \in \Theta,\ i = 1, ..., p, \tag{8.13}$$

where

$$z^i(\theta) = \int_{\mathcal{X}}\int_{\mathcal{X}} G^i(y,x;\theta)p(y,x;\theta_*)\mu(dy)P^\circ_{\theta_*}(dx);$$

$$n^{-1}(\dot{\mathbb{Z}}_n^{i,j}(\theta) - \dot{Z}_n^{i,j}(\theta)) \xrightarrow{\mathsf{P}_{\theta_*}} 0, \quad \forall\theta \in \Theta,\ (i,j) \in \{1, ..., p\}^2, \tag{8.14}$$

$$n^{-1}\dot{Z}_n^{i,j}(\theta) - (-I^{i,j}(\theta)) \xrightarrow{\mathsf{P}_{\theta_*}} 0, \quad \forall\theta \in \Theta,\ (i,j) \in \{1, ..., p\}^2, \tag{8.15}$$

where

$$-I^{i,j}(\theta) = \int_{\mathcal{X}}\int_{\mathcal{X}} H^{i,j}(y,x;\theta)p(y,x;\theta_*)\mu(dy)P^\circ_{\theta_*}(dx).$$

Indeed, (8.12) and (8.14) can be proved by using the corollary to Lenglart's inequality (Corollary 4.3.3 (i)), while (8.13) and (8.15) are immediate from the ergodicity. Moreover, under some more conditions (see the remark below), we can extend the above assertions up to:

$$\sup_{\theta\in\Theta} |n^{-1}\mathbb{Z}_n^i(\theta) - z^i(\theta)| \xrightarrow{\mathsf{P}_{\theta_*}} 0, \quad i = 1, ..., p; \tag{8.16}$$

$$\sup_{\theta\in N} |n^{-1}\dot{\mathbb{Z}}_n^{i,j}(\theta) - (-I^{i,j}(\theta))| \xrightarrow{\mathsf{P}_{\theta_*}} 0, \quad (i,j) \in \{1, ..., p\}^2, \tag{8.17}$$

where N is a neighborhood of θ_*. The latter yields that, once we have that $\theta \mapsto I(\theta)$ is continuous on N, it holds for any sequence of Θ-valued random elements $\widetilde{\theta}_n$ converging in outer-probability to θ_* that

$$n^{-1}\dot{\mathbb{Z}}_n(\widetilde{\theta}_n) \xrightarrow{\mathsf{P}^*_{\theta_*}} -I(\theta_*), \tag{8.18}$$

where $I(\theta) = (I^{i,j}(\theta))_{(i,j)\in\{1,...,p\}^2}$ for every $\theta \in \Theta$.

Remark. Recalling Theorem 7.3.5, some possible regularity conditions under which (8.16) and (8.17) hold true are the following: the set Θ is bounded, and there exist a constant $\alpha \in (0,1]$ and some measurable functions $\hat{G}(y,x)$ and $\hat{H}(y,x)$ that are integrable with respect to $p(y,x;\theta_*)\mu(dy)P^\circ_{\theta_*}(dx)$ such that

$$||G(y,x;\theta) - G(y,x;\theta')|| \le \hat{G}(y,x)||\theta - \theta'||^\alpha, \quad \forall\theta,\theta' \in \Theta,$$

$$||H(y,x;\theta) - H(y,x;\theta')|| \le \hat{H}(y,x)||\theta - \theta'||^\alpha, \quad \forall\theta,\theta' \in N,$$

where N is a bounded neighborhood of θ_*.

Indeed, recalling Exercise 6.6.2 (ii), we have that

$$\frac{1}{n}\sum_{k=1}^{n} \widehat{G}(X_k, X_{k-1}) = O_{\mathsf{P}_{\theta_*}}(1),$$

because its "predictable compensator"

$$\frac{1}{n}\sum_{k=1}^{n} \mathsf{E}_{\theta_*}[\widehat{G}(X_k,X_{k-1})|\mathcal{F}_{k-1}] = \frac{1}{n}\sum_{k=1}^{n}\int_{\mathcal{X}} \widehat{G}(y,X_{k-1})p(y,X_{k-1};\theta_*)\mu(dy)$$

converges in probability to a (finite) limit, namely $\int_{\mathcal{X}}\int_{\mathcal{X}} \widehat{G}(y,x)p(y,x;\theta_*)\mu(dy)P^{\circ}_{\theta_*}$, and thus it is $O_{\mathsf{P}_{\theta_*}}(1)$. It is clear that $\theta \mapsto z^i(\theta)$ is α-Hölder continuous. Hence Theorem 7.3.5 yields (8.16). Noting also that $\theta \mapsto I(\theta)$ is α-Hölder continuous on N, the proof for (8.17) is the same as that for (8.16).

Discussion 8.3.4 (Consequences from the "standard regularity condition") If the "standard regularity condition" to guarantee that

$$\int_{\mathcal{X}} \partial_i p(y,x;\theta_*)\mu(dy) = \partial_i \int_{\mathcal{X}} p(y,x;\theta_*)\mu(dy) = 0, \quad \forall x \in \mathcal{X}, \quad i = 1,...,p,$$

is satisfied, then it holds that

$$Z_n^i(\theta_*) = 0, \text{ a.s.,} \quad \text{and} \quad z^i(\theta_*) = 0, \quad i = 1,...,p, \tag{8.19}$$

and that

$$I^{i,j}(\theta_*) = \int_{\mathcal{X}}\int_{\mathcal{X}} \frac{\partial_i p(y,x;\theta_*)\partial_j p(y,x;\theta_*)}{p(y,x;\theta_*)}\mu(dy)P^{\circ}_{\theta_*}(dx), \quad (i,j) \in \{1,...,p\}^2.$$

The first equality in (8.19) implies that $\mathbb{Z}_n(\theta_*)$ is the terminal variable of a discrete-time martingale. In order to apply the corresponding CLT (Theorem 7.1.3) with Lyapunov's condition rather than Lindeberg's one, we assume that $||G(y,x;\theta_*)||^{2+\delta}$ is $p(y,x;\theta_*)\mu(dy)P^{\circ}_{\theta_*}(dx)$-integrable for some $\delta > 0$; then we obtain that

$$n^{-1/2}\mathbb{Z}_n(\theta_*) \xrightarrow{\mathsf{P}_{\theta_*}} \mathcal{N}_p(0, I(\theta_*)) \quad \text{in } \mathbb{R}^p. \tag{8.20}$$

The second equality in (8.19) is a key point to prove the consistency. Finally, it should be remarked that the Fisher information matrix $I(\theta_*)$ has also been reduced to the well-known form as above, owing to the "standard regularity condition".

Based on the discussions so far, we may conclude the following:

Proposition 8.3.5 (i) Under the regularity conditions for (8.16), if the condition **[C2]** in Theorem 8.2.1 is satisfied for $z(\theta) = (z^1(\theta),...,z^p(\theta))^{\text{tr}}$, then for any sequence of Θ-valued random variables $\widehat{\theta}_n$ satisfying

$$n^{-1}\mathbb{Z}_n(\widehat{\theta}_n) = o_{\mathsf{P}_{\theta_*}}(1),$$

it holds that $\widehat{\theta}_n \xrightarrow{\mathsf{P}_{\theta_*}} \theta_*$.

(ii) Under the regularity conditions for (8.20) and (8.18), if the Fisher information matrix $I(\theta_*)$ is positive definite, then for any sequence of Θ-valued random variables $\widehat{\theta}_n$ satisfying

$$n^{-1/2}\mathbb{Z}_n(\widehat{\theta}_n) = o_{\mathsf{P}_{\theta_*}}(1) \quad \text{and} \quad \widehat{\theta}_n \xrightarrow{\mathsf{P}_{\theta_*}} \theta_*,$$

it holds that

$$\begin{aligned}\sqrt{n}(\widehat{\theta}_n - \theta_*) &= \frac{1}{\sqrt{n}} I(\theta_*)^{-1}\mathbb{Z}_n(\theta_*) + o_{\mathsf{P}_{\theta_*}}(1) \\ &\stackrel{\mathsf{P}_{\theta_*}}{\Longrightarrow} \mathcal{N}_p(0, I(\theta_*)^{-1}) \quad \text{in } \mathbb{R}^p.\end{aligned}$$

In Proposition 8.3.5, we *assumed* (8.16), (8.20) and (8.18) to apply the general theorems; some sufficient conditions for them were given in Discussions 8.3.3 and 8.3.4. However, this style of description is not the only way to give proofs of the consistency and the asymptotic normality of MLEs in the current (and any statistical) models, as we will observe below.

Discussion 8.3.6 (General theory (sometimes) merely gives a guideline) Here, let us observe an example of Markov chain models, for which a direct calculation is already sufficient to derive the asymptotic normality of MLEs. Our conclusion in this discussion will be that the role of a general theory is sometimes merely providing a guideline to solve a given concrete problem.

Consider the autoregressive process

$$X_k = \theta f(X_{k-1}) + \varepsilon_k, \quad k = 1, 2, ...,$$

where $\theta \in \Theta \subset \mathbb{R}$ is a parameter of interest, f is a known function, and ε_k's are i.i.d. random variables distributed to $\mathcal{N}(0, \sigma^2)$ with σ being a unknown, nuisance parameter in which we are not interested. Suppose that the process $(X_k)_{k=0,1,2,...}$ is ergodic with the invariant measure $P^{\circ}_{\theta,\sigma}$.

To estimate the true value θ_* for θ, consider the likelihood function $L_n(\theta)$ based on $X_0, X_1, ..., X_n$, constructed as if the true value σ_* for σ were known to statisticians, given by

$$L_n(\theta) = \prod_{k=1}^{n} p(X_k, X_{k-1}; \theta), \quad \theta \in \Theta,$$

where

$$p(X_k, X_{k-1}; \theta) = \frac{1}{\sqrt{2\pi\sigma_*^2}} \exp\left(-\frac{(X_k - \theta f(X_{k-1}))^2}{2\sigma_*^2}\right).$$

Then, the estimating function $\mathbb{Z}_n(\theta)$, defined by $\frac{\partial}{\partial\theta}\log L_n(\theta)$, is given by

$$\mathbb{Z}_n(\theta) = \sum_{k=1}^{n} \frac{f(X_{k-1})}{\sigma_*^2}\{X_k - \theta f(X_{k-1})\}.$$

While one can easily see that

$$n^{-1/2}\mathbb{Z}_n(\theta_*) = \frac{1}{\sqrt{n}} \sum_{k=1}^{n} \frac{f(X_{k-1})}{\sigma_*^2} \varepsilon_k \stackrel{\mathsf{P}_{\theta_*,\sigma_*}}{\Longrightarrow} \mathcal{N}(0, I(\theta_*)) \quad \text{in } \mathbb{R},$$

where

$$I(\theta_*) = \frac{\int_{\mathbb{R}} f(x)^2 P^\circ_{\theta_*,\sigma_*}(dx)}{\sigma_*^2},$$

the estimating equation $\mathbb{Z}_n(\theta) = 0$ in this model has a unique, explicit solution

$$\begin{aligned}\widehat{\theta}_n &= \frac{\sum_{k=1}^n f(X_{k-1})X_k}{\sum_{k=1}^n f(X_{k-1})^2} \quad \text{(this does not involve } \sigma_*!\text{)} \\ &= \theta_* + \frac{\sum_{k=1}^n f(X_{k-1})\varepsilon_k}{\sum_{k=1}^n f(X_{k-1})^2},\end{aligned}$$

and therefore we obtain that

$$\begin{aligned}\sqrt{n}(\widehat{\theta}_n - \theta_*) &= \frac{\sigma_*^2}{n^{-1}\sum_{k=1}^n f(X_{k-1})^2} \cdot n^{-1/2}\sum_{k=1}^n \frac{f(X_{k-1})}{\sigma_*^2}\varepsilon_k \\ &= I(\theta_*)^{-1} \cdot n^{-1/2}\mathbb{Z}_n(\theta_*) + o_{P_{\theta_*,\sigma_*}}(1) \\ &\stackrel{P_{\theta_*,\sigma_*}}{\Longrightarrow} \mathcal{N}(0, I(\theta_*)^{-1}) \quad \text{in } \mathbb{R}.\end{aligned}$$

We have directly obtained the asymptotic representation of $\sqrt{n}(\widehat{\theta}_n - \theta_*)$ in the current example, not via our general theorem (i.e., Theorem 8.2.2). At the same time, however, it should be remarked that getting familiar with the general theories must be helpful for us to find the most economical ways to solve many concrete problems.

8.4 Interim Summary for Approach Overview

As we have studied the above example of Markov chain models, now is time to summarise our approach to the consistency and the asymptotic normality of Z-estimators, based on some tools of the martingale methods.

8.4.1 Consistency

Recalling the prototype of our approach, given as our discussion for Markov chain models, let us observe the following two charts.

In general cases:

$z(\theta)$	$\longleftarrow$	$s_n^{-1}\mathbb{Z}_n(\theta)$	$\dashleftarrow$	$s_n^{-1}\mathbb{Z}_n(\theta)$
deteministic limit		predictable compensator		estimating function

Notice that, in the above chart, the notation "$X_n \leftarrow\!\!-\!- \mathbb{X}_n$" may be interpreted as "X_n is the projection of $\mathbb{X}_n$", and the notation "$x \longleftarrow X_n$" means that "x is the limit of X_n as $n \to \infty$.

In the cases of i.i.d. data:				
$z(\theta)$	$=$	$s_n^{-1} Z_n(\theta)$	$\leftarrow\!\!-\!-$	$s_n^{-1} \mathbb{Z}_n(\theta)$
deterministic limit	$=$	predictable compensator		estimating function

It is often possible to show that $s_n^{-1}\mathbb{Z}_n(\theta)$ is asymptotically equivalent to $s_n^{-1} Z_n(\theta)$ for every $\theta \in \Theta$, in the sense that the difference of these two random sequences converges in probability to zero, by some martingale tools. It is also possible to have that $s_n^{-1} Z_n(\theta)$ converges in probability to $z(\theta)$ for every $\theta \in \Theta$, by some ergodic theorems. These two convergences in hands, with the help also of Theorem 7.3.5, we can (often) verify the condition **[C1]** in Theorem 8.2.1. Therefore, by assuming the identifiability condition **[C2]** in the same theorem for $z(\theta)$, we can prove the consistency of Z-estimators.

8.4.2 Asymptotic normality

It often holds that $\mathbb{Z}_n(\theta) - Z_n(\theta)$ is (the terminal variable of) a local martingale by the construction of $Z_n(\theta)$. It should hold also that

$$Z_n(\theta_*) = 0, \quad \text{if the estimating function } \mathbb{Z}_n(\theta) \text{ is well introduced!}$$

Thus, it is true that

$$\mathbb{Z}_n(\theta_*) \quad \text{is often (the terminal variable of) a local martingale.}$$

Hence, in order to obtain the asymptotic normality of Z-estimators, a good idea is to apply the asymptotic representation theorem (Theorem 8.2.2) to have that

$$r_n(\widehat{\theta}_n - \theta_*) = J(\theta_*)^{-1}(q_n^{-1}\mathbb{Z}_n(\theta_*)) + o_P(1),$$

and then apply a martingale CLT to $q_n^{-1}\mathbb{Z}_n(\theta_*)$ to prove its weak convergence to a Gaussian limit.

8.5 Examples, I-2 (Advanced Topics)

8.5.1 Method of moment estimators

Let $X, X_1, X_2, \ldots$ be a sequence of i.i.d. random variables from a distribution P_θ on a measurable space $(\mathcal{X}, \mathcal{A})$, where θ is from an open, convex subset Θ of $\mathbb{R}^p$. Let

$\psi = (\psi^1, ..., \psi^p)^{\rm tr}$ be a given vector of measurable functions on $(\mathcal{X}, \mathcal{A})$. Assuming that $E_\theta[||\psi(X)||] < \infty$ for every $\theta \in \Theta$, define

$$\begin{aligned} \mathbb{Z}_n(\theta) &= \sum_{k=1}^{n} (\psi(X_k) - e(\theta)) \\ &= \sum_{k=1}^{n} (\psi^1(X_k) - e^1(\theta), ..., \psi^p(X_k) - e^p(\theta))^{\rm tr}, \end{aligned}$$

where

$$e(\theta) = (E_\theta[\psi^1(X)], ..., E_\theta[\psi^p(X)])^{\rm tr}.$$

The (approximate) solution $\widehat{\theta}_n$ to the estimating equation $\mathbb{Z}_n(\theta) = 0$ is called the *method of moment estimator*. If e is one-to-one, then the estimator is explicitly determined as $\widehat{\theta}_n = e^{-1}(\frac{1}{n}\sum_{k=1}^{n} \psi(X_k))$.

If $\theta \mapsto e(\theta)$ is continuously differentiable and $E_{\theta_*}[||\psi(X)||^2] < \infty$, then both of the general theorems in Section 8.2 can be applied to this estimating function $\mathbb{Z}_n(\theta)$ with the rate sequences $q_n = r_n = \sqrt{n}$ and $s_n = n$, and the derivative matrices $\dot{\mathbb{Z}}_n(\theta) = (\dot{\mathbb{Z}}_n^{i,j}(\theta))_{(i,j)\in\{1,...,p\}^2}$, where $\dot{\mathbb{Z}}_n^{i,j}(\theta) = \partial_j \mathbb{Z}_n^i(\theta) = -n\partial_j e^i(\theta)$.

To prove the consistency, it is sufficient to check the conditions **[C1]**[4] and **[C2]** in Theorem 8.2.1 for $z(\theta) = e(\theta_*) - e(\theta)$. However, notice in particular that the argument to prove the consistency is more straightforward, even in the case where the set Θ is unbounded, if e is one-to-one and e^{-1} is continuous; just apply the law of large numbers to $\frac{1}{n}\sum_{k=1}^{n} \psi(X_k)$ and use the continuous mapping theorem for the function $e^{-1}(\cdot)$.

As for the asymptotic normality, if the matrix $J(\theta_*) = (\partial_j e^i(\theta_*))_{(i,j)\in\{1,...,p\}^2}$ is regular, then Theorem 8.2.2 yields that

$$\begin{aligned} \sqrt{n}(\widehat{\theta}_n - \theta_*) &= J(\theta_*)^{-1}(n^{-1/2}\mathbb{Z}_n(\theta_*)) + o_{P_{\theta_*}}(1) \\ &\stackrel{P_{\theta_*}}{\Longrightarrow} J(\theta_*)^{-1}\mathcal{N}_p(0, \Sigma(\theta_*)) \quad \text{in } \mathbb{R}^p \\ &\stackrel{d}{=} \mathcal{N}_p(0, J(\theta_*)^{-1}\Sigma(\theta_*)(J(\theta_*)^{-1})^{\rm tr}), \end{aligned}$$

where $\Sigma(\theta_*) = E_{\theta_*}[(\psi(X) - e(\theta_*))(\psi(X) - e(\theta_*))^{\rm tr}]$.

Remark. The matrix $J(\theta_*)$ appearing above may not be symmetric.

8.5.2 Quasi-likelihood for drifts in ergodic diffusion models

Let us consider the 1-dimensional stochastic differential equation

$$X_t = X_0 + \int_0^t \beta(X_s; \theta) ds + \int_0^t \sigma(X_s) dW_s,$$

[4]This is verified very easily in the current example, at least when the set Θ is bounded. In fact, it suffices just to check that $\theta \mapsto e(\theta)$ is α-Hölder continuous for some $\alpha \in (0, 1]$; see Theorem 7.3.5.

where $s \rightsquigarrow W_s$ is a standard Wiener process. In this model, the drift coefficient involves the unknown parameter θ of interest, which belong to an open, convex subset Θ of $\mathbb{R}^p$, while the diffusion coefficient σ is assumed to be a *known* function.

Suppose that the process X is observable only at finitely many time points $0 = t_0^n < t_1^n < \cdots < t_n^n$, and put $\Delta_n = \max_{1 \le k \le n} |t_k^n - t_{k-1}^n|$. Throughout this subsection, we assume that

$$t_n^n \to \infty \quad \text{and} \quad \Delta_n = o((t_n^n)^{-1}) \quad \text{as } n \to \infty. \tag{8.21}$$

The latter is satisfied if $n\Delta_n^2 \to 0$.

Under this sampling scheme, it would be natural to adopt the contrast function, called the *quasi-likelihood*, given by

$$\begin{aligned} L_n(\theta) &= \prod_{k=1}^n \frac{1}{\sqrt{2\pi\sigma(X_{t_{k-1}^n})^2 |t_k^n - t_{k-1}^n|}} \\ &\quad \cdot \exp\left(-\frac{(X_{t_k^n} - X_{t_{k-1}^n} - \beta(X_{t_{k-1}^n};\theta)|t_k^n - t_{k-1}^n|)^2}{2\sigma(X_{t_{k-1}^n})^2 |t_k^n - t_{k-1}^n|} \right). \end{aligned}$$

Denote the gradient vector and the Hessian matrix of $\log L_n(\theta)$ by $\mathbb{Z}_n(\theta)$ and $\dot{\mathbb{Z}}_n(\theta)$, respectively:

$$\begin{aligned} \mathbb{Z}_n^i(\theta) &= \sum_{k=1}^n \frac{\partial_i \beta(X_{t_{k-1}^n};\theta)}{\sigma(X_{t_{k-1}^n})^2} \{X_{t_k^n} - X_{t_{k-1}^n} - \beta(X_{t_{k-1}^n};\theta)|t_k^n - t_{k-1}^n|\}; \\ \dot{\mathbb{Z}}_n^{i,j}(\theta) &= \sum_{k=1}^n \left(\frac{\partial_{i,j}\beta(X_{t_{k-1}^n};\theta)}{\sigma(X_{t_{k-1}^n})^2} \{X_{t_k^n} - X_{t_{k-1}^n} - \beta(X_{t_{k-1}^n};\theta)|t_k^n - t_{k-1}^n|\} \right. \\ &\qquad \left. - \frac{\partial_i \beta(X_{t_{k-1}^n};\theta)\partial_j \beta(X_{t_{k-1}^n};\theta)}{\sigma(X_{t_{k-1}^n})^2} |t_k^n - t_{k-1}^n| \right). \end{aligned}$$

Here, we have used the notations such as $\partial_i = \frac{\partial}{\partial \theta_i}, \partial_{i,j} = \frac{\partial^2}{\partial \theta_i \partial \theta_j}$, as in the previous subsections.

In our arguments for the statistical analysis in this model, the fact that the term

$$X_{t_k^n} - X_{t_{k-1}^n}$$

appearing in the definitions of $\mathbb{Z}_n(\theta)$ and $\dot{\mathbb{Z}}_n(\theta)$ can be well approximated by

$$\beta(X_{t_{k-1}^n};\theta_*)|t_k^n - t_{k-1}^n| + \sigma(X_{t_{k-1}^n})(W_{t_k^n} - W_{t_{k-1}^n})$$

will be important. Let us state it as a lemma below.

Lemma 8.5.1 Suppose that $x \mapsto S(x;\theta_*)$ and $x \mapsto \sigma(x)$ are Lipschitz continuous, and that $\sup_{t\in[0,\infty)} \mathsf{E}[|X_t|^q] < \infty$ for every $q \ge 1$. Under the sampling scheme (8.21) it holds for any measurable function $g:\mathbb{R}\to\mathbb{R}$ with polynomial growth (see Definition 7.4.2) that

$$\begin{aligned}
&\frac{1}{t_n^n}\sum_{k=1}^n g(X_{t_{k-1}^n})(X_{t_k^n}-X_{t_{k-1}^n}) \\
&= \frac{1}{t_n^n}\sum_{k=1}^n g(X_{t_{k-1}^n})\{\beta(X_{t_{k-1}^n};\theta_*)|t_k^n-t_{k-1}^n|+\sigma(X_{t_{k-1}^n})(W_{t_k^n}-W_{t_{k-1}^n})\} \\
&\quad +o_{\mathsf{P}}((t_n^n)^{-1/2}) \\
&= \frac{1}{t_n^n}\sum_{k=1}^n g(X_{t_{k-1}^n})\beta(X_{t_{k-1}^n};\theta_*)|t_k^n-t_{k-1}^n|+o_{\mathsf{P}}(1).
\end{aligned}$$

Proof. Let us prove the first equation. Consider the decomposition

$$X_{t_k^n}-X_{t_{k-1}^n}=A_{n,k}^a+A_{n,k}^b+A_{n,k}^c,$$

where

$$\begin{aligned}
A_{n,k}^a &= \int_{t_{k-1}^n}^{t_k^n}\beta(X_{t_{k-1}^n};\theta_*)ds=\beta(X_{t_{k-1}^n};\theta_*)|t_k^n-t_{k-1}^n|, \\
A_{n,k}^b &= \int_{t_{k-1}^n}^{t_k^n}(\beta(X_s;\theta_*)-\beta(X_{t_{k-1}^n};\theta_*))ds, \\
A_{n,k}^c &= \int_{t_{k-1}^n}^{t_k^n}\sigma(X_{t_{k-1}^n})dW_s=\sigma(X_{t_{k-1}^n})(W_{t_k^n}-W_{t_{k-1}^n}), \\
A_{n,k}^d &= \int_{t_{k-1}^n}^{t_k^n}(\sigma(X_s)-\sigma(X_{t_{k-1}^n}))dW_s.
\end{aligned}$$

By the first inequality of Theorem 7.4.1, there exists a constant $C>0$ depending only on the Lipschitz coefficient K of $\beta(\cdot;\theta_*)$ such that

$$\begin{aligned}
\mathsf{E}[|A_{n,k}^b||\mathcal{F}_{t_{k-1}^n}] &\le \int_{t_{k-1}^n}^{t_k^n}K\mathsf{E}[|X_s-X_{t_{k-1}^n}||\mathcal{F}_{t_{k-1}^n}]ds \\
&\le C|t_k^n-t_{k-1}^n|^{3/2}(1+|X_{t_{k-1}^n}|).
\end{aligned}$$

Hence we have that

$$\begin{aligned}
&\frac{1}{t_n^n}\sum_{k=1}^n|g(X_{t_{k-1}^n})|\mathsf{E}[|A_{n,k}^b||\mathcal{F}_{t_{k-1}^n}] \\
&\le \frac{1}{t_n^n}\sum_{k=1}^n|g(X_{t_{k-1}^n})|\cdot C|t_k^n-t_{k-1}^n|(1+|X_{t_{k-1}^n}|)\cdot\Delta_n^{1/2},
\end{aligned}$$

and the right-hand side is actually $O_{\mathsf{P}}(\Delta_n^{1/2})$ by Exercise 2.3.4, and thus it is $o_{\mathsf{P}}((t_n^n)^{-1/2})$ by the assumption (8.21).

On the other hand, it follows the corollary to Lenglart's inequality (Corollary 4.3.3 (i)) that

$$\frac{1}{\sqrt{t_n^n}}\sum_{k=1}^{n} g(X_{t_{k-1}^n})A_{n,k}^d \xrightarrow{\mathsf{P}} 0,$$

because the predictable quadratic variation is bounded by

$$\begin{aligned}
&\frac{1}{t_n^n}\sum_{k=1}^{n} g(X_{t_{k-1}^n})^2 \int_{t_{k-1}^n}^{t_k^n} \mathsf{E}[(\sigma(X_s)-\sigma(X_{t_{k-1}^n}))^2|\mathcal{F}_{t_{k-1}^n}]ds \\
&\leq C\frac{1}{t_n^n}\sum_{k=1}^{n} g(X_{t_{k-1}^n})^2(1+|X_{t_{k-1}^n}|)^2|t_k^n - t_{k-1}^n|^2,
\end{aligned}$$

where $C>0$ is a constant depending only on the Lipschitz coefficient of σ, and this is actually evaluated as $O_{\mathsf{P}}(\Delta_n)$ and thus as $o_{\mathsf{P}}(1)$. The proof of the first equation is finished.

The second equation is proved by using the corollary to Lenglart's inequality because the terminal value of the predictable quadratic variation of the discrete-time martingale

$$\frac{1}{t_n^n}\sum_{k=1}^{m} g(X_{t_{k-1}^n})\sigma(X_{t_{k-1}^n})(W_{t_k^n}-W_{t_{k-1}^n}), \quad m=1,2,\dots,n$$

is given by

$$\frac{1}{(t_n^n)^2}\sum_{k=1}^{n}(g(X_{t_{k-1}^n})\sigma(X_{t_{k-1}^n}))^2|t_k^n - t_{k-1}^n|,$$

which converges in probability to zero. The proof is complete. □

Now, let us start our discussion on statistical analysis in the model. During the rest part of this subsection, we suppose that the process $t \rightsquigarrow X_t$ is ergodic with the invariant measure P°; then, it holds for any P°-integrable function f that

$$\frac{1}{T}\int_0^T f(X_t)dt \xrightarrow{\mathsf{P}} \int_{\mathbb{R}} f(x)P^\circ(dx) \quad \text{as } T\to\infty.$$

Since the invariant measure P° depends on the true value θ_*, we denote it by $P_{\theta_*}^\circ$ when we should emphasize what is the true value θ_*. Moreover, we suppose also that

$$\inf_{x\in\mathbb{R}} \sigma(x) > 0$$

and that[5]

$$\sup_{t\in[0,\infty)} \mathsf{E}[|X_t|^q] < \infty, \quad \forall q \geq 1.$$

[5]Recall the footnote remark to Theorem 7.4.1 for the validity of this assumptions.

Lemma 8.5.2 Suppose that $x \mapsto \beta(x;\theta_*)$ and $x \mapsto \sigma(x)$ are Lipschitz continuous. Suppose also that $\theta \mapsto \beta(x;\theta)$ is three times continuously differentiable on Θ and that $\beta(x;\theta)$, $\partial_i\beta(x;\theta)$, $\partial_{i,j}\beta(x;\theta)$, $\partial_{i,j,k}\beta(x;\theta)$ are differentiable with respect to x and their derivatives are bounded by a function with polynomial growth not depending on θ.

(i) If the set Θ is bounded, then it holds that

$$\sup_{\theta\in\Theta} ||(t_n^n)^{-1}\mathbb{Z}_n(\theta) - z(\theta)|| \xrightarrow{\mathsf{P}} 0,$$

where

$$z^i(\theta) = \int_{\mathbb{R}} \frac{\partial_i\beta(x;\theta)}{\sigma(x)^2}(\beta(x;\theta_*) - \beta(x;\theta))P^\circ_{\theta_*}(dx), \quad i = 1,\dots,p.$$

(ii) It holds that

$$(t_n^n)^{-1/2}\mathbb{Z}_n(\theta_*) \stackrel{\mathsf{P}}{\Longrightarrow} \mathcal{N}_p(0, J(\theta_*)) \quad \text{in } \mathbb{R}^p,$$

where the matrix $J(\theta_*) = (J^{i,j}(\theta_*))_{(i,j)\in\{1,\dots,p\}^2}$ given by

$$J^{i,j}(\theta_*) = \int_{\mathbb{R}} \frac{\partial_i\beta(x;\theta_*)\partial_j\beta(x;\theta_*)}{\sigma(x)^2}P^\circ_{\theta_*}(dx).$$

Moreover, for any sequence of Θ-valued random element $\widetilde{\theta}_n$ converging in outer-probability to θ_*, it holds that

$$(t_n^n)^{-1}\dot{\mathbb{Z}}_n(\widetilde{\theta}_n) \xrightarrow{\mathsf{P}^*} -J(\theta_*),$$

The above lemma combined with Theorems 8.2.1 and 8.2.2 yields the following results.

Theorem 8.5.3 (i) Under the situation of Lemma 8.5.2 (i), if the condition **[C2]** in Theorem 8.2.1 is satisfied for $z(\theta)$ given above, then, for any sequence of Θ-valued random variables $\widehat{\theta}_n$ satisfying

$$(t_n^n)^{-1}\mathbb{Z}_n(\widehat{\theta}_n) = o_{\mathsf{P}_{\theta_*}}(1),$$

it holds that $\widehat{\theta}_n \xrightarrow{\mathsf{P}} \theta_*$.

(ii) Under the situation of Lemma 8.5.2 (ii), if the matrix $J(\theta_*)$ given above is positive definite, then, for any sequence of Θ -valued random variables $\widehat{\theta}_n$ satisfying

$$(t_n^n)^{-1/2}\mathbb{Z}_n(\widehat{\theta}_n) = o_{\mathsf{P}}(1) \quad \text{and} \quad \widehat{\theta}_n \xrightarrow{\mathsf{P}} \theta_*,$$

it holds that

$$\begin{aligned}\sqrt{t_n^n}(\widehat{\theta}_n - \theta_*) &= J(\theta_*)^{-1}((t_n^n)^{-1/2}\mathbb{Z}_n(\theta_*)) + o_{\mathsf{P}}(1) \\ &\stackrel{\mathsf{P}}{\Longrightarrow} \mathcal{N}_p(0, J(\theta_*)^{-1}) \quad \text{in } \mathbb{R}^p.\end{aligned}$$

Proof of Lemma 8.5.2 (i). In order to apply Theorem 7.3.5 (ii), let us first prove that $(t_n^n)^{-1}\mathbb{Z}_n(\theta) \xrightarrow{\mathsf{P}} z(\theta)$ for every $\theta \in \Theta$. By the second equation of Lemma 8.5.1, it holds that

$$(t_n^n)^{-1}(\mathbb{Z}_n^i(\theta) - Z_n^i(\theta)) \xrightarrow{\mathsf{P}} 0, \quad i = 1, ..., p,$$

where

$$Z_n^i(\theta) = \sum_{k=1}^n \frac{\partial_i \beta(X_{t_{k-1}^n};\theta)}{\sigma(X_{t_{k-1}^n})^2}(\beta(X_{t_{k-1}^n};\theta_*) - \beta(X_{t_{k-1}^n};\theta))|t_k^n - t_{k-1}^n|,$$

while it follows from Theorem 7.4.4 (ii) that

$$(t_n^n)^{-1}Z_n^i(\theta) \xrightarrow{\mathsf{P}} z^i(\theta), \quad i = 1, ..., p.$$

Next, in order to check the condition (7.25) in Theorem 7.3.5, let us write

$$(t_n^n)^{-1}\partial_j \mathbb{Z}_n^i(\theta) = A_n(\theta) + \frac{1}{t_n^n}\sum_{k=1}^n \frac{\partial_{i,j}\beta(X_{t_{k-1}^n};\theta)}{\sigma(X_{t_{k-1}^n})^2}\int_{t_{k-1}^n}^{t_k^n} \sigma(X_s)dW_s.$$

It is easy to see that $\sup_{\theta\in\Theta}|A_n(\theta)| = O_{\mathsf{P}}(1)$. By using Theorem 7.3.5, it is proved that the second term on the right-hand side converges, uniformly in θ, to zero in probability. This completes the proof. □

Proof of Lemma 8.5.2 (ii). The first claim is proved by the CLT for discrete-time martingales (Theorem 7.1.3) with the help of the first equation in Lemma 8.5.1.

To show the second claim, by applying the second equation of Lemma 8.5.1 to

$$g_\theta(x) = \frac{\partial_{i,j}\beta(x;\theta)}{\sigma(x)^2},$$

we can prove that $(t_n^n)^{-1}\dot{\mathbb{Z}}_n^{i,j}(\theta)$ is asymptotically equivalent to $(t_n^n)^{-1}\dot{Z}_n^{i,j}(\theta)$, where

$$\begin{aligned}\dot{Z}_n^{i,j}(\theta) &= \sum_{k=1}^n g_\theta(X_{t_{k-1}^n})(\beta(X_{t_{k-1}^n};\theta_*) - \beta(X_{t_{k-1}^n};\theta))|t_k^n - t_{k-1}^n| \\ &\quad - \sum_{k=1}^n \frac{\partial_i\beta(X_{t_{k-1}^n};\theta)\partial_j\beta(X_{t_{k-1}^n};\theta)}{\sigma(X_{t_{k-1}^n})^2}|t_k^n - t_{k-1}^n|.\end{aligned}$$

It follows from Theorem 7.4.4 (ii) that the latter converges in probability to

$$-J^{i,j}(\theta) = \int_{\mathbb{R}}\left\{g_\theta(x)(\beta(x;\theta_*) - \beta(x;\theta)) - \frac{\partial_i\beta(x;\theta)\partial_j\beta(x;\theta)}{\sigma(x)^2}\right\}P_{\theta_*}^\circ(dx).$$

At this moment, we have obtained only the convergences in probability, for every θ. However, the stronger claim that

$$\sup_{\theta\in N}|(t_n^n)^{-1}\dot{\mathbb{Z}}_n^{ij}(\theta) - (-J^{i,j}(\theta))| \xrightarrow{\mathsf{P}} 0, \quad (i,j) \in \{1, ..., p\}^2,$$

where N is a bounded neighborhood of θ_*, can be proved by checking the condition (7.25) in Theorem 7.3.5. Since $\theta \mapsto J(\theta)$ is continuous, for any sequence of random elements $\widetilde{\theta}_n$ converging in outer-probability to θ_*, it holds that $\dot{\mathbb{Z}}_n^{i,j}(\widetilde{\theta}_n) \xrightarrow{\mathsf{P}^*} -J(\theta_*)$. The proof is finished. □

8.5.3 Quasi-likelihood for volatilities in ergodic diffusion models

Let us consider the 1-dimensional stochastic differential equation

$$X_t = X_0 + \int_0^t \beta(X_s)ds + \int_0^t \sigma(X_s;\theta)dW_s,$$

where $s \rightsquigarrow W_s$ is a standard Wiener process. In contrast with the model considered in the preceding subsection, in the current model the diffusion coefficient involves the unknown parameter θ of interest, which belong to an open, convex subset Θ of $\mathbb{R}^p$, while the drift coefficient β is regarded as a nuisance parameter that is assumed to be *unknown* to statisticians.

As in the previous subsection, suppose that the process $t \rightsquigarrow X_t$ is observable only at finitely many time points $0 = t_0^n < t_1^n < \cdots < t_n^n$, and put $\Delta_n = \max_{1\le k\le n} |t_k^n - t_{k-1}^n|$. To develop an asymptotic theory, we will assume that

$$t_n^n \to \infty \quad \text{and} \quad n\Delta_n^2 \to 0, \quad \text{as } n \to \infty, \tag{8.22}$$

and moreover that

$$\sum_{k=1}^n \left| \frac{|t_k^n - t_{k-1}^n|}{t_n^n} - \frac{1}{n} \right| \to 0, \quad \text{as } n \to \infty. \tag{8.23}$$

The latter assumption (8.23) is satisfied for the *equidistant sampling scheme* $t_k^n = k\Delta_n$.

The *quasi-likelihood* in this model is given by

$$\begin{aligned} L_n(\beta,\theta) &= \prod_{k=1}^n \frac{1}{\sqrt{2\pi\sigma(X_{t_{k-1}^n};\theta)^2|t_k^n - t_{k-1}^n|}} \\ &\quad \cdot \exp\left(-\frac{(X_{t_k^n} - X_{t_{k-1}^n} - \beta(X_{t_{k-1}^n})|t_k^n - t_{k-1}^n|)^2}{2\sigma(X_{t_{k-1}^n};\theta)^2|t_k^n - t_{k-1}^n|} \right). \end{aligned}$$

As it will be seen from now on, the effect from the drift coefficient β is negligible under the situation where $n\Delta_n^2 \to 0$. Based on this fact, we set $\beta = 0$ to introduce the estimating function $\mathbb{Z}_n(\theta)$ given by

$$\begin{aligned} \mathbb{Z}_n^i(\theta) &= \frac{\partial}{\partial\theta_i} \log L_n(0,\theta) \\ &= \sum_{k=1}^n \frac{\partial_i\sigma(X_{t_{k-1}^n};\theta)}{\sigma(X_{t_{k-1}^n};\theta)^3} \left\{ \frac{(X_{t_k^n} - X_{t_{k-1}^n})^2}{|t_k^n - t_{k-1}^n|} - \sigma(X_{t_{k-1}^n};\theta)^2 \right\}. \end{aligned}$$

As in the previous subsections, we set

$$\dot{\mathbb{Z}}_n^{i,j}(\theta) = \frac{\partial^2}{\partial\theta_i\partial\theta_j} \log L_n(0,\theta).$$

Again, we use the notations such as $\partial_i = \frac{\partial}{\partial\theta_i}$, $\partial_{i,j} = \frac{\partial^2}{\partial\theta_i\partial\theta_j}$.

The key point in our statistical analysis is that the term $(X_{t_k^n} - X_{t_{k-1}^n})^2$ appearing in the definition of $\mathbb{Z}_n(\theta)$ can be well approximated by $\sigma(X_{t_{k-1}^n};\theta_*)^2(W_{t_k^n} - W_{t_{k-1}^n})^2$. We state this fact as a lemma below.

Lemma 8.5.4 Suppose that $x \mapsto \beta(x)$ and $x \mapsto \sigma(x;\theta_*)$ are Lipschitz continuous, and moreover that the latter is two times continuously differentiable with respect to x and its derivatives are bounded by a function with polynomial growth (see Definition 7.4.2). Suppose also that $\sup_{t\in[0,\infty)} \mathsf{E}[|X_t|^q] < \infty$ for any $q \geq 1$ and that the condition (8.22) is satisfied. Then, for any measurable function $g : \mathbb{R} \to \mathbb{R}$ with polynomial growth it holds that

$$\begin{aligned}
&\frac{1}{n}\sum_{k=1}^{n} g(X_{t_{k-1}^n})\frac{(X_{t_k^n} - X_{t_{k-1}^n})^2}{|t_k^n - t_{k-1}^n|} \\
&= \frac{1}{n}\sum_{k=1}^{n} g(X_{t_{k-1}^n})\frac{\sigma(X_{t_{k-1}^n};\theta_*)^2(W_{t_k^n} - W_{t_{k-1}^n})^2}{|t_k^n - t_{k-1}^n|} + o_{\mathsf{P}}(n^{-1/2}) \\
&= \frac{1}{n}\sum_{k=1}^{n} g(X_{t_{k-1}^n})\sigma(X_{t_{k-1}^n};\theta_*)^2 + o_{\mathsf{P}}(1).
\end{aligned}$$

Proof. To prove the first equation, consider the decomposition

$$(X_{t_k^n} - X_{t_{k-1}^n})^2 = A_{n,k}^{aa} + A_{n,k}^{bb} + A_{n,k}^{cc} + 2A_{n,k}^{ab} + 2A_{n,k}^{bc} + 2A_{n,k}^{ca},$$

where

$$\begin{aligned}
A_{n,k}^{aa} &= \left(\int_{t_{k-1}^n}^{t_k^n} \beta(X_s)ds\right)^2, \\
A_{n,k}^{bb} &= \left(\int_{t_{k-1}^n}^{t_k^n} (\sigma(X_s;\theta_*) - \sigma(X_{t_{k-1}^n};\theta_*)dW_s\right)^2, \\
A_{n,k}^{cc} &= \left(\int_{t_{k-1}^n}^{t_k^n} \sigma(X_{t_{k-1}^n};\theta_*)dW_s\right)^2 \\
&= \sigma(X_{t_{k-1}^n};\theta_*)^2(W_{t_k^n} - W_{t_{k-1}^n})^2, \\
A_{n,k}^{ab} &= \int_{t_{k-1}^n}^{t_k^n} \beta(X_s)ds \int_{t_{k-1}^n}^{t_k^n} (\sigma(X_s;\theta_*) - \sigma(X_{t_{k-1}^n};\theta_*))dW_s, \\
A_{n,k}^{bc} &= \int_{t_{k-1}^n}^{t_k^n} (\sigma(X_s;\theta_*) - \sigma(X_{t_{k-1}^n};\theta_*))dW_s \int_{t_{k-1}^n}^{t_k^n} \sigma(X_{t_{k-1}^n};\theta_*)dW_s, \\
A_{n,k}^{ca} &= \int_{t_{k-1}^n}^{t_k^n} \sigma(X_{t_{k-1}^n};\theta_*)dW_s \int_{t_{k-1}^n}^{t_k^n} \beta(X_s)ds.
\end{aligned}$$

In the rest part of this proof, for given non-negative $\mathcal{F}_{t_{k-1}^n}$-measurable random variable Y we denote

$$Y \lesssim |t_k^n - t_{k-1}^n|^2$$

when there exist some constants $C, q \geq 1$ depending only on β, σ such that

$$Y \leq C|t_k^n - t_{k-1}^n|^2(1 + |X_{t_{k-1}^n}|)^q.$$

From now on, for all terms except for $A_{n,k}^{cc}$ we will prove that either that

$$\mathsf{E}[|A_{n,k}^{\cdot\cdot}||\mathcal{F}_{t_{k-1}^n}] \lesssim |t_k^n - t_{k-1}^n|^2, \tag{8.24}$$

or that there exists a constant $\gamma > 0$ such that

$$|\mathsf{E}[A_{n,k}^{\cdot\cdot}|\mathcal{F}_{t_{k-1}^n}]| \lesssim |t_k^n - t_{k-1}^n|^2 \quad \text{and} \quad \mathsf{E}[(A_{n,k}^{\cdot\cdot})^2|\mathcal{F}_{t_{k-1}^n}]| \lesssim |t_k^n - t_{k-1}^n|^{2+\gamma}, \tag{8.25}$$

where the notation "$A_{n,k}^{\cdot\cdot}$" should be read as "$A_{n,k}^{aa}$", "$A_{n,k}^{ab}$", and so on. The former implies that

$$\frac{1}{n}\sum_{k=1}^n |g(X_{t_{k-1}^n})|\mathsf{E}\left[\frac{|A_{n,k}^{\cdot\cdot}|}{|t_k^n - t_{k-1}^n|}\middle|\mathcal{F}_{t_{k-1}^n}\right] = O_{\mathsf{P}}(\Delta_n) = o_{\mathsf{P}}(n^{-1/2}),$$

while the latter yields that

$$\sum_{k=1}^n \zeta_k^n = \sum_{k=1}^n \mathsf{E}[\zeta_k^n|\mathcal{F}_{t_{k-1}^n}] + \sum_{k=1}^n (\zeta_k^n - \mathsf{E}[\zeta_k^n|\mathcal{F}_{t_{k-1}^n}]) = o_{\mathsf{P}}(1) + o_{\mathsf{P}}(1),$$

where

$$\zeta_k^n = \frac{1}{\sqrt{n}}g(X_{t_{k-1}^n})\frac{A_{n,k}^{\cdot\cdot}}{|t_k^n - t_{k-1}^n|}.$$

The assertion of the lemma follows from the combination of the evaluations for all these terms.

First, by Hölder's inequality and the second inequality in Theorem 7.4.1 we have

$$\mathsf{E}[|A_{n,k}^{aa}||\mathcal{F}_{t_{k-1}^n}] \leq \int_{t_{k-1}^n}^{t_k^n} ds \int_{t_{k-1}^n}^{t_k^n} \mathsf{E}[\beta(X_s)^2|\mathcal{F}_{t_{k-1}^n}]ds \lesssim |t_k^n - t_{k-1}^n|^2.$$

Secondly, by the first inequality in Theorem 7.4.1 we have

$$\mathsf{E}[|A_{n,k}^{bb}||\mathcal{F}_{t_{k-1}^n}] \leq \int_{t_{k-1}^n}^{t_k^n} \mathsf{E}[(\sigma(X_s;\theta_*) - \sigma(X_{t_{k-1}^n};\theta_*))^2|\mathcal{F}_{t_{k-1}^n}]ds \lesssim |t_k^n - t_{k-1}^n|^2.$$

Thirdly, by these results and the Cauchy-Schwarz inequality we have

$$\mathsf{E}[|A_{n,k}^{ab}||\mathcal{F}_{t_{k-1}^n}] \leq \sqrt{\mathsf{E}[A_{n,k}^{aa}|\mathcal{F}_{t_{k-1}^n}]}\sqrt{\mathsf{E}[A_{n,k}^{bb}|\mathcal{F}_{t_{k-1}^n}]} \lesssim |t_k^n - t_{k-1}^n|^2.$$

As for the other two terms, note that we have to compute conditional expectations before taking the absolute value when we check the first condition in (8.25). First, since it holds that

$$\mathsf{E}[A_{n,k}^{bc}|\mathcal{F}_{t_{k-1}^n}] = \int_{t_{k-1}^n}^{t_k^n} \mathsf{E}[\sigma(X_s;\theta_*) - \sigma(X_{t_{k-1}^n};\theta_*)|\mathcal{F}_{t_{k-1}^n}]\sigma(X_{t_{k-1}^n};\theta_*)ds,$$

Theorem 7.4.3 yields that

$$|\mathsf{E}[A^{bc}_{n,k}|\mathcal{F}_{t^n_{k-1}}]| \lesssim |t^n_k - t^n_{k-1}|^2.$$

On the other hand, since

$$\begin{aligned}\mathsf{E}[A^{ca}_{n,k}|\mathcal{F}_{t^n_{k-1}}]\\ = \mathsf{E}\left[\int_{t^n_{k-1}}^{t^n_k} \sigma(X_{t^n_{k-1}};\theta_*)dW_s \int_{t^n_{k-1}}^{t^n_k} (\beta(X_s) - \beta(X_{t^n_{k-1}}))ds \middle| \mathcal{F}_{t^n_{k-1}}\right],\end{aligned}$$

by using some techniques similar to the ones that we have used so far, we can prove that

$$|\mathsf{E}[A^{ca}_{n,k}|\mathcal{F}_{t^n_{k-1}}]| \lesssim |t^n_k - t^n_{k-1}|^2.$$

Checking that $A^{bc}_{n,k}$ and $A^{ca}_{n,k}$ satisfy the second condition in (8.25) is easy; use the Cauchy-Schwarz and extended Burkholder-Davis-Gundy's inequalities. The proof of the first equation of the lemma is finished.

The second equation can be proved by using the facts that

$$\mathsf{E}[(W_{t^n_k} - W_{t^n_{k-1}})^2|\mathcal{F}_{t^n_{k-1}}] = |t^n_k - t^n_{k-1}| \tag{8.26}$$

and that

$$\mathsf{E}[((W_{t^n_k} - W_{t^n_{k-1}})^2 - |t^n_k - t^n_{k-1}|)^2|\mathcal{F}_{t^n_{k-1}}] = 2|t^n_k - t^n_{k-1}|^2 \tag{8.27}$$

combined with the corollary to Lenglart's inequality. □

Now, let us start our discussion on statistical analysis in the model. During the rest part of this subsection, we suppose that the process $t \rightsquigarrow X_t$ is ergodic with the invariant measure P°; then, it holds for any P°-integrable function f that

$$\frac{1}{T}\int_0^T f(X_t)dt \xrightarrow{\mathsf{P}} \int_{\mathbb{R}} f(x)P^\circ(dx) \quad \text{as } T \to \infty.$$

Again, since the invariant measure P° depends on the true value θ_*, we often denote it by $P^\circ_{\theta_*}$. Moreover, we suppose also that

$$\inf_{x\in\mathbb{R}} \sigma(x) > 0 \quad \text{and} \quad \sup_{t\in[0,\infty)} \mathsf{E}[|X_t|^q] < \infty, \quad \forall q \geq 1.$$

Recall also that the conditions (8.22) and (8.23) for the sampling scheme are always assumed for our statistical analysis throughout this subsection.

Lemma 8.5.5 Suppose the following conditions (a) and (b) are satisfied.

(a) $x \mapsto S(x)$ and $x \mapsto \sigma(x;\theta_*)$ are Lipschitz continuous. Moreover, the latter is two times continuously differentiable and its derivatives are functions with polynomial growth.

(b) $\theta \mapsto \sigma(x;\theta)$ is three times continuously differentiable on Θ, for every $x \in \mathbb{R}$.

Moreover, $\sigma(x;\theta)$ and the derivatives $\partial_i\sigma(x;\theta)$, $\partial_{i,j}\sigma(x;\theta)$, $\partial_{i,j,k}\sigma(x;\theta)$ are differentiable with respect to x and their derivatives are bounded by a function with polynomial growth not depending on θ.

(i) If the set Θ is bounded, then it holds that

$$\sup_{\theta\in\Theta}||n^{-1}\mathbb{Z}_n(\theta)-z(\theta)||\xrightarrow{\mathsf{P}}0,$$

where

$$z^i(\theta)=\int_{\mathbb{R}}\frac{\partial_i\sigma(x;\theta)}{\sigma(x;\theta)^3}\{\sigma(x;\theta_*)^2-\sigma(x;\theta)^2\}P^{\circ}_{\theta_*}(dx),\quad i=1,...,p.$$

(ii) It holds that

$$n^{-1/2}\mathbb{Z}_n(\theta_*)\overset{\mathsf{P}}{\Longrightarrow}\mathcal{N}_p(0,J(\theta_*))\quad\text{in }\mathbb{R}^p,$$

where the matrix $J(\theta_*)=(J^{ij}(\theta_*))_{(i,j)\in\{1,...,p\}^2}$ is given by

$$J^{ij}(\theta_*)=2\int_{\mathbb{R}}\frac{\partial_i\sigma(x;\theta_*)\partial_j\sigma(x;\theta_*)}{\sigma(x;\theta_*)^2}P^{\circ}_{\theta_*}(dx).$$

Moreover, for any sequence of Θ-valued random elements $\widetilde{\theta}_n$ converging in outer-probability to θ_*, it holds that $n^{-1}\dot{\mathbb{Z}}_n(\widetilde{\theta}_n)\xrightarrow{\mathsf{P}^*}-J(\theta_*)$.

The following theorem is a consequence from the above lemma with the help of Theorems 8.2.1 and 8.2.2.

Theorem 8.5.6 (i) Under the same situation of Lemma 8.5.5 (i), if the condition **[C2]** in Theorem 8.2.1 is satisfied for $z(\theta)$, then for any sequence of random variables $\widehat{\theta}_n$ satisfying

$$n^{-1}\mathbb{Z}_n(\widehat{\theta}_n)=o_{\mathsf{P}}(1),$$

it holds that $\widehat{\theta}_n\xrightarrow{\mathsf{P}}\theta_*$.

(ii) Under the same situation of Lemma 8.5.5 (ii), if the matrix $J(\theta_*)$ is positive definite, then for any sequence of random variables $\widehat{\theta}_n$ satisfying

$$n^{-1/2}\mathbb{Z}_n(\widehat{\theta}_n)=o_{\mathsf{P}}(1)\quad\text{and}\quad\widehat{\theta}_n\xrightarrow{\mathsf{P}}\theta_*,$$

it holds that

$$\begin{aligned}\sqrt{n}(\widehat{\theta}_n-\theta_*)&=J(\theta_*)^{-1}(n^{-1/2}\mathbb{Z}_n(\theta_*))+o_{\mathsf{P}}(1)\\&\overset{\mathsf{P}}{\Longrightarrow}\mathcal{N}_p(0,J(\theta_*)^{-1})\quad\text{in }\mathbb{R}^p.\end{aligned}$$

Proof of Lemma 8.5.5 (i). It follows from Lemma 8.5.4 that for every $\theta\in\Theta$, $n^{-1}\mathbb{Z}_n^i(\theta)$ is asymptotically equivalent to $n^{-1}Z_n^i(\theta)$, where

$$Z_n^i(\theta)=\sum_{k=1}^n\frac{\partial_i\sigma(X_{t^n_{k-1}};\theta)}{\sigma(X_{t^n_{k-1}};\theta)^3}\{\sigma(X_{t^n_{k-1}};\theta_*)^2-\sigma(X_{t^n_{k-1}};\theta)^2\}.$$

Thus, by Theorem 7.4.4 (ii) we obtain that $n^{-1}\mathbb{Z}_n(\theta) \xrightarrow{\mathsf{P}} z(\theta)$. Although this convergence is pointwise at this moment, we can extend it up to the convergence uniformly in $\theta \in \Theta$ by checking the condition (7.26) of Theorem 7.3.5. □

Proof of Lemma 8.5.5 (ii). To prove the first claim, first observe that $n^{-1/2}\mathbb{Z}_n(\theta_*)$ is asymptotically equivalent to

$$\frac{1}{\sqrt{n}}\sum_{k=1}^{n}\frac{\partial_i\sigma(X_{t_{k-1}^n};\theta_*)}{\sigma(X_{t_{k-1}^n};\theta_*)^3}\left\{\frac{\sigma(X_{t_{k-1}^n};\theta_*)^2(W_{t_k^n}-W_{t_{k-1}^n})^2}{|t_k^n-t_{k-1}^n|}-\sigma(X_{t_{k-1}^n};\theta_*)^2\right\}$$

using also the first equality in Lemma 8.5.4. Thus with the help of (8.26) and (8.27) the claim follows from the CLT for discrete-time martingales (Theorem 7.1.3).

To prove the second claim, first note that we can write

$$\begin{aligned} n^{-1}\dot{\mathbb{Z}}_n^{ij}(\theta) &= \frac{1}{n}\sum_{k=1}^{n} g_\theta(X_{t_{k-1}^n})\left\{\frac{(X_{t_k^n}-X_{t_{k-1}^n})^2}{|t_k^n-t_{k-1}^n|}-\sigma(X_{t_{k-1}^n};\theta)^2\right\} \\ &\quad -\frac{2}{n}\sum_{k=1}^{n}\frac{\partial_i\sigma(X_{t_{k-1}^n};\theta)}{\sigma(X_{t_{k-1}^n};\theta)^3}\sigma(X_{t_{k-1}^n};\theta)\partial_j\sigma(X_{t_{k-1}^n};\theta), \end{aligned}$$

where

$$g_\theta(x) = \frac{\partial}{\partial\theta_j}\frac{\partial_i\sigma(x;\theta)}{\sigma(x;\theta)^3}.$$

Since the first term on the right-hand side is proved to be asymptotically equivalent, pointwise, to

$$\frac{1}{n}\sum_{k=1}^{n} g_\theta(X_{t_{k-1}^n})\{\sigma(X_{t_{k-1}^n};\theta_{\theta_*})^2-\sigma(X_{t_{k-1}^n};\theta)^2\}$$

by the second equation of Lemma 8.5.4, it follows from Lemma 7.4.4 (ii) that $n^{-1}\dot{\mathbb{Z}}_n^{ij}(\theta)$ converges in probability to

$$\dot{z}^{ij}(\theta) = \int_{\mathbb{R}}\left(g_\theta(x)\{\sigma(x;\theta_*)^2-\sigma(x;\theta)^2\}-2\frac{\partial_i\sigma(x;\theta)\partial_j\sigma(x;\theta)}{\sigma(x;\theta)^2}\right)P_{\theta_*}^{\circ}(dx).$$

Although this convergence is pointwise at this moment, we can extend it to

$$\sup_{\theta\in N}|n^{-1}\dot{\mathbb{Z}}_n^{ij}(\theta)-\dot{z}^{ij}(\theta)| \xrightarrow{\mathsf{P}} 0,$$

where N is a bounded neighborhood of θ_*, by checking the condition (7.26) in Theorem 7.3.5 and the fact that $\theta \mapsto \dot{z}(\theta)$ is Lipschitz continuous. Therefore, for any sequence of Θ-valued random elements $\widetilde{\theta}_n$ converging in outer-probability to θ_*, it holds that $n^{-1}\dot{\mathbb{Z}}_n(\widetilde{\theta}_n) \xrightarrow{\mathsf{P}^*} \dot{z}(\theta_*) = -J(\theta_*)$. The proof is finished. □

8.5.4 Partial-likelihood for Cox's regression models

Let $t \rightsquigarrow N_t^k$, $k = 1, 2, ...$, be counting processes which do not have simultaneous jumps. Suppose that each N^k admits the intensity process

$$\lambda_t^{k,\theta} = Y_t^k e^{\theta^{\mathrm{tr}} Z_t^k} \alpha(t), \quad t \in [0, \infty),$$

where $t \mapsto \alpha(t)$ is a deterministic, non-negative measurable function not depending on k, called the *baseline hazard function*, $t \rightsquigarrow Y_t^k$ is a $\{0,1\}$-valued, predictable process, and $t \rightsquigarrow Z_t^k$ is an $\mathbb{R}^p$-valued, predictable process called the *covariate process*. The baseline hazard function represents the rate of risk common for all individual k's, in which we are not so much interested. The individual k is observable whenever $Y_t^k = 1$. We are interested in estimating the unknown parameter θ appearing as the linear coefficient for the covariate process which represents some characteristics that are different over individual k's, like age, sex, the amount of medicines given to each individual, etc. Our statistical analysis is based on the data

$$(N_t^k, Y_t^k) \text{ on } [0, T], \quad \text{and} \quad Z_t^k \text{ on } \{t \in [0, T] : Y_t^k = 1\}, \quad k = 1, ..., n,$$

where $T > 0$ is a fixed time. Note that the counting process $t \rightsquigarrow N_t^k$ never jumps at any time point t in the set $\{t \in [0, T] : Y_t^k = 1\}^c$.

Supposing that θ belongs to an open, convex subset Θ of $\mathbb{R}^p$, let us introduce some notations:

$$\begin{aligned}
S_t^{n,0}(\theta) &= \sum_{k=1}^n Y_t^k e^{\theta^{\mathrm{tr}} Z_t^k}, \\
S_t^{n,1}(\theta) &= (\partial_1 S_t^{n,0}(\theta), ..., \partial_p S_t^{n,0}(\theta))^{\mathrm{tr}} \\
&= \sum_{k=1}^n Z_t^k Y_t^k e^{\theta^{\mathrm{tr}} Z_t^k}, \\
S_t^{n,2}(\theta) &= (\partial_{i,j} S_t^{n,0}(\theta))_{(i,j) \in \{1,...,p\}^2} \\
&= \sum_{k=1}^n Z_t^k (Z_t^k)^{\mathrm{tr}} Y_t^k e^{\theta^{\mathrm{tr}} Z_t^k}.
\end{aligned}$$

The logarithm of *Cox's partial-likelihood* is given by

$$\mathbb{M}_n(\theta) = \sum_{k=1}^n \int_0^T (\theta^{\mathrm{tr}} Z_t^k - \log S_t^{n,0}(\theta)) dN_t^k,$$

and we will use the estimating function

$$\begin{aligned}
\mathbb{Z}_n(\theta) &= (\partial_1 \mathbb{M}_n(\theta), ..., \partial_p \mathbb{M}_n(\theta))^{\mathrm{tr}} \\
&= \sum_{k=1}^n \int_0^T \left(Z_t^k - \frac{S_t^{n,1}(\theta)}{S_t^{n,0}(\theta)} \right) dN_t^k.
\end{aligned}$$

Let us discuss the validity of this estimating function. In order to explain the key idea, let us first consider the case where the baseline hazard function α were "known" to statisticians for a while. In this case, the (true) log-likelihood function should be given by

$$\begin{aligned}\ell_n(\theta) &= \sum_{k=1}^{n}\int_0^T \log\lambda_t^{k,\theta}dN_t^k - \sum_{k=1}^{n}\int_0^T \lambda_t^{k,\theta}dt \\ &= \sum_{k=1}^{n}\int_0^T (\theta^{\mathrm{tr}}Z_t^k + \log(Y_t^k\alpha(t)))dN_t^k - \sum_{k=1}^{n}\int_0^T Y_t^k e^{\theta^{\mathrm{tr}}Z_t^k}\alpha(t)dt,\end{aligned}$$

(see Subsection 6.5.4), and thus it would be natural to introduce the estimating function by differentiating $\ell_n(\theta)$ with respect to θ:

$$\sum_{k=1}^{n}\left\{\int_0^T Z_t^k dN_t^k - \int_0^T Z_t^k Y_t^k e^{\theta^{\mathrm{tr}}Z_t^k}\alpha(t)dt\right\}. \tag{8.28}$$

However, since we actually would like to assume that the baseline hazard function α is *unknown* to statisticians, we should remove it from the second term. To do it, noting that the predictable compensator for the total sum process $\bar{N}_t = \sum_{k=1}^{n} N_t^k$ is given by

$$\int_0^t S_s^{n,0}(\theta)\alpha(s)ds,$$

we shall estimate

$$\alpha(s)ds$$

by the *Breslow estimator*[6]

$$\frac{1}{S_s^{n,0}(\theta)}d\bar{N}_s.$$

By replacing the second term of (8.28) with $\int_0^T \frac{S_t^{n,1}(\theta)}{S_t^{n,0}(\theta)}d\bar{N}_t$, we obtain

$$\sum_{k=1}^{n}\int_0^T Z_t^k dN_t^k - \int_0^T \frac{S_t^{n,1}(\theta)}{S_t^{n,0}(\theta)}d\bar{N}_t,$$

which coincides with Cox's estimating function $\mathbb{Z}_n(\theta)$.

From mathematical point of view, it is important that if we substitute the true value $\theta = \theta_*$ into the estimating function $\mathbb{Z}_n(\theta)$, then the resulting $\mathbb{Z}_n(\theta_*)$ is the terminal variable of a locally square-integrable martingale. In fact, it holds that

$$\begin{aligned}\mathbb{Z}_n(\theta_*) &= \sum_{k=1}^{n}\int_0^T \left(Z_t^k - \frac{S_t^{n,1}(\theta_*)}{S_t^{n,0}(\theta_*)}\right)dN_t^k \\ &= \sum_{k=1}^{n}\int_0^T \left(Z_t^k - \frac{S_t^{n,1}(\theta_*)}{S_t^{n,0}(\theta_*)}\right)(dN_t^k - Y_t^k e^{\theta_*^{\mathrm{tr}}Z_t^k}\alpha(t)dt),\end{aligned}$$

[6]Note that $S_s^{n,0}(\theta) = 0$ implies $Y_s^k = 0$ for all k; in this case, the process $\bar{N}$ never jumps at s.

because

$$\sum_{k=1}^{n}\int_0^T \left(Z_t^k - \frac{S_t^{n,1}(\theta_*)}{S_t^{n,0}(\theta_*)} \right) Y_t^k e^{\theta_*^{\mathrm{tr}} Z_t^k} \alpha(t)dt = 0.$$

In other words, the "predictable compensator" $Z_n(\theta_*)$ for $\mathbb{Z}_n(\theta_*)$ is zero, as we expected.

In order to apply Theorem 8.2.2, note that the Taylor expansion of $\mathbb{Z}_n(\theta)$ around $\theta = \theta_*$ is given by $\mathbb{Z}_n(\theta) = \mathbb{Z}_n(\theta_*) + \dot{\mathbb{Z}}_n(\widetilde{\theta})(\theta - \theta_*)$, where

$$\dot{\mathbb{Z}}_n(\theta) = -\sum_{k=1}^{n}\int_0^T \frac{S_t^{n,2}(\theta)S_t^{n,0}(\theta) - S_t^{n,1}(\theta)S_t^{n,1}(\theta)^{\mathrm{tr}}}{S_t^{n,0}(\theta)^2} dN_t^k$$

and $\widetilde{\theta}$ is a point on the segment connecting θ and θ_*.

Now, let us introduce a set of regularity conditions to develop the asymptotic analysis in Cox's regression model.

Condition 8.5.7 (a) The function α is integrable: $\int_0^T \alpha(t)dt < \infty$.

(b) The set Θ is bounded, and the processes Z^k's are uniformly bounded.

(c) There exist an $\mathbb{R}$-valued function $s^0(t;\theta)$, an $\mathbb{R}^p$-valued function $s^1(t;\theta)$ and a $(p \times p)$-matrix valued function $s^2(t;\theta)$ such that

$$\sup_{t\in[0,T]} \left\| \frac{1}{n} S_t^{n,l}(\theta) - s^l(t;\theta) \right\| \xrightarrow{\mathsf{P}} 0, \quad \forall \theta \in \Theta, \quad l = 0,1,2.$$

Moreover, it holds that

$$\inf_{\theta\in\Theta} \inf_{t\in[0,T]} s^0(t;\theta) > 0.$$

(d_1) All entries of the $\mathbb{R}^p$-valued function

$$\theta \mapsto z(\theta) = \int_0^T \left(\frac{s^1(t;\theta_*)}{s^0(t;\theta_*)} - \frac{s^1(t;\theta)}{s^0(t;\theta)} \right) s^0(t;\theta_*)\alpha(t)dt$$

are Lipschitz continuous on Θ.

(d_2) All entries of the $(p \times p)$-matrix valued function

$$\theta \mapsto J(\theta) = \int_0^T \frac{s^2(t;\theta)s^0(t;\theta) - s^1(t;\theta)s^1(t;\theta)^{\mathrm{tr}}}{s^0(t;\theta)^2} s^0(t;\theta_*)\alpha(t)dt$$

are Lipschitz continuous on a neighborhood of θ_*.

(e) The $(p \times p)$-matrix

$$J(\theta_*) = \int_0^T \left(s^2(t;\theta_*) - \frac{s^1(t;\theta_*)s^1(t;\theta_*)^{\mathrm{tr}}}{s^0(t;\theta_*)} \right) \alpha(t)dt$$

is positive definite.

Remark. Condition 8.5.7 (b) can be weakened. However, this assumption, which is not a real restriction in practice, considerably simplifies the proofs of the theorems below.

We are now ready to prove the consistency and the asymptotic normality of maximum partial-likelihood estimators.

Theorem 8.5.8 Suppose that (a), (b), (c) for $l = 0, 1$, and (d_1) in Condition 8.5.7 are satisfied. If the condition **[C2]** in Theorem 8.2.2 is satisfied for $\theta \mapsto z(\theta)$, then, for any sequence of Θ-valued random variables $\widehat{\theta}_n$ satisfying

$$n^{-1}\mathbb{Z}_n(\widehat{\theta}_n) = o_{\mathsf{P}}(1),$$

it holds that $\widehat{\theta}_n \xrightarrow{\mathsf{P}} \theta_*$.

Proof. By the corollary to Lenglart's inequality, it holds for every $\theta \in \Theta$ that $n^{-1}\mathbb{Z}_n(\theta)$ is asymptotically equivalent to $n^{-1}Z_n(\theta)$, where

$$Z_n(\theta) = \sum_{k=1}^{n} \int_0^T \left(\frac{S_t^{n,1}(\theta_*)}{S_t^{n,0}(\theta_*)} - \frac{S_t^{n,1}(\theta)}{S_t^{n,0}(\theta)} \right) S_t^{n,0}(\theta_*)\alpha(t)dt.$$

Furthermore, $n^{-1}Z_n(\theta)$ converges in probability to $z(\theta)$ by Condition 8.5.7 (c) for $l = 0, 1$. Hence, we have proved that $||n^{-1}\mathbb{Z}_n(\theta) - z(\theta)|| \xrightarrow{\mathsf{P}} 0$ for every $\theta \in \Theta$.

Due to Condition 8.5.7 (d_1) and the fact

$$\sup_{\theta \in \Theta} |n^{-1}\partial_j \mathbb{Z}_n^i(\theta)| = O_{\mathsf{P}}(1), \quad (i, j) \in \{1, ..., p\}^2,$$

we can apply Theorem 7.3.5 to conclude that the pointwise convergence is extended up to

$$\sup_{\theta \in \Theta} ||n^{-1}\mathbb{Z}_n(\theta) - z(\theta)|| \xrightarrow{\mathsf{P}} 0.$$

Thus Theorem 8.2.1 yields the assertion. □

Theorem 8.5.9 Assume Condition 8.5.7. Suppose also that the condition **[C2]** in Theorem 8.2.1 is satisfied for $\theta \mapsto z(\theta)$ given in Condition 8.5.7 (d_1). Then, for any sequence of Θ-valued random variables $\widehat{\theta}_n$ satisfying

$$n^{-1/2}\mathbb{Z}_n(\widehat{\theta}_n) = o_{\mathsf{P}}(1),$$

it holds that

$$\sqrt{n}(\widehat{\theta}_n - \theta_*) \xrightarrow{\mathsf{P}} \mathcal{N}_p(0, J(\theta_*)^{-1}).$$

Proof. The consistency is a consequence from the previous theorem. By Theorem 8.2.2 it is sufficient to prove that

$$n^{-1/2}\mathbb{Z}_n(\theta_*) \xrightarrow{\mathsf{P}} \mathcal{N}_p(0, J(\theta_*)) \quad \text{in } \mathbb{R}^p,$$

and that it holds for any sequence of Θ-valued random elements $\widetilde{\theta}_n$ converging in outer-probability to θ_* that

$$n^{-1}\dot{\mathbb{Z}}_n(\widetilde{\theta}_n) \stackrel{\mathsf{P}}{\longrightarrow} -J(\theta_*).$$

As it has already seen above, $n^{-1/2}\mathbb{Z}_n(\theta_*)$ is the terminal variable of the locally square-integrable martingale

$$t \rightsquigarrow M_t^n = \frac{1}{\sqrt{n}}\sum_{k=1}^n \int_0^t \left(Z_u^k - \frac{S_u^{n,1}(\theta_*)}{S_u^{n,0}(\theta_*)} \right) (dN_u^k - Y_u^k e^{\theta_*^{\mathrm{tr}} Z_u^k} \alpha(u)du).$$

We shall apply the CLT for stochastic integrals (Theorem 7.1.6). The convergence of the predictable quadratic co-variation matrices $(\langle M^{n,i}, M^{n,j}\rangle_T)_{(i,j)\in\{1,\ldots,p\}^2}$ is verified as follows:

$$\begin{aligned}
&\frac{1}{n}\sum_{k=1}^n \int_0^T \left(Z_t^k - \frac{S_t^{n,1}(\theta_*)}{S_t^{n,0}(\theta_*)} \right) \left(Z_t^k - \frac{S_t^{n,1}(\theta_*)}{S_t^{n,0}(\theta_*)} \right)^{\mathrm{tr}} Y_t^k e^{\theta_*^{\mathrm{tr}} Z_t^k} \alpha(t)dt \\
&\quad = \frac{1}{n}\sum_{k=1}^n \int_0^T \left(S_t^{n,2}(\theta_*) - \frac{S_t^{n,1}(\theta_*) S_t^{n,1}(\theta_*)^{\mathrm{tr}}}{S_t^{n,0}(\theta_*)} \right) \alpha(t)dt \\
&\quad \stackrel{\mathsf{P}}{\longrightarrow} \int_0^T \left(s^2(t;\theta_*) - \frac{s^1(t;\theta_*) s^1(t;\theta_*)^{\mathrm{tr}}}{s^0(t;\theta_*)} \right) \alpha(t)dt \\
&\quad = J(\theta_*).
\end{aligned}$$

It is easy to show that Lyapunov's condition is satisfied, due to Condition 8.5.7 (b). Thus we have that $n^{-1/2}\mathbb{Z}_n(\theta_*) = M_T^n \stackrel{\mathsf{P}}{\Longrightarrow} \mathcal{N}_p(0, J(\theta_*))$ in $\mathbb{R}^p$.

On the other hand, it is easy to prove that $n^{-1}\dot{\mathbb{Z}}_n(\theta)$ converges in probability to $-J(\theta)$ for every $\theta \in \Theta$, and that this pointwise convergence can be extended up to

$$\sup_{\theta\in N} ||n^{-1}\dot{\mathbb{Z}}_n(\theta) - (-J(\theta))|| \stackrel{\mathsf{P}}{\longrightarrow} 0,$$

where N is a neighborhood of θ_*, by the same argument as the previous proofs. Noting again that $\theta \mapsto J(\theta)$ is continuous by assumption, we obtain that $n^{-1}\dot{\mathbb{Z}}_n(\widetilde{\theta}_n)$ converges in outer-probability to $-J(\theta_*)$ for any sequence of Θ-valued random elements $\widetilde{\theta}_n$ converging in outer probability to θ_*. The proof is finished. □

8.6 *More* General Theory for Z-estimators

In this section, two advanced theorems for Z-estimators are discussed; they are viewed as extensions of the two theorems presented in Section 8.2, respectively.

8.6.1 Consistency of Z-estimators, II

This subsection presents a slight extension of Theorem 8.2.1, which gives a set of sufficient conditions for the consistency of Z-estimators, to the case where the rate sequence $(s_n)_{n=1,2,...}$ appearing there is replaced by a sequence $(S_n)_{n=1,2,...}$ of $(p \times p)$-diagonal matrices, which is typically set to be $S_n = nI_p$ where I_p denotes the $(p \times p)$-identity matrix.

Theorem 8.6.1 (Consistency of Z-estimators) Consider the set-up given at the beginning of Section 8.2. Suppose that for the true value θ_* of the unknown parameter $\theta \in \Theta$ there exist a sequence $(S_n)_{n=1,2,...}$ of $(p \times p)$-diagonal, positive definite matrices and an $\mathbb{R}^p$-valued (possibly,) random function $\theta \rightsquigarrow \zeta(\theta)$ satisfying the following conditions **[C1*]** and **[C2*]**.

[C1*] $\sup_{\theta \in \Theta} ||S_n^{-1}\mathbb{Z}_n(\theta) - \zeta(\theta)|| \xrightarrow{\mathsf{P}} 0$.

[C2*] It holds that

$$\inf_{\theta:||\theta-\theta_*||>\varepsilon} ||\zeta(\theta)|| > 0 = ||\zeta(\theta_*)||, \quad \forall \varepsilon > 0, \quad \text{almost surely.}$$

Then, for any sequence of Θ-valued random variables $\widehat{\theta}_n$ satisfying

$$S_n^{-1}\mathbb{Z}_n(\widehat{\theta}_n) = o_{\mathsf{P}}(1),$$

it holds that $\widehat{\theta}_n \xrightarrow{\mathsf{P}} \theta_*$.

The proof is exactly the same as that of Theorem 8.2.1, so it is omitted.

8.6.2 Asymptotic representation of Z-estimators, II

This subsection provides an extension of Theorem 8.2.2, which derives the asymptotic representation of Z-estimators, to the case where the rate sequences $(q_n)_{n=1,2,...}$ and $(r_n)_{n=1,2,...}$ appearing there are replaced by two sequences $(Q_n)_{n=1,2,...}$ and $(R_n)_{n=1,2,...}$ of $(p \times p)$-diagonal matrices, which are typically set to be $Q_n = R_n = \sqrt{n}I_p$ where I_p denotes the $(p \times p)$-identity matrix.

Theorem 8.6.2 (Asymptotic representation of Z-estimators) Consider the set-up given at the beginning of Section 8.2. Suppose that for the true value θ_* of the unknown parameter $\theta \in \Theta$ the following conditions **[AR1*]**, **[AR2*]** and **[ARJ]** are satisfied.

[AR1*] There exists a sequence of $(p \times p)$-diagonal, positive definite matrices Q_n such that $Q_n^{-1}\mathbb{Z}_n(\theta_*)$ converges weakly in $\mathbb{R}^p$ to a random variable $L(\theta_*)$.

[AR2*] There also exists a sequence of $(p \times p)$-diagonal, positive definite matrices R_n such that, for any sequence of Θ-valued random elements $\widetilde{\theta}_n$ converging in outer-probability to θ_*, it holds that $Q_n^{-1}\dot{\mathbb{Z}}_n(\widetilde{\theta}_n)R_n^{-1}$ converges in outer-probability to a (possibly,) random $(p \times p)$-matrix $-\mathcal{J}(\theta_*)$, which is regular almost surely.

[ARJ] The matrix $\mathcal{J}(\theta_*)$ is deterministic, or more generally, the random sequences $Q_n^{-1}\mathbb{Z}_n(\theta_*)$ and $Q_n^{-1}\dot{\mathbb{Z}}_n(\theta_*)R_n^{-1}$ converge in distribution to $L(\theta_*)$ and $-\mathcal{J}(\theta_*)$, jointly[7].

Then, for any sequence of Θ-valued random variables $\widehat{\theta}_n$ satisfying

$$Q_n^{-1}\mathbb{Z}_n(\widehat{\theta}_n) = o_{\mathsf{P}}(1) \quad \text{and} \quad \widehat{\theta}_n \xrightarrow{\mathsf{P}} \theta_*,$$

it holds that

$$R_n(\widehat{\theta}_n - \theta_*) = \mathcal{J}(\theta_*)^{-1}Q_n^{-1}\mathbb{Z}_n(\theta_*) + o_{\mathsf{P}}(1).$$

In particular, it holds also that

$$R_n(\widehat{\theta}_n - \theta_*) \stackrel{\mathsf{P}}{\Longrightarrow} \mathcal{J}(\theta_*)^{-1}L(\theta_*).$$

Remark. In view of **[AR2*]**, when the random matrices $\dot{\mathbb{Z}}_n(\theta)$ are symmetric, the rate matrices Q_n and R_n should be taken to be *equal*.

Proof. Just replace (8.4) in the proof of Theorem 8.2.2 with

$$Q_n^{-1}\mathbb{Z}_n(\widehat{\theta}_n) = Q_n^{-1}\mathbb{Z}_n(\theta_*) + (Q_n^{-1}\dot{\mathbb{Z}}_n(\widetilde{\theta}_n)R_n^{-1})(R_n(\widehat{\theta}_n - \theta_*)),$$

and repeat the same argument. □

8.7 Example, II (*More* Advanced Topic: Different Rates of Convergence)

8.7.1 Quasi-likelihood for ergodic diffusion models

Let us consider the 1-dimensional diffusion process $t \rightsquigarrow X_t$ which is the unique strong solution to the stochastic differential equation

$$X_t = X_0 + \int_0^t \beta(X_s;\theta_1)ds + \int_0^t \sigma(X_s;\theta_2)dW_s,$$

where $s \rightsquigarrow W_s$ is a standard Wiener process. For illustration, the parameters are assumed to come from $\theta_1 \in \Theta_1 \subset \mathbb{R}$ and $\theta_2 \in \Theta_2 \subset \mathbb{R}$, and we denote $\theta = (\theta_1,\theta_2)^{\mathrm{tr}} \in \Theta = \{(\theta_1,\theta_2)^{\mathrm{tr}};(\theta_1,\theta_2) \in \Theta_1 \times \Theta_2\} \subset \mathbb{R}^2$. We assume that the process $t \rightsquigarrow X_t$ is ergodic with the invariant measure $P^\circ_{\theta_*}$ under the true value $\theta_* = (\theta_{*1},\theta_{*2})^{\mathrm{tr}}$ of the unknown parameter θ. Suppose that, we are able to observe the process $t \rightsquigarrow X_t$ at discrete time grids $0 = t_0^n < t_1^n < \cdots < t_n^n$, and we shall consider the sampling scheme

$$t_n^n \to \infty, \quad n\Delta_n^2 \to 0 \quad \text{and} \quad \sum_{k=1}^n \left| \frac{|t_k^n - t_{k-1}^n|}{t_n^n} - \frac{1}{n} \right| \to 0,$$

[7] Recall Slutsky's lemma (Lemma 2.3.7).

as $n \to \infty$, where $\Delta_n = \max_{1 \le k \le n} |t_k^n - t_{k-1}^n|$.

Now, introduce

$$\mathbb{Z}_n(\theta) = (\partial_1 \mathbb{M}_n(\theta), \partial_2 \mathbb{M}_n(\theta))^{\mathrm{tr}},$$

where

$$\mathbb{M}_n(\theta) = \\ -\sum_{k: t_{k-1}^n \le t_n^n} \left\{ \log \sigma(X_{t_{k-1}^n}; \theta_2) + \frac{(X_{t_k^n} - X_{t_{k-1}^n} - \beta(X_{t_{k-1}^n}; \theta_1)|t_k^n - t_{k-1}^n|)^2}{2\sigma(X_{t_{k-1}^n}; \theta_2)^2 |t_k^n - t_{k-1}^n|} \right\}.$$

The right choice of the rate matrices would be

$$Q_n = R_n = \begin{pmatrix} \sqrt{t_n^n} & 0 \\ 0 & \sqrt{n} \end{pmatrix} \quad \text{and} \quad S_n = \begin{pmatrix} t_n^n & 0 \\ 0 & n \end{pmatrix}.$$

Under some regularity conditions, which we do not dare to write down here explicitly, we can show that the 2-dimensional random vectors $\mathbb{Z}_n(\theta)$ and the (2×2)-random matrices $\dot{\mathbb{Z}}_n(\theta)$, namely,

$$\begin{aligned} \mathbb{Z}_n(\theta) &= (\mathbb{Z}_n^1(\theta), \mathbb{Z}_n^2(\theta))^{\mathrm{tr}}, \\ \dot{\mathbb{Z}}_n(\theta) &= \begin{pmatrix} \dot{\mathbb{Z}}_n^{1,1}(\theta) & \dot{\mathbb{Z}}_n^{1,2}(\theta) \\ \dot{\mathbb{Z}}_n^{2,1}(\theta) & \dot{\mathbb{Z}}_n^{2,2}(\theta) \end{pmatrix}, \end{aligned}$$

satisfy the following, respectively (since both of the limits $\zeta(\theta)$ and $\mathcal{J}(\theta_*)$ are non-random in the current case, we denote them by $z(\theta)$ and $J(\theta_*)$, respectively):

$$\sup_{\theta \in \Theta} ||S_n^{-1} \mathbb{Z}_n(\theta) - z(\theta)|| = o_{\mathsf{P}}(1);$$

$$R_n^{-1} \dot{\mathbb{Z}}_n(\widetilde{\theta}_n) R_n^{-1} \xrightarrow{\mathsf{P}^*} -J(\theta_*),$$

where $\widetilde{\theta}_n$ is any Θ-valued random elements converging in outer-probability to θ_*. Here, the limits are given by $z(\theta) = (z^1(\theta), z^2(\theta))^{\mathrm{tr}}$, where

$$z^1(\theta) = \int_{\mathbb{R}} \frac{\partial_1 \beta(x; \theta_1)}{\sigma(x; \theta_2)} \{\beta(x; \theta_{*1}) - \beta(x; \theta_1)\} P_{\theta_*}^{\circ}(dx),$$

$$z^2(\theta) = \int_{\mathbb{R}} \frac{\partial_2 \sigma(x; \theta_2)}{\sigma(x; \theta_2)^3} \{\sigma(x; \theta_{*2})^2 - \sigma(x; \theta_2)^2\} P_{\theta_*}^{\circ}(dx),$$

and

$$J(\theta_*) = \begin{pmatrix} J^{1,1}(\theta_*) & 0 \\ 0 & J^{2,2}(\theta_*) \end{pmatrix},$$

where

$$J^{1,1}(\theta_*) = \int_{\mathbb{R}} \frac{(\partial_1 \beta(x; \theta_{*1}))^2}{\sigma(x; \theta_{*2})^2} P_{\theta_*}^{\circ}(dx), \quad J^{2,2}(\theta_*) = 2 \int_{\mathbb{R}} \frac{(\partial_2 \sigma(x; \theta_{*2}))^2}{\sigma(x; \theta_{*2})^2} P_{\theta_*}^{\circ}(dx).$$

Moreover, it holds that

$$R_n^{-1}\mathbb{Z}_n(\theta_*) \stackrel{\mathsf{P}}{\Longrightarrow} \mathcal{N}_2(0, J(\theta_*)) \quad \text{in } \mathbb{R}^2.$$

Therefore, for any sequence of Θ-valued random variables $\widehat{\theta}_n$ satisfying

$$S_n^{-1}\mathbb{Z}_n(\widehat{\theta}_n) = o_{\mathsf{P}}(1),$$

it holds that $\widehat{\theta}_n \stackrel{\mathsf{P}}{\longrightarrow} \theta_*$. Also, for any sequence of Θ-valued random variables $\widehat{\theta}_n$ satisfying

$$R_n^{-1}\mathbb{Z}_n(\widehat{\theta}_n) = o_{\mathsf{P}_{\theta_*}}(1) \quad \text{and} \quad \widehat{\theta}_n \stackrel{\mathsf{P}}{\longrightarrow} \theta_*,$$

it holds that

$$\begin{aligned} R_n(\widehat{\theta}_n - \theta_*) &= J(\theta_*)^{-1}R_n^{-1}\mathbb{Z}_n(\theta_*) + o_{\mathsf{P}}(1) \\ &\stackrel{\mathsf{P}}{\Longrightarrow} \mathcal{N}_2(0, J(\theta_*)^{-1}) \quad \text{in } \mathbb{R}^2. \end{aligned}$$

Exercise 8.7.1 Complete the calculation for this example. In particular, prove that $(t_n^n n)^{-1/2}\mathbb{Z}_n^{1,2}(\widetilde{\theta}_n) = (t_n^n n)^{-1/2}\mathbb{Z}_n^{2,1}(\widetilde{\theta}_n) = o_{\mathsf{P}^*}(1)$ for any Θ-valued random elements $\widetilde{\theta}_n$ converging in outer-probability to θ_*.

9

Optimal Inference in Finite-Dimensional LAN Models

This chapter presents a unified theory to show the asymptotic efficiency of the maximum likelihood estimators (and some other estimators based on the likelihood functions) in any parametric models where the asymptotic behavior of the log-likelihood function can be fully analyzed. The theory is based on the concept of *local asymptotic normality (LAN)* of sequences of statistical experiments. The main results are Hájek-Inagaki's convolution theorem and Hájek-Le Cam's asymptotic minimax theorem, both established in the early 1970's. The asymptotic representation theorem (Theorems 8.2.2 and 8.6.2) studied in the previous chapter plays an important role in the process of applying the LAN theory with the help of Le Cam's third lemma.

Throughout this chapter, the limit operations are always taken as $n \to \infty$.

9.1 Local Asymptotic Normality

The main object which we treat in this chapter is the log-likelihood function in a finite-dimensional parametric model, say, $\ell_n(\theta)$, where $\theta \in \Theta \subset \mathbb{R}^p$ for an integer $p \geq 1$. For example, in the case of i.i.d. model, it has been given as

$$\ell_n(\theta) = \sum_{k=1}^{n} \log p(X_k;\theta), \quad \theta \in \Theta.$$

In general, our discussion in the previous chapter was performed typically by setting $\mathbb{Z}_n(\theta)$ to be the gradient vector and $\dot{\mathbb{Z}}_n(\theta)$ the Hessian matrix, respectively, of $\ell_n(\theta)$:

$$\mathbb{Z}_n(\theta) = \dot{\ell}_n(\theta) = \left(\tfrac{\partial}{\partial\theta_1}\ell_n(\theta), \ldots, \tfrac{\partial}{\partial\theta_p}\ell_n(\theta)\right)^{\mathrm{tr}};$$

$$\dot{\mathbb{Z}}_n(\theta) = \ddot{\ell}_n(\theta) = \left(\tfrac{\partial^2}{\partial\theta_i\partial\theta_j}\ell_n(\theta)\right)_{(i,j)\in\{1,\ldots,p\}^2}.$$

Introducing the two matrices $Q_n = R_n = \sqrt{n}I_p$, where I_p denotes the $(p\times p)$-identity matrix, let us consider the following Taylor expansion:

$$\begin{aligned}
\ell_n(\theta_* + R_n^{-1}h) - \ell_n(\theta_*) &= (R_n^{-1}h)^{\mathrm{tr}}\dot{\ell}_n(\theta_*) + \frac{1}{2}(R_n^{-1}h)^{\mathrm{tr}}\ddot{\ell}_n(\widetilde{\theta}_n)R_n^{-1}h \\
&= h^{\mathrm{tr}}(R_n^{-1}\mathbb{Z}_n(\theta_*)) + \frac{1}{2}h^{\mathrm{tr}}R_n^{-1}\dot{\mathbb{Z}}_n(\widetilde{\theta}_n)R_n^{-1}h,
\end{aligned}$$

DOI: 10.1201/9781315117768-9

where $\widetilde{\theta}_n$ is a random point on the segment connecting θ_* and $\theta_* + R_n^{-1}h$. As we have already proved in many examples, we can often obtain that

$$R_n^{-1}\mathbb{Z}_n(\theta_*) \stackrel{\mathsf{P}_{\theta_*}}{\Longrightarrow} \mathcal{N}_p(0, I(\theta_*)) \quad \text{in } \mathbb{R}^p \tag{9.1}$$

and that

$$R_n^{-1}\dot{\mathbb{Z}}_n(\widetilde{\theta}_n)R_n^{-1} = -I(\theta_*) + o_{\mathsf{P}_{\theta_*}}(1). \tag{9.2}$$

Remark. Since $\dot{\mathbb{Z}}_n(\theta)$ is a symmetric matrix in the current situation, we shall always introduce the two sequences of matrices Q_n and R_n which are *equal* during this chapter.

Since (9.1) and (9.2) have always been the key points in the course of proving the asymptotic normality of Z-estimators, especially, maximum likelihood estimators (MLEs), extracting these two points we introduce the definition of *local asymptotic normality*, by setting

$$R_n^{-1}\mathbb{Z}_n(\theta_*) = \Delta_n(\theta_*) \quad \text{and} \quad R_n^{-1}\dot{\mathbb{Z}}_n(\widetilde{\theta}_n)R_n^{-1} = -I(\theta_*) + o_{\mathsf{P}_{\theta_*}}(1).$$

Definition 9.1.1 (Local asymptotic normality (LAN)) Let Θ be an open, convex subset of $\mathbb{R}^p$. For every $n \in \mathbb{N}$, let $(\mathcal{X}_n, \mathcal{A}_n)$ be a measurable space on which a family $\{P_{n,\theta}; \theta \in \Theta\}$ of probability measures are commonly defined, and let R_n be a diagonal, positive definite matrix.

The sequence of *statistical experiments* $(\mathcal{X}_n, \mathcal{A}_n, \{P_{n,\theta}; \theta \in \Theta\})$ is said to be *locally asymptotically normal (LAN)* at $\theta_* \in \Theta$ if there exists a positive definite matrix $I(\theta_*)$ such that

$$\log \frac{dP_{n,\theta_*+R_n^{-1}h}}{dP_{n,\theta_*}} = h^{\mathrm{tr}}\Delta_n(\theta_*) - \frac{1}{2}h^{\mathrm{tr}}I(\theta_*)h + o_{P_{n,\theta_*}}(1), \quad \forall h \in \mathbb{R}^p,$$

and that

$$\Delta_n(\theta_*) \stackrel{P_{n,\theta_*}}{\Longrightarrow} \mathcal{N}_p(0, I(\theta_*)) \quad \text{in } \mathbb{R}^p.$$

9.2 Asymptotic Efficiency

There are two theorems that provide the characterizations of asymptotically efficient estimators in the LAN models.

Theorem 9.2.1 (Hájek-Inagaki's convolution theorem) Let a sequence of statistical experiments $(\mathcal{X}_n, \mathcal{A}_n, \{P_{n,\theta}; \theta \in \Theta\})$ be locally asymptotically normal at $\theta_* \in \Theta$. Suppose that an estimator T_n for θ_*, which is assumed to be $\mathcal{A}_n$-measurable, is *regular*, in the sense that there exists a Borel law $\mathcal{L}(\theta_*)$, not depending on h, such that

$$R_n(T_n - (\theta_* + R_n^{-1}h)) \stackrel{P_{n,\theta_*+R_n^{-1}h}}{\Longrightarrow} \mathcal{L}(\theta_*) \quad \text{in } \mathbb{R}^p, \quad \forall h \in \mathbb{R}^p. \tag{9.3}$$

Then, the limit $\mathcal{L}(\theta_*)$ is the distribution of the sum of two independent random variables, namely, $G(\theta_*) + H(\theta_*)$, such that

$$G(\theta_*) \text{ is distributed to } \mathcal{N}_p(0, I(\theta_*)^{-1}).$$

In order to state the next theorem, let us introduce a definition of a class of loss functions.

Definition 9.2.2 (Subconvex function) A non-negative measurable function ℓ on $\mathbb{R}^p$ is said to be of *subconvex* if the set $\{x : \ell(x) \leq a\}$ is closed, convex and symmetric for any $a \geq 0$.

Theorem 9.2.3 (Hàjek-Le Cam's asymptotic minimax theorem) Let a sequence of statistical experiments $(\mathcal{X}_n, \mathcal{A}_n, \{P_{n,\theta}; \theta \in \Theta\})$ be locally asymptotically normal at $\theta_* \in \Theta$. Let ℓ be a non-negative measurable function which is subconvex. Then, for any estimator T_n, which is assumed to be $\mathcal{A}_n$-measurable, it holds that

$$\sup_{I \subset \mathbb{R}^p} \liminf_{n \to \infty} \max_{h \in I} E_{n,\theta_* + R_n^{-1}h} \left[\ell(T_n - (\theta_* + R_n^{-1}h)) \right] \geq E[\ell(G(\theta_*))], \tag{9.4}$$

where the first supremum is taken over all finite subsets I of $\mathbb{R}^p$, and

$$G(\theta_*) \text{ is distributed to } \mathcal{N}_p(0, I(\theta_*)^{-1}).$$

See Chapter 3.11 of van der Vaart and Wellner (1996) for the proofs of these two theorems.

9.3 How to Apply

In the i.i.d. and Markov chain models which we studied in Subsections 8.3.1 and 8.3.2, respectively, we have already proved that the sequences of statistical experiments are locally asymptotically normal; that is, (8.10) and (8.11) for the former model, and (8.20) and (8.18) for the latter one. Hence, in order to establish the asymptotic efficiency in the sense of the convolution theorem presented above, it suffices to show that the estimator T_n under consideration meets (9.3) with $\mathcal{L}(\theta_*) \stackrel{d}{=} G(\theta_*)$; that is,

$$R_n(T_n - (\theta_* + R_n^{-1}h)) \stackrel{P_{n,\theta_* + R_n^{-1}h}}{\Longrightarrow} G(\theta_*) \quad \text{in } \mathbb{R}^p, \quad \forall h \in \mathbb{R}^p. \tag{9.5}$$

Also, once (9.5) is established, we may conclude that the estimator T_n is asymptotically efficient also in the sense of the asymptotic minimax theorem, at least for any *bounded, continuous* non-negative function ℓ which is subconvex. Indeed, it follows from the definition of the convergence in law that

$$\lim_{n \to \infty} E_{\theta_* + R_n^{-1}h} \left[\ell(R_n(T_n - (\theta_* + R_n^{-1}h))) \right] = E[\ell(G(\theta_*))],$$

and moreover, noting that the maximum operations over finitely many h's does not affect the limit operation "$\lim_{n\to\infty}$", we have that for any finite subset I of $\mathbb{R}^p$,

$$\lim_{n\to\infty}\max_{h\in I} E_{\theta_*+R_n^{-1}h}\left[\ell(R_n(T_n-(\theta_*+R_n^{-1}h)))\right]=E[\ell(G(\theta_*))];$$

consequently, the equality holds true in (9.4).

Therefore, it remains only to establish (9.5). This can be proved in the cases where the asymptotic representation

$$\begin{aligned} R_n(T_n-\theta_*) &= R_n(\widehat{\theta}_n-\theta_*) \\ &= J(\theta_*)^{-1}R_n^{-1}\mathbb{Z}_n(\theta_*)+o_{P_{n,\theta_*}}(1) \\ &= I(\theta_*)^{-1}\Delta_n(\theta_*)+o_{P_{n,\theta_*}}(1) \end{aligned}$$

holds true, as in Subsections 8.3.1 and 8.3.2. Let us first prepare a lemma.

Lemma 9.3.1 (Le Cam's third lemma) Given sequences of $\mathbb{R}^p$-valued random variables X_n and probability measures P_n and Q_n, if

$$\left(X_n^{\mathrm{tr}},\log\frac{dQ_n}{dP_n}\right)^{\mathrm{tr}} \stackrel{P_n}{\Longrightarrow} \mathcal{N}_{p+1}\left(\left(\begin{array}{c}\mu\\ -\frac{1}{2}\sigma^2\end{array}\right),\left(\begin{array}{cc}\Sigma^2 & \tau\\ \tau^{\mathrm{tr}} & \sigma^2\end{array}\right)\right) \quad \text{in } \mathbb{R}^{p+1},$$

then it holds that

$$X_n \stackrel{Q_n}{\Longrightarrow} \mathcal{N}_p(\mu+\tau,\Sigma) \quad \text{in } \mathbb{R}^p.$$

See Example 3.10.8 of van der Vaart and Wellner (1996) for a proof.

In order to apply this lemma, observe that, for any $h\in\mathbb{R}^p$,

$$\begin{aligned} &\left((R_n(T_n-\theta_*))^{\mathrm{tr}},\log\frac{dQ_n}{dP_n}\right)^{\mathrm{tr}} \\ &= \left((I(\theta_*)^{-1}\Delta_n(\theta_*))^{\mathrm{tr}},h^{\mathrm{tr}}\Delta_n(\theta_*)-\frac{1}{2}h^{\mathrm{tr}}I(\theta_*)h\right)^{\mathrm{tr}}+o_{P_{n,\theta_*}}(1) \\ &\stackrel{P_{n,\theta_*}}{\Longrightarrow} \mathcal{N}_{p+1}\left(\left(\begin{array}{c}0\\ -\frac{1}{2}h^{\mathrm{tr}}I(\theta_*)h\end{array}\right),\left(\begin{array}{cc}I(\theta_*)^{-1} & h\\ h^{\mathrm{tr}} & h^{\mathrm{tr}}I(\theta_*)h\end{array}\right)\right) \quad \text{in } \mathbb{R}^{p+1}. \end{aligned}$$

Thus, Le Cam's third lemma implies that

$$R_n(T_n-\theta_*) \stackrel{P_{n,\theta_*+R_n^{-1}h}}{\Longrightarrow} \mathcal{N}_p(h,I(\theta_*)^{-1}) \quad \text{in } \mathbb{R}^p, \quad \forall h\in\mathbb{R}^p,$$

which yields (9.5).

10

Z-Process Method for Change Point Problems

This chapter develops a general, unified approach based on a partial sum process of estimating function, which we call "Z-process", to some change point problems in mathematical statistics. The asymptotic representation theorem for Z-estimators and the functional CLTs for martingales, both of which were established in the previous chapters, will play the key roles to analyze our test statistic defined based on the Z-process. The limit distribution of the test statistic under the null hypothesis that there is no change point is the supremum of standard Brownian bridges. The consistency of the test under the alternative is also proved in a general way. Some applications to Markov chain models and diffusion process models are discussed.

Throughout this chapter, the limit operations are taken as $n \to \infty$, unless otherwise stated.

10.1 Illustrations with Independent Random Sequences

Let us give an illustration by an example of a parametric model for independent random sequence. Suppose that, under the null hypothesis that there is no change point, the data $X_1, ..., X_n$ comes from a distribution in the parametric family $\{p(x;\theta); \theta \in \Theta\}$ of probability densities with respect to a σ-finite measure μ on a measurable space $(\mathcal{X}, \mathcal{A})$. We introduce the partial sum process

$$\mathbb{M}_n(u,\theta) = \sum_{k=1}^{[un]} \log p(X_k;\theta), \quad \forall u \in [0,1],$$

and the gradient vectors $\dot{\mathbb{M}}_n(u,\theta)$ of $\mathbb{M}_n(u,\theta)$ with respect to θ. In a similar way to the previous chapters, we set $\mathbb{Z}_n(u,\theta) = \dot{\mathbb{M}}_n(u,\theta)$. Let $\widehat{\theta}_n$ be the MLE for the full data $X_1, ..., X_n$ as a special case of Z-estimators; that is, $\widehat{\theta}_n$ is a solution to the estimating equation

$$\mathbb{Z}_n(1,\theta) = 0.$$

The fact that the sequence of stochastic processes

$$u \rightsquigarrow n^{-1/2}\mathbb{Z}_n(u,\theta_*) \quad \text{converges weakly to} \quad u \rightsquigarrow I(\theta_*)^{1/2}B(u)$$

in the Skorokhod space $D[0,1]$, where $u \rightsquigarrow B(u)$ is a vector of independent standard Brownian motions, is immediate from Donsker's theorem (Corollary 7.2.4). However, it does not seem so well-known that the sequence of stochastic processes

$$u \rightsquigarrow n^{-1/2}\mathbb{Z}_n(u,\widehat{\theta}_n) \quad \text{converges weakly to} \quad u \rightsquigarrow I(\theta_*)^{1/2}B^\circ(u) \qquad (10.1)$$

DOI: 10.1201/9781315117768-10

in $D[0,1]$, where $u \rightsquigarrow B^\circ(u)$ is a vector of independent standard Brownian bridges. Horváth and Parzen (1994) is apparently the first to have introduced the statistic

$$\mathcal{T}_n = n^{-1} \sup_{u\in[0,1]} \mathbb{Z}_n(u,\widehat{\theta}_n)^{\mathrm{tr}}\widehat{I}_n^{-1}\mathbb{Z}_n(u,\widehat{\theta}_n)$$

for change point problems, where $\widehat{I}_n$ is a consistent estimator for the Fisher information matrix $I(\theta_*)$. It follows from (10.1) and the continuous mapping theorem that

$$\mathcal{T}_n \stackrel{\mathsf{P}}{\Longrightarrow} \sup_{u\in[0,1]} ||B^\circ(u)||^2 \quad \text{in } \mathbb{R}.$$

Let us call this approach, which was pioneered by the innovative work of Horváth and Parzen (1994), the *Z-process method*[1]. We will present a general theory with the help of asymptotic representation of Z-estimators. We will also prove the consistency of the test under the alternatives in a general way.

To close this section, let us get a preview of deriving the weak convergence (10.1). First, observe the following equation obtained by the Taylor expansion:

$$n^{-1/2}\mathbb{Z}_n(u,\widehat{\theta}_n) = n^{-1/2}\mathbb{Z}_n(u,\theta_*) + (n^{-1}\dot{\mathbb{Z}}_n(u,\widetilde{\theta}_n))(n^{1/2}(\widehat{\theta}_n-\theta_*)),$$

where $\widetilde{\theta}_n$ is a Θ-valued random element conversing in outer-probability to θ_*. Since the asymptotic representation theorem for Z-estimators says that

$$n^{1/2}(\widehat{\theta}_n-\theta_*) = I(\theta_*)^{-1}n^{-1/2}\mathbb{Z}_n(1,\theta_*) + o_{\mathsf{P}}(1),$$

and since the ergodic theorem implies that

$$n^{-1}\dot{\mathbb{Z}}_n(u,\widetilde{\theta}_n) = n^{-1}\dot{\mathbb{Z}}_n(u,\theta_*) + o_{\mathsf{P}}(1) = -uI(\theta_*) + o_{\mathsf{P}}(1),$$

we can continue the above computation as follows:

$$\begin{aligned}
&n^{-1/2}\mathbb{Z}_n(u,\widehat{\theta}_n)\\
&= n^{-1/2}\mathbb{Z}_n(u,\theta_*) + (n^{-1}\dot{\mathbb{Z}}_n(u,\widetilde{\theta}_n))(n^{1/2}(\widehat{\theta}_n-\theta_*))\\
&= n^{-1/2}\mathbb{Z}_n(u,\theta_*) + (-uI(\theta_*))(I(\theta_*)^{-1}n^{-1/2}\mathbb{Z}_n(1,\theta_*)) + o_{\mathsf{P}}(1)\\
&= n^{-1/2}\mathbb{Z}_n(u,\theta_*) - u(n^{-1/2}\mathbb{Z}_n(1,\theta_*)) + o_{\mathsf{P}}(1)\\
&\stackrel{\mathsf{P}}{\Longrightarrow} I(\theta_*)^{1/2}B(u) - u(I(\theta_*)^{1/2}B(1))\\
&= I(\theta_*)^{1/2}(B(u)-uB(1))\\
&\stackrel{d}{=} I(\theta_*)^{1/2}B^\circ(u).
\end{aligned}$$

Hence, the key idea of the approach presented in this chapter is to substitute the result of asymptotic representation of Z-estimators into the Taylor expansion, and then to use the joint convergence of $n^{-1/2}\mathbb{Z}_n(u,\theta_*)$ and $n^{-1/2}\mathbb{Z}_n(1,\theta_*)$. The latter is possible because we have the functional weak convergence of martingales in hands.

[1] Horváth and Parzen (1994) originally gave the name "Fisher-score change process" for the stochastic process $u \rightsquigarrow n^{-1/2}\mathbb{Z}_n(u,\widehat{\theta}_n)$ appearing above. The reason why we introduce a new name here is that what we can treat are not only the gradient vectors of the (true) log-likelihood functions but also general estimating functions.

10.2 Z-Process Method: General Theorem

Let $D[0,1]$ be the Skorokhod space; that is, the space of càdlàg functions defined on $[0,1]$ taking values in a finite-dimensional Euclidean space. We equip this space with either the uniform metric[2] or the Skorokhod metric. Throughout this section, all stochastic processes, denoted such as $u \rightsquigarrow X(u)$, are assumed to take values in $D[0,1]$.

The general set-up for the Z-process method is the following.

(General set-up.) Let Θ be an open, convex subset of $\mathbb{R}^p$. For every $n \in \mathbb{N}$ and $\theta \in \Theta$, let $u \rightsquigarrow \mathbb{Z}_n(u,\theta)$ be an $\mathbb{R}^p$-valued càdlàg process, defined on a probability space $(\Omega^n, \mathcal{F}^n, \mathrm{P}^n)$. We suppose that for every $u \in [0,1]$ and $\omega \in \Omega^n$ the $\mathbb{R}^p$-valued function $\theta \mapsto \mathbb{Z}_n(u,\theta)(\omega)$ is continuously differentiable with the derivative matrix $\dot{\mathbb{Z}}_n(u,\theta)(\omega)$. Introduce three sequences of diagonal, positive definite matrices Q_n, R_n amd S_n, with right orders of growth for our later purposes, such that $S_n - Q_n$ is non-negative definite.

Typically, the rate matrices are given as $Q_n = R_n = \sqrt{n}I_p$ and $S_n = nI_p$, where I_p denotes the $(p \times p)$-identity matrix.

We consider the following testing problem:

(Change Point Problem, CPP.) Test the hypothesis

H_0: the true value $\theta_* \in \Theta$ does not change during $u \in [0,1]$

versus

H_1: there exists a constant $u_\star \in (0,1)$ such that the true value is $\theta_* \in \Theta$ for $u \in [0,u_\star]$, and $\theta_{**} \in \Theta$ for $u \in (u_\star, 1]$, where $\theta_* \neq \theta_{**}$.

Remark. The true value(s) θ_* (and θ_{**}) and the constant $u_\star$ are assumed to be *unknown* to statisticians.

We shall require the following properties for the (deterministic) "limits", namely $z_{\theta_*}(\theta)$ and $\bar{z}(u,\theta)$, of the sequence of random vectors $S_n^{-1}\mathbb{Z}_n(1,\theta)$ and $S_n^{-1}\mathbb{Z}_n(u,\theta)$, under H_0 and H_1, respectively.

Condition 10.2.1 [N] Under H_0, it holds that

$$\sup_{\theta\in\Theta} ||S_n^{-1}\mathbb{Z}_n(1,\theta) - z_{\theta_*}(\theta)|| = o_{\mathrm{P}^n}(1), \tag{10.2}$$

where the (deterministic) limits $z_{\theta_*}(\theta)$ satisfies that

$$\inf_{\theta:||\theta-\theta_*||>\varepsilon} ||z_{\theta_*}(\theta)|| > 0 = ||z_{\theta_*}(\theta_*)||, \quad \forall\varepsilon > 0. \tag{10.3}$$

[2] Recall that $D[0,1]$ (or, sometimes denoted by $D([0,1],\mathbb{R}^p)$) may be regarded as a subset of $\ell^\infty([0,1]\times\{1,...,p\})$.

[A] Under H_1, it holds that

$$\sup_{\theta\in\Theta} ||S_n^{-1}\mathbb{Z}_n(u,\theta) - \bar{z}(u,\theta)|| = o_{\mathsf{P}^n}(1), \quad \forall u \in [0,1], \tag{10.4}$$

where

$$\bar{z}(u,\theta) = (u \wedge u_\star) z_{\theta_*}(\theta) + (u - (u \wedge u_\star)) z_{\theta_{**}}(\theta).$$

Moreover, the limits $\bar{z}(u,\theta)$'s satisfy that there exists a $\theta_\star \in \Theta$ such that

$$\inf_{\theta:||\theta-\theta_\star||>\varepsilon} ||\bar{z}(1,\theta)|| > 0 = ||\bar{z}(1,\theta_\star)||, \quad \forall \varepsilon > 0. \tag{10.5}$$

Remark. The conditions (10.2), (10.3), (10.4) and (10.5) are natural in many cases. Notice here that these conditions imply that

$$\begin{aligned}
\sup_{u\in(0,1)} ||\bar{z}(u,\theta_\star)|| &\geq ||\bar{z}(u_\star,\theta_\star)|| \\
&= ||\bar{z}(u_\star,\theta_\star) - u_\star \bar{z}(1,\theta_\star)|| \\
&= u_\star(1-u_\star)||z_{\theta_*}(\theta_\star) - z_{\theta_{**}}(\theta_\star)||,
\end{aligned}$$

which is *positive*; if this were zero, the condition (10.5) should imply that $z_{\theta_*}(\theta_\star) = z_{\theta_{**}}(\theta_\star) = 0$, which contradicts with $\theta_* \neq \theta_{**}$. This positive value is closely related to the power of our test under H_1.

Now, we prepare a lemma to prove the consistency of a sequence of Z-estimators. This lemma can be proved exactly in the same way as Theorems 8.2.1 and 8.6.1, so the proof is omitted.

Lemma 10.2.2 (i) Under **[N]**, for any sequence of Θ-valued random variables $\widehat{\theta}_n$ such that $S_n^{-1}\mathbb{Z}_n(1,\widehat{\theta}_n) = o_{\mathsf{P}^n}(1)$, it holds that $\widehat{\theta}_n \xrightarrow{\mathsf{P}^n} \theta_*$.

(ii) Under **[A]**, for any sequence of Θ-valued random variables $\widehat{\theta}_n$ such that $S_n^{-1}\mathbb{Z}_n(1,\widehat{\theta}_n) = o_{\mathsf{P}^n}(1)$, it holds that $\widehat{\theta}_n \xrightarrow{\mathsf{P}^n} \theta_\star$.

We are ready to present our main result of this section.

Theorem 10.2.3 Consider the CPP with the general set-up described at the beginning of this section. Let $(\widehat{\theta}_n)_{n=1,2,\ldots}$ be any sequence of Θ-valued random variables such that $Q_n^{-1}\mathbb{Z}_n(1,\widehat{\theta}_n) = o_{\mathsf{P}^n}(1)$. Let $(\widehat{J}_n)_{n=1,2,\ldots}$ be any sequence of consistent estimators for $J(\theta_*)$, under H_0, where $J(\theta_*)$ is a positive definite matrix appearing in (i) below, and each $\widehat{J}_n$ is a $(p\times p)$-matrix valued random variable that are positive definite almost surely. Introduce the test statistic

$$\mathcal{T}_n = \sup_{u\in(0,1]} (Q_n^{-1}\mathbb{Z}_n(u,\widehat{\theta}_n))^{\mathrm{tr}} \widehat{J}_n^{-1} Q_n^{-1}\mathbb{Z}_n(u,\widehat{\theta}_n).$$

(i) Under **[N]**, suppose that there exists a positive definite matrix $J(\theta_*)$ such that, for any random elements $\widetilde{\theta}_n(u)$ on the segment connecting θ_* and $\widehat{\theta}_n$, it holds that

$$\sup_{u\in[0,1]} ||Q_n^{-1}\dot{\mathbb{Z}}_n(u,\widetilde{\theta}_n(u))R_n^{-1} + uJ(\theta_*)|| = o_{\mathsf{P}^{n*}}(1). \tag{10.6}$$

Suppose also that

$$Q_n^{-1}\mathbb{Z}_n(u,\theta_*) \stackrel{\mathsf{P}^n}{\Longrightarrow} J(\theta_*)^{1/2}B(u), \quad \text{in } D[0,1], \tag{10.7}$$

where $u \rightsquigarrow B(u)$ is a vector of independent standard Brownian motions, and that $\widehat{J}_n \stackrel{\mathsf{P}^n}{\longrightarrow} J(\theta_*)$. Then, it holds that

$$\mathcal{T}_n \stackrel{\mathsf{P}^n}{\Longrightarrow} \sup_{u\in[0,1]} ||B(u)-uB(1)||^2 \quad \text{in } \mathbb{R}.$$

Therefore the test is asymptotically distribution free[3].

(ii) Under **[A]**, it holds that

$$\mathcal{T}_n \geq \lambda(S_nQ_n^{-1}\widehat{J}_n^{-1}Q_n^{-1}S_n)\left\{||\bar{z}(u_\star,\theta_\star)||^2 + o_{\mathsf{P}^n}(1)\right\},$$

where $\lambda(A)$ denotes the smallest eigenvalue of the random matrix A. Recalling that $||\bar{z}(u_\star,\theta_\star)|| > 0$, if $\lambda(S_nQ_n^{-1}\widehat{J}_n^{-1}Q_n^{-1}S_n)$ tends to ∞ in probability, then the test is consistent.

Proof. First let us prove (i). It follows the Taylor expansion and the asymptotic representation theorem for Z-estimators (Theorem 8.6.2) that

$$\begin{aligned}
&Q_n^{-1}\mathbb{Z}_n(u,\widehat{\theta}_n)\\
&= Q_n^{-1}\mathbb{Z}_n(u,\theta_*) + Q_n^{-1}\dot{\mathbb{Z}}_n(u,\widetilde{\theta}_n(u))R_n^{-1}R_n(\widehat{\theta}_n-\theta_*)\\
&= Q_n^{-1}\mathbb{Z}_n(u,\theta_*) - (uJ(\theta_*))J(\theta_*)^{-1}Q_n^{-1}\mathbb{Z}_n(1,\theta_*) + e_n(u)\\
&\stackrel{\mathsf{P}^n}{\Longrightarrow} J(\theta_*)^{1/2}(B(u)-uB(1)), \quad \text{in } D[0,1],
\end{aligned}$$

where $\widetilde{\theta}_n(u)$ is a random element on the segment connecting θ_* and $\widehat{\theta}_n$, and the reminder terms $e_n(u)$ satisfy that $\sup_{u\in[0,1]}||e_n(u)|| = o_{\mathsf{P}^{n*}}(1)$. Consequently, the claim (i) follows from the continuous mapping theorem.

The inequality in (ii) is proved as follows:

$$\begin{aligned}
\mathcal{T}_n &\geq (Q_n^{-1}\mathbb{Z}_n(u_\star,\widehat{\theta}_n))^{\mathrm{tr}}\widehat{J}_n^{-1}Q_n^{-1}\mathbb{Z}_n(u_\star,\widehat{\theta}_n)\\
&= (S_n^{-1}\mathbb{Z}_n(u_\star,\widehat{\theta}_n))^{\mathrm{tr}}(S_nQ_n^{-1}\widehat{J}_n^{-1}Q_n^{-1}S_n)S_n^{-1}\mathbb{Z}_n(u_\star,\widehat{\theta}_n)\\
&\geq \lambda(S_nQ_n^{-1}\widehat{J}_n^{-1}Q_n^{-1}S_n)||S_n^{-1}\mathbb{Z}_n(u_\star,\widehat{\theta}_n)||^2\\
&= \lambda(S_nQ_n^{-1}\widehat{J}_n^{-1}Q_n^{-1}S_n)\left\{||\bar{z}(u_\star,\theta_\star)||^2 + o_{\mathsf{P}^n}(1)\right\}.
\end{aligned}$$

All claims have been proved. □

[3]The distribution of the limit, namely $\sup_{u\in[0,1]}||B(u)-uB(1)||^2 = \sup_{u\in[0,1]}||B^\circ(u)||^2$, where $u \rightsquigarrow B^\circ(u)$ is the vector of independent standard Brownian bridges, can be computed analytically, at least when $p=1$.

10.3 Examples

10.3.1 Rigorous arguments for independent random sequences

In this subsection, we shall give some rigorous arguments concerning the change point problem for independent random sequences discussed intuitively in Section 10.1.

Let $(\mathcal{X},\mathcal{A},\mu)$ be a σ-finite measure space. Let $X_1,X_2,...$ be an independent sequence of $\mathcal{X}$-valued random variables. As candidates for the true densities of the distribution of the random sequence, let us consider a family $\{p(\cdot;\theta);\theta\in\Theta\}$ of probability densities with respect to μ indexed by elements θ of an open, convex subset Θ of $\mathbb{R}^p$.

Let us consider the CPP with the set-up given below. Assuming that the densities $p(x;\theta)$ are differentiable with respect to θ on Θ, we introduce the estimating function, which is the first derivatives of the log-likelihood function based on the data $X_1,...,X_n$, given by

$$\mathbb{Z}_n(\theta)=\sum_{k=1}^{n}(\partial_1\log p(X_k;\theta),...,\partial_p\log p(X_k;\theta))^{\mathrm{tr}}.$$

Here and in the sequel, the derivatives with respect to θ are denoted such as $\partial_i=\frac{\partial}{\partial\theta_i}$, $\partial_{i,j}=\frac{\partial^2}{\partial\theta_i\partial\theta_j}$.

Discussion 10.3.1 (The null hypothesis H_0) Since our approach is based on the asymptotic representation of the MLEs as a special case of Z-estimators, which are defined as any random variables $\widehat{\theta}_n$ satisfying

$$n^{-1}\mathbb{Z}_n(1,\widehat{\theta}_n)=o_{\mathrm{P}}(1),$$

we assume all the conditions, including that the Fisher information matrix $I(\theta_*)=(I^{i,j}(\theta_*)_{(i,j)\in\{1,...,p\}^2}$, where

$$I^{i,j}(\theta_*)=\int_{\mathcal{X}}\frac{\partial_i p(x;\theta_*)\partial_j p(x;\theta_*)}{p(x;\theta_*)}\mu(dx),\quad (i,j)\in\{1,...,p\}^2,$$

is positive definite (see Subsection 8.3.1 for the other conditions), to ensure that

$$\begin{aligned}\sqrt{n}(\widehat{\theta}_n-\theta_*) &= I(\theta_*)^{-1}(n^{-1/2}\mathbb{Z}_n(1,\theta_*))+o_{\mathrm{P}}(1)\\ &\stackrel{\mathrm{P}}{\Longrightarrow} I(\theta_*)^{-1}\mathcal{N}_p(0,I(\theta_*))\quad\text{in }\mathbb{R}^p\\ &\stackrel{d}{=} \mathcal{N}_p(0,I(\theta_*)^{-1}).\end{aligned}$$

We remark that the condition **[N]** in Section 10.2 was imposed for the consistency of Z-estimators under H_0, and that it is included in the conditions in Subsection 8.3.1. It is natural to introduce $\widehat{I}_n = (\widehat{I}_n^{i,j})_{(i,j)\in\{1,...,p\}^2}$ given by

$$\widehat{I}_n^{i,j} = \frac{1}{n}\sum_{k=1}^{n} \frac{\partial_i p(X_k;\widehat{\theta}_n)\partial_j p(X_k;\widehat{\theta}_n)}{p(X_k;\widehat{\theta}_n)^2}, \quad (i,j)\in\{1,...,p\}^2,$$

as a consistent estimator for $I(\theta_*)$ under H_0.

The condition (10.7) in Theorem 10.2.3 is immediate from Donsker's theorem (Corollary 7.2.4):

$$n^{-1/2}\mathbb{Z}_n(u,\theta_*) \xrightarrow{\mathsf{P}} I(\theta_*)^{1/2}B(u), \quad \text{in } D[0,1].$$

Finally, let us check the condition (10.6). Since we have assumed (8.8), it holds that

$$\begin{aligned}
&\sup_{u\in[0,1]} |n^{-1}\dot{\mathbb{Z}}_n^{i,j}(u,\widetilde{\theta}_n(u)) - u(-I^{i,j}(\theta_*))| \\
\leq\ &\sup_{u\in[0,1]} |n^{-1}\dot{\mathbb{Z}}_n^{i,j}(u,\theta_*) + uI^{i,j}(\theta_*)| + \frac{1}{n}\sum_{k=1}^{n} K(X_k)\cdot||\widehat{\theta}_n - \theta_*||^{\alpha},
\end{aligned}$$

on the set $\bigcap_{u\in[0,1]}\{\widetilde{\theta}_n(u)\in N\} \supset \{\widehat{\theta}_n \in N\}$, where N is a neighborhood of θ_* appearing in (8.8). The first term on the right-hand side converges in probability to zero by Theorem 7.3.4, while the second term is $O_{\mathsf{P}}(1)\cdot o_{\mathsf{P}}(1) = o_{\mathsf{P}}(1)$. Since $\mathsf{P}(\widehat{\theta}_n \in N^c) = o(1)$, the condition (10.6) is satisfied. All the conditions for Theorem 10.2.3 (i) have been established.

Discussion 10.3.2 (The alternative H_1) Recall that we have proved $||\bar{z}(u_\star,\theta_\star)|| > 0$, and notice that the matrix $S_n Q_n^{-1}\widehat{V}_n^{-1}Q_n^{-1}S_n$ is reduced to $n\widehat{I}_n^{-1}$, where

$$\begin{aligned}
\widehat{I}_n^{i,j} &= \frac{1}{n}\sum_{k=1}^{n} \frac{\partial_i p(X_k;\widehat{\theta}_n)\partial_j p(X_k;\widehat{\theta}_n)}{p(X_k;\widehat{\theta}_n)^2} \\
&\xrightarrow{\mathsf{P}} \int_{\mathcal{X}} \frac{\partial_i p(x;\theta_\star)\partial_j p(x;\theta_\star)}{p(x;\theta_\star)^2}\{u_\star p(x;\theta_*) + (1-u_\star)p(x;\theta_{**})\}\mu(dx) \\
&=: I^{i,j}(\theta_\star).
\end{aligned}$$

If we assume that the limit matrix $I(\theta_\star) = (I^{i,j}(\theta_\star))_{(i,j)\in\{1,...,p\}^2}$ is positive definite, we actually have $\lambda(n\widehat{I}_n^{-1}) = n\{\lambda(I(\theta_\star)^{-1}) + o_{\mathsf{P}}(1)\} \to \infty$ in probability. Therefore, the test is consistent.

10.3.2 Markov chain models

Let $(\mathcal{X},\mathcal{A},\mu)$ be a σ-finite measure space. Let $X_0,X_1,X_2,...$ be a Markov chain with the state space $\mathcal{X}$, the initial density q, and the parametric family of transition density $\{p(\cdot,\cdot;\theta);\theta\in\Theta\}$, where Θ is an open, convex subset of $\mathbb{R}^p$, defined on a parametric family of probability space $(\Omega,\mathcal{F},\{\mathsf{P}_\theta;\theta\in\Theta\})$; that is, for any $\theta\in\Theta$,

$$\mathsf{P}_\theta(X_0\in A) = \int_A q(x)\mu(dx), \quad \forall A\in\mathcal{A},$$

$$\mathsf{P}_\theta(X_k \in A|X_{k-1} = x) = \int_A p(y,x;\theta)\mu(dy), \quad \forall A \in \mathcal{A}, \quad \forall x \in \mathcal{X}, \quad k = 1,2,....$$

Suppose that, under the probability measure P_θ, the Markov chain is ergodic with the invariant measure P_θ°; then, for any P_θ°-integrable function f on $\mathcal{X}$, it holds that

$$\frac{1}{n}\sum_{k=1}^{n} f(X_k) \xrightarrow{\mathsf{P}_\theta} \int_{\mathcal{X}} f(x)P_\theta^\circ(dx).$$

Let us consider the CPP with the set-up given below. Assuming that the function $\theta \mapsto p(y,x;\theta)$ is two times continuously differentiable, we introduce the sequence of Z-processes $\mathbb{Z}_n(u,\theta)$ and its derivative matrices $\dot{\mathbb{Z}}_n(u,\theta)$ by

$$\mathbb{Z}_n^i(u,\theta) = \sum_{k=1}^{[un]} G^i(X_k, X_{k-1};\theta), \quad i = 1,...,p,$$

and

$$\dot{\mathbb{Z}}_n^{i,j}(u,\theta) = \sum_{k=1}^{[un]} H^{i,j}(X_k, X_{k-1};\theta), \quad (i,j) \in \{1,...,p\}^2,$$

where

$$G^i(y,x;\theta) = \frac{\partial_i p(y,x;\theta)}{p(y,x;\theta)}$$

and

$$H^{i,j}(y,x;\theta) = \frac{p(y,x;\theta)\partial_{i,j}p(y,x;\theta) - \partial_i p(y,x;\theta)\partial_j p(y,x;\theta)}{p(y,x;\theta)^2},$$

with the notations $\partial_i = \frac{\partial}{\partial\theta_i}$ and $\partial_{i,j} = \frac{\partial^2}{\partial\theta_i\partial\theta_j}$.

Let us start our discussion under the null hypothesis H_0. Again, our approach is based on the asymptotic representation of the MLEs as a special case of Z-estimators, which are defined as any random variables $\widehat{\theta}_n$ satisfying

$$n^{-1}\mathbb{Z}_n(1,\widehat{\theta}_n) = o_{\mathsf{P}}(1);$$

thus we assume all the conditions, including that the Fisher information matrix $I(\theta_*) = (I^{i,j}(\theta_*))_{(i,j)\in\{1,...,p\}^2}$, where

$$I^{i,j}(\theta_*) = \int_{\mathcal{X}}\int_{\mathcal{X}} \frac{\partial_i p(y,x;\theta_*)\partial_j p(y,x;\theta_*)}{p(y,x;\theta_*)}\mu(dy)P_{\theta_*}^\circ(dx), \quad (i,j) \in \{1,...,p\}^2,$$

is positive definite (see Section 8.3.2 for the other conditions), to ensure that

$$\begin{aligned}\sqrt{n}(\widehat{\theta}_n - \theta_*) &= I(\theta_*)^{-1}(n^{-1/2}\mathbb{Z}_n(1,\theta_*)) + o_{\mathsf{P}}(1)\\ &\xrightarrow{\mathsf{P}} I(\theta_*)^{-1}\mathcal{N}_p(0, I(\theta_*)) \quad \text{in } \mathbb{R}^p\\ &\stackrel{d}{=} \mathcal{N}_p(0, I(\theta_*)^{-1}).\end{aligned}$$

It is natural to introduce $\widehat{I}_n = (\widehat{I}_n^{i,j})_{(i,j)\in\{1,\dots,p\}^2}$ given by

$$\widehat{I}_n^{i,j} = \frac{1}{n}\sum_{k=1}^{n} \frac{\partial_i p(X_k,X_{k-1};\widehat{\theta}_n)\partial_j p(X_k,X_{k-1};\widehat{\theta}_n)}{p(X_k,X_{k-1};\widehat{\theta}_n)^2}, \quad (i,j)\in\{1,\dots,p\}^2,$$

as a consistent estimator for $I(\theta_*)$ under H_0.

The condition (10.7) in Theorem 10.2.3 is immediate from the functional CLT for discrete-time martingales (Theorem 7.2.3):

$$n^{-1/2}\mathbb{Z}_n(u,\theta_*) \xrightarrow{\mathsf{P}} I(\theta_*)^{1/2}B(u), \quad \text{in } D[0,1].$$

Finally, let us check the condition (10.6). As in Subsection 8.3.2, we assume that there exist a constant $\alpha\in(0,1]$ and a measurable function $\widehat{H}(y,x)$ that is integrable with respect to $p(y,x;\theta_*)\mu(dy)P_{\theta_*}^{\circ}$ such that

$$||H(y,x;\theta)-H(y,x;\theta')|| \le \widehat{H}(y,x)||\theta-\theta'||^{\alpha}, \quad \forall\theta,\theta'\in\Theta^4.$$

Thus, it holds that

$$\begin{aligned}
&\sup_{u\in[0,1]} |n^{-1}\dot{\mathbb{Z}}_n^{i,j}(u,\widetilde{\theta}_n)+uI^{i,j}(\theta_*)| \\
&\le \sup_{u\in[0,1]} |n^{-1}\dot{\mathbb{Z}}_n^{i,j}(u,\theta_*)+uI^{i,j}(\theta_*)| + \frac{1}{n}\sum_{k=1}^{n}\widehat{H}(X_k,X_{k-1})\cdot||\widehat{\theta}_n-\theta_*||^{\alpha}.
\end{aligned}$$

The second term on the right-hand side is $O_{\mathsf{P}}(1)\cdot o_{\mathsf{P}}(1)=o_{\mathsf{P}}(1)$ (apply Exercise 6.6.2 (ii) to deduce that $n^{-1}\sum_{k=1}^{n}\widehat{H}(X_k,X_{k-1})=O_{\mathsf{P}}(1)$). Regarding the first term, observe first that

$$\begin{aligned}
n^{-1}\dot{\mathbb{Z}}_n^{i,j}(u,\theta_*) &= \frac{1}{n}\sum_{k=1}^{[un]}\{H^{i,j}(X_k,X_{k-1};\theta_*)-\mathsf{E}_{\theta_*}[H^{i,j}(X_k,X_{k-1};\theta_*)|\mathcal{F}_{k-1}]\} \\
&\quad + \frac{1}{n}\sum_{k=1}^{[un]}\mathsf{E}_{\theta_*}[H^{i,j}(X_k,X_{k-1};\theta_*)|\mathcal{F}_{k-1}].
\end{aligned}$$

The supremum with respect to $u\in[0,1]$ of the absolute value of the first term on the right-hand side converges in probability to zero by the corollary to Lenglart's inequality (Corollary 4.3.3) if we assume that $\mathsf{E}_{\theta_*}[(H^{i,j}(X_k,X_{k-1}))^2]<\infty$ for every k. So we have that

$$\begin{aligned}
&\sup_{u\in[0,1]} |n^{-1}\dot{\mathbb{Z}}_n^{i,j}(u,\theta_*)+uI^{i,j}(\theta_*)| \\
&\le \sup_{u\in[0,1]} \left|\frac{1}{n}\sum_{k=1}^{[un]}\widetilde{H}^{i,j}(X_{k-1}) - u\int_{\mathcal{X}}\widetilde{H}^{i,j}(x)P_{\theta_*}^{\circ}(dx)\right| + o_{\mathsf{P}}(1),
\end{aligned}$$

[4] This assumption may be replaced by that for "$\theta,\theta'\in N$, where N is a neighborhood of θ_*".

where $\widetilde{H}^{i,j}(x) = \int_{\mathcal{X}} H^{i,j}(y,x;\theta_*)p(y,x;\theta_*)\mu(dy)$, and the right-hand side converges in probability to zero by Theorem 7.3.4. Hence, the condition (10.6) is satisfied. All the conditions for Theorem 10.2.3 (i) have been established.

The argument to prove the consistency of the test under H_1 is similar to that for the independent random sequences in Subsection 10.3.1, and it is omitted.

10.3.3 Final exercises: three models of ergodic diffusions

Consider a 1-dimensional diffusion process $t \rightsquigarrow X_t$ which is the unique strong solution to the stochastic differential equation

$$X_t = X_0 + \int_0^t \beta(X_s;\theta)ds + \int_0^t \sigma(X_s;\theta)dW_s,$$

where $s \rightsquigarrow W_s$ is a standard Wiener process, and the drift and/or diffusion coefficients involve unknown parameter θ of interest, which is from a set $\Theta \subset \mathbb{R}^p$ for an integer $p \geq 1$.

Let us suppose that, the diffusion process $t \rightsquigarrow X_t$ is ergodic under P_θ with the invariant measure P_θ°; then, it holds for any P_θ°-measurable function f and any $0 \leq u < v \leq 1$ that

$$\begin{aligned} \frac{1}{T}\int_{uT}^{vT} f(X_t)dt &= v \cdot \frac{1}{vT}\int_0^{vT} f(X_t)dt - u \cdot \frac{1}{uT}\int_0^{uT} f(X_t)dt \\ &\xrightarrow{\mathsf{P}_\theta} (v-u)\int_{\mathbb{R}} f(x)P_\theta^\circ(dx), \quad \text{as } T \to \infty. \end{aligned}$$

Suppose that, we are able to observe the diffusion process $t \rightsquigarrow X_t$ at discrete time grids $0 = t_0^n < t_1^n < \cdots < t_n^n$, and we shall consider the sampling scheme

$$t_n^n \to \infty, \quad n\Delta_n^2 \to 0 \quad \text{and} \quad \sum_{k=1}^n \left| \frac{|t_k^n - t_{k-1}^n|}{t_n^n} - \frac{1}{n} \right| \to 0,$$

as $n \to \infty$, where $\Delta_n = \max_{1 \leq k \leq n} |t_k^n - t_{k-1}^n|$.

Let us consider the CPPs for *three kinds of diffusion models* given in Exercises 10.3.1, 10.3.2 and 10.3.3 given below.

In order to construct a test statistic for each model, we shall first introduce an appropriate partial sum process of contrast function $\mathbb{M}_n(u,\theta)$ and its gradient vector $\mathbb{Z}_n(u,\theta) = (\partial_1\mathbb{M}_n(u,\theta), \ldots, \partial_p\mathbb{M}_n(u,\theta))^{\mathrm{tr}}$. Next find some appropriate sequences of $(p \times p)$-diagonal, positive definite matrices $Q_n (= R_n)$ and S_n, and a $(p \times p)$-matrix $J(\theta_*)$ which is positive definite, such that

$$u \rightsquigarrow Q_n^{-1}\mathbb{Z}_n(u,\theta_*) \quad \text{converges weakly in } D[0,1] \text{ to} \quad u \rightsquigarrow J(\theta_*)^{1/2}B^\circ(u),$$

where $u \rightsquigarrow B(u)$ is a standard Brownian motion. To estimate the matrix $J(\theta_*)$, introduce an appropriate estimator $\widehat{J}_n$ which is positive definite almost surely, and then propose the statistic

$$\mathcal{T}_n = \sup_{u \in [0,1]} (Q_n^{-1}\mathbb{Z}_n(u,\widehat{\theta}_n))^{\mathrm{tr}} \widehat{J}_n^{-1} Q_n^{-1}\mathbb{Z}_n(u,\widehat{\theta}_n), \tag{10.8}$$

where $\widehat{\theta}_n$ is any Θ-valued random variable such that

$$Q_n^{-1}\mathbb{Z}_n(1,\widehat{\theta}_n) = o_{\mathsf{P}}(1) \quad \text{and} \quad \widehat{\theta}_n \xrightarrow{\mathsf{P}} \theta_* \quad \text{under } H_0.$$

For the discussions in the following exercises, you may assume some (reasonable) conditions, including:

- $\theta \mapsto \beta(x;\theta)$ and $\theta \mapsto \sigma(x;\theta)$ are three times continuously differentiable with derivatives which are, as a function of x, bounded by a function of polynomial growth;
- $\inf_{x\in\mathbb{R}} \sigma(x;\theta) > 0$ for any $\theta \in \Theta$;
- $\sup_{t\in[0,\infty)} \mathsf{E}_\theta[|X_t|^q] < \infty$ for any $q \geq 1$ and any $\theta \in \Theta$.

Exercise 10.3.1 Consider the model discussed in Subsection 8.5.2; that is, the drift coefficient involves the unknown parameter of interest, say $\beta(x;\theta)$, but the diffusion coefficient is assumed to be a *known* function, say $\sigma(x)$. To test the hypotheses H_0 versus H_1, consider the partial sum processes of contrast functions

$$\mathbb{M}_n(u,\theta) = -\sum_{k:t_{k-1}^n \leq ut_n^n} \frac{(X_{t_k^n} - X_{t_{k-1}^n} - \beta(X_{t_{k-1}^n};\theta)|t_k^n - t_{k-1}^n|)^2}{2\sigma(X_{t_{k-1}^n})^2|t_k^n - t_{k-1}^n|}.$$

Construct the corresponding test statistic $\mathcal{T}_n$ by (10.8) and derive its asymptotic properties under H_0 and H_1, following the next steps.

(i) Find some appropriate rate matrices $Q_n (= R_n)$ and S_n.

(ii) Find the matrix $J(\theta_*)$, and construct an appropriate estimator $\widehat{J}_n$ for $J(\theta_*)$.

(iii) Under H_0, prove that $\mathcal{T}_n \overset{\mathsf{P}_{\theta_*}}{\Longrightarrow} \sup_{u\in[0,1]} ||B^\circ(u)||^2$ in $\mathbb{R}$, where $u \rightsquigarrow B^\circ(u)$ is a p-dimensional vector of independent, standard Brownian bridges.

(iv) Prove the consistency of the test under H_1.

Exercise 10.3.2 Consider the model discussed in Subsection 8.5.3; that is, the diffusion coefficient involves the unknown parameter of interest, say $\sigma(x;\theta)$, and the drift coefficient $\beta(x)$ is regarded as an unknown *nuisance function*. To test the hypotheses H_0 versus H_1, consider the partial sum processes of contrast functions

$$\mathbb{M}_n(u,\theta) = -\sum_{k:t_{k-1}^n \leq ut_n^n} \left\{ \log\sigma(X_{t_{k-1}^n};\theta) + \frac{(X_{t_k^n} - X_{t_{k-1}^n})^2}{2\sigma(X_{t_{k-1}^n};\theta)^2|t_k^n - t_{k-1}^n|} \right\}.$$

Construct the corresponding test statistic $\mathcal{T}_n$ by (10.8) and derive its asymptotic properties under H_0 and H_1, following the steps as in Exercise 10.3.1.

Exercise 10.3.3 Consider the model discussed in Subsection 8.7.1; that is, both the drift and diffusion coefficients involve 1-dimensional unknown parameter of interest, respectively, say $\beta(x;\theta_1)$ and $\sigma(x;\theta_2)$, where $\theta = (\theta_1,\theta_2)^{\mathrm{tr}}$ is from a subset of $\mathbb{R}^2$.

To test the hypotheses H_0 versus H_1, consider the partial sum processes of contrast functions

$$\mathbb{M}_n(u,\theta) = - \sum_{k:t_{k-1}^n \le ut_n^n} \left\{ \log\sigma(X_{t_{k-1}^n};\theta_2) + \frac{(X_{t_k^n} - X_{t_{k-1}^n} - \beta(X_{t_{k-1}^n};\theta_1)|t_k^n - t_{k-1}^n|)^2}{2\sigma(X_{t_{k-1}^n};\theta_2)^2|t_k^n - t_{k-1}^n|} \right\}.$$

Construct the corresponding test statistic $\mathcal{T}_n$ by (10.8) and derive its asymptotic properties under H_0 and H_1, following the steps as in Exercise 10.3.1 with $p = 2$.

Part A

Appendices

A1

Supplements

The first section of this chapter is devoted to providing some (hopefully) useful tools for the $p \gg n$ problems (i.e., the so-called "short, fat data" problems) appearing in high-dimensional statistical models.

First, a new inequality, which may be called *stochastic maximal inequality*, for the maxima of finite-dimensional martingales is presented. This naming comes from the fact that the both sides of the inequality are some stochastic processes, where a reminder term of a local martingale starting from zero is added to the right-hand side; any deterministic values of expectations or probabilities are not involved.

Next, the new inequality is applied to obtain various maximal inequalities, where the both sides consist of some deterministic values in terms of expectations and/or probabilities. Those inequalities may possibly bring us an alternative approach to high-dimensional statistics taking a route somewhat different from the ones based on the classical inequalities of Bernstein, Hoeffding, and others.

The second section of this chapter serves some supplementary tools for the main text.

A1.1 A Stochastic Maximal Inequality and Its Applications

It has been well understood that obtaining a good bound for

$$\mathsf{E}\left[\max_{1\leq i\leq p}|X_i|\right]$$

is one of the essential parts in high-dimensional statistics. Regarding this issue, the so-called maximal inequalities based on Orlicz norms, which are well explained in Chapter 2.2 of van der Vaart and Wellner (1996), have been playing a key role to derive some "oracle inequalities". As one of the fruits of such approaches, the following result has already been well-known: by combining Bernsten's inequality (Theorem 6.6.5) with Lemma 2.2.10 of van der Vaart and Wellner (1996), it holds for any locally square-integrable martingales $M^1,...,M^p$ satisfying $\max_{1\leq i\leq p}|\Delta M^i|\leq a$ for a constant $a\geq 0$ and any stopping time T that

$$\left(\mathsf{E}\left[\max_{1\leq i\leq p}\sup_{t\in[0,T]}|M_t^i|^q 1\left\{\max_{1\leq i\leq p}\langle M^i\rangle_T\leq\sigma^2\right\}\right]\right)^{1/q}$$
$$\leq K_q\left(a\log(1+p)+\sigma\sqrt{\log(1+p)}\right),\quad \forall\sigma>0,\quad \forall q\geq 1,$$

DOI: 10.1201/9781315117768-A1

where $K_q > 0$ is a constant depending only on q.

The assumption that $\max_i |\Delta M^i| \leq a$ for a constant $a \geq 0$ is sometimes an untractable restriction, but it can be replaced by a higher order moment condition on $\max_i |\Delta M^i|$ if we use the inequality of van de Geer (1995, 2000) instead of the (usual) Bernstein inequality. On the other hand, this section provides an alternative approach to this problem.

A1.1.1 Continuous-time case

Throughout this subsection, we shall consider a p-dimensional local martingale $M = (M^1, ..., M^p)^{\text{tr}}$ starting from zero, defined on a stochastic basis $\mathbf{B}$, such that

$$\langle M^{i,c}, M^{j,c} \rangle_t = \int_0^t c_s^{i,j} ds, \quad \forall t \in [0, \infty), \quad (i,j) \in \{1, ..., p\}^2,$$

where $M^{i,c}$ denotes the continuous martingale part of M^i. We introduce the notation

$$[M]_t = J_t + \int_0^t \max_{1 \leq i \leq p} c_s^{i,i} ds, \quad \text{where} \quad J_t = \sum_{s \leq t} \max_{1 \leq i \leq p} (\Delta M_s^i)^2, \quad \forall t \in [0, \infty).$$

In particular, when M is a p-dimensional locally square-integrable martingale, we also use the notation

$$\langle M \rangle_t = J_t^p + \int_0^t \max_{1 \leq i \leq p} c_s^{i,i} ds, \quad \forall t \in [0, \infty),$$

where J^p denotes the predictable compensator for the locally integrable, increasing process J. These notations[1] $[M]$ and $\langle M \rangle$ coincide with the standard ones when $p = 1$.

In the sequel, let a stochastic basis $\mathbf{B}$, on which M and other objects are defined, be given (except the last corollary, where a sequence of stochastic bases $\mathbf{B}^n$ should be introduced).

Lemma A1.1.1 (Stochastic maximal inequality) Let p be any positive integer.

(i) When M is a p-dimensional local martingale starting from zero, for any constant $\sigma > 0$ there exists a (1-dimensional) local martingale M' starting from zero such that

$$\max_{1 \leq i \leq p} |M_t^i| \leq \left(2\sigma + \frac{[M]_t}{\sigma} \right) \sqrt{\log(1+p)} + M_t', \quad \forall t \in [0, \infty), \quad \text{a.s.}$$

(ii) When M is a p-dimensional locally square-integrable martingale starting from zero, for any constant $\sigma > 0$ there exists a (1-dimensional) local martingale M'' starting from zero such that

$$\max_{1 \leq i \leq p} |M_t^i| \leq \left(2\sigma + \frac{\langle M \rangle_t}{\sigma} \right) \sqrt{\log(1+p)} + M_t'', \quad \forall t \in [0, \infty), \quad \text{a.s.}$$

[1] Some authors might have used the notations $[M]$ and $\langle M \rangle$ for the two matrix valued stochastic processes $([M^i, M^j])_{(i,j) \in \{1,...,p\}^2}$ and $(\langle M^i, M^j \rangle)_{(i,j) \in \{1,...,p\}^2}$, respectively. However, since these notations for the matrices are not used in this monograph, there would be no danger of confusion.

Proof. Introducing the degenerate local martingale $M^0 \equiv 0$, we shall consider the $(p+1)$-dimensional local martingale $\widehat{M} = (M^0, M^1, ..., M^p)^{\mathrm{tr}}$, for which it holds that[2]

$$\max_{0\le i\le p} M_t^i \ge 0, \quad \forall t \in [0,\infty).$$

Note that the process $[\widehat{M}]$ for the augmented $(p+1)$-dimensional local martingale $\widehat{M}$ coincides with the process $[M]$ for the original p-dimensional local martingale M.

For any constant $a > 0$, it follows from Itô's formula that

$$\begin{aligned}
&\frac{\max_{0\le i\le p} M_t^i}{a} \\
&\le \log\left(\sum_{i=0}^{p} \exp(M_t^i/a)\right) \\
&= \log(1+p) + \sum_{i=0}^{p} \int_0^t \frac{\exp(M_{s-}^i/a)}{a\sum_{\iota=0}^{p}\exp(M_{s-}^\iota/a)} dM_s^i \\
&\quad + \frac{1}{2}\sum_{i=0}^{p}\int_0^t \frac{\exp(M_{s-}^i/a)}{a^2\sum_{\iota=0}^{p}\exp(M_{s-}^\iota/a)} c_s^{i,i} ds \\
&\quad - \frac{1}{2}\sum_{i=0}^{p}\sum_{j=0}^{p}\int_0^t \frac{\exp(M_{s-}^i/a)\exp(M_{s-}^j/a)}{a^2(\sum_{\iota=0}^{p}\exp(M_{s-}^\iota/a))^2} c_s^{i,j} ds \\
&\quad + \frac{1}{2}\sum_{i=0}^{p}\sum_{s\le t} \frac{\exp(\widetilde{M}_s^i/a)}{a^2\sum_{\iota=0}^{p}\exp(\widetilde{M}_s^\iota/a)} (\Delta M_s^i)^2 \\
&\quad - \frac{1}{2}\sum_{i=0}^{p}\sum_{j=0}^{p}\sum_{s\le t} \frac{\exp(\widetilde{M}_s^i/a)\exp(\widetilde{M}_s^j/a)}{a^2(\sum_{\iota=0}^{p}\exp(\widetilde{M}_s^\iota/a))^2} \Delta M_s^i \Delta M_s^j \\
&= \log(1+p) + M_t^{(1)} \\
&\quad + \frac{1}{2}\sum_{i=0}^{p}\int_0^t \frac{v_{s-}^i}{a^2\sum_{\iota=0}^{p} v_{s-}^\iota} c_s^{i,i} ds - \frac{1}{2}\int_0^t \frac{v_{s-}^{\mathrm{tr}} c_s v_{s-}}{a^2(\sum_{\iota=0}^{p} v_{s-}^\iota)^2} ds \\
&\quad + \frac{1}{2}\sum_{i=0}^{p}\sum_{s\le t} \frac{\widetilde{v}_s^i}{a^2\sum_{\iota=0}^{p}\widetilde{v}_s^\iota}(\Delta M_s^i)^2 - \frac{1}{2}\sum_{s\le t}\frac{(\sum_{i=0}^{p}\widetilde{v}_s^i \Delta M_s^i)^2}{a^2(\sum_{\iota=0}^{p}\widetilde{v}_s^\iota)^2} \\
&\le \log(1+p) + \frac{1}{2}\frac{[\widehat{M}]_t}{a^2} + M_t^{(1)} \\
&= \log(1+p) + \frac{1}{2}\frac{[M]_t}{a^2} + M_t^{(1)},
\end{aligned}$$

where each $\widetilde{M}_s^i$ is a point on the segment connecting M_{s-}^i and M_s^i,

$$M^{(1)} = \sum_{i=0}^{p}\int_0^{\cdot} \frac{\exp(M_{s-}^i/a)}{a\sum_{\iota=0}^{p}\exp(M_{s-}^\iota/a)} dM_s^i$$

---

[2]Our argument is based on an inequality of the form "$\max_i |M_t^i| \le \max_i M_t^i + \max_i(-M_t^i)$", which is true if both of the two terms on the right-hand side are non-negative. This is why we consider the augmented $(p+1)$-dimensional local martingale $\widehat{M}$.

is a local martingale starting from zero, the matrix $c_s = (c_s^{i,j})_{(i,j)\in\{0,1,...,p\}^2}$ is non-negative definite,

$$v_{s-} = (v_{s-}^0, v_{s-}^1, ..., v_{s-}^p)^{\text{tr}} = (\exp(M_{s-}^0/a), \exp(M_{s-}^1/a), ..., \exp(M_{s-}^p/a))^{\text{tr}},$$

and

$$\widetilde{v}_s = (\widetilde{v}_s^0, \widetilde{v}_s^1, ..., \widetilde{v}_s^p)^{\text{tr}} = (\exp(\widetilde{M}_s^0/a), \exp(\widetilde{M}_s^1/a), ..., \exp(\widetilde{M}_s^p/a))^{\text{tr}}.$$

Since a similar inequality holds also for $-\widehat{M}$, we have that

$$\begin{aligned}\frac{\max_{1\le i\le p}|M_t^i|}{a} &\le \frac{\max_{0\le i\le p} M_t^i + \max_{0\le i\le p}(-M_t^i)}{a} \\ &\le 2\log(1+p) + \frac{[M]_t}{a^2} + M_t^{(2)},\end{aligned}$$

where $M^{(2)}$ is a local martingale starting from zero. By setting $a = \sigma/\sqrt{\log(1+p)}$, we obtain that

$$\begin{aligned}\max_{1\le i\le p}|M_t^i| &\le 2a\log(1+p) + \frac{[M]_t}{a} + aM_t^{(2)} \\ &= \left(2\sigma + \frac{[M]_t}{\sigma}\right)\sqrt{\log(1+p)} + aM_t^{(2)}.\end{aligned}$$

The proof of (i) is finished.

The claim (ii) is immediate from (i), because $[M] - \langle M\rangle$ is a local martingale. □

Theorem A1.1.2 (Maximal inequality) Let p be any positive integer.

(i) When M is a p-dimensional local martingale starting from zero, it holds for any finite stopping time T that

$$\mathsf{E}\left[\max_{1\le i\le p}|M_T^i|\right] \le 2\sqrt{2}\sqrt{\mathsf{E}[[M]_T]}\sqrt{\log(1+p)}.$$

(ii) When M is a p-dimensional locally square-integrable martingale starting from zero, it holds for any finite stopping time T that

$$\mathsf{E}\left[\max_{1\le i\le p}|M_T^i|\right] \le 2\sqrt{2}\sqrt{\mathsf{E}[\langle M\rangle_T]}\sqrt{\log(1+p)}.$$

Proof. We shall apply Lemma A1.1.1 (i); introduce a localizing sequence (T_n) of stopping times for which $M_\cdot^{\prime,T_n} = M'_{\cdot\wedge T_n}$ is a uniformly integrable martingale. Put $\sigma = \sqrt{\mathsf{E}[[M]_T]/2}$, which may be assumed to be finite; otherwise, the desired inequality is trivial. When a finite stopping time T is given, apply the optional sampling theorem to the bounded stopping time $T\wedge n$ to deduce that

$$\begin{aligned}\mathsf{E}\left[\max_{1\le i\le p}|M_{T\wedge n}^{i,T_n}|\right] &\le \left(2\sigma + \frac{\mathsf{E}[[M]_{(T\wedge n)\wedge T_n}]}{\sigma}\right)\sqrt{\log(1+p)} \\ &\le 2\sqrt{2}\sqrt{\mathsf{E}[[M]_T]}\sqrt{\log(1+p)}.\end{aligned}$$

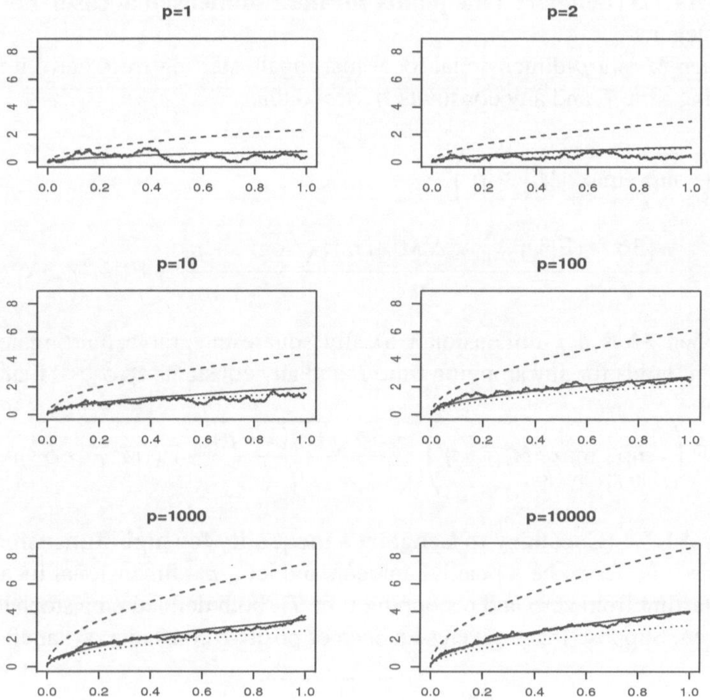

FIGURE A1.1
A path of $t \rightsquigarrow \max_{1\le i\le p}|W_t^i|$, where $W^1,\ldots,W^p$ are independent standard Wiener processes, the average of 1000 iterative, independent paths of $t \rightsquigarrow \max_{1\le i\le p}|W_t^i|$ (solid), the curve $t \mapsto 2\sqrt{2}\sqrt{t\log(1+p)}$ (dashed), as well as the curve $t \mapsto \sqrt{t\log(1+p)}$ (dotted) for a reference, for $p = 1, 2, 10, 100, 1000$ and 10000.

We thus have that, for any constant $K > 0$,

$$\mathsf{E}\left[\max_{1\le i\le p}|M_{T\wedge n}^{i,T_n}|\wedge K\right] \le 2\sqrt{2}\sqrt{\mathsf{E}[[M]_T]}\sqrt{\log(1+p)}.$$

Letting $n \to \infty$, the bounded convergence theorem yields that

$$\mathsf{E}\left[\max_{1\le i\le p}|M_T^i|\wedge K\right] \le 2\sqrt{2}\sqrt{\mathsf{E}[[M]_T]}\sqrt{\log(1+p)}.$$

Finally, let $K \to \infty$ to apply the monotone convergence theorem. The proof of (i) is finished.

The proof of (ii) is exactly the same as that of (i), just changing the choice of $\sigma = \sqrt{\mathsf{E}[\langle M\rangle_T]/2}$. □

Theorem A1.1.3 (Lenglart's inequality for multi-dimensional case) Let p be any positive integer.

(i) When M is a p-dimensional local martingale starting from zero, it holds for any stopping time T and any constants $\eta, \sigma > 0$ that

$$\begin{aligned} &\mathsf{P}\left(\sup_{t\in[0,T]} \max_{1\leq i\leq p} |M_t^i| \geq \eta \right) \\ &\leq \frac{\{3\sigma + (\mathsf{E}[\sup_{t\in[0,T]} \Delta[M]_t]/\sigma)\}\sqrt{\log(1+p)}}{\eta} + \mathsf{P}([M]_T > \sigma^2). \end{aligned}$$

(ii) When M is a p-dimensional locally square-integrable martingale starting from zero, it holds for any stopping time T and any constants $\eta, \sigma > 0$ that

$$\mathsf{P}\left(\sup_{t\in[0,T]} \max_{1\leq i\leq p} |M_t^i| \geq \eta \right) \leq \frac{3\sigma\sqrt{\log(1+p)}}{\eta} + \mathsf{P}(\langle M\rangle_T > \sigma^2).$$

Corollary A1.1.4 (Corollary to Lenglart's inequality for high-dimensional case) For every $n \in \mathbb{N}$, let p_n be a positive integer, and let a p_n-dimensional local martingale M^n starting from zero and a stopping time T_n, both defined on a stochastic basis $\mathbf{B}^n$, be given. Suppose that a given sequence of positive constants σ_n satisfies that

$$\lim_{n\to\infty} \sigma_n \sqrt{\log(1+p_n)} = 0.$$

(i) If

$$\lim_{n\to\infty} \mathsf{E}^n\left[\sup_{t\in[0,T_n]} \Delta[M^n]_t \right] \frac{\sqrt{\log(1+p_n)}}{\sigma_n} = 0 \quad \text{and} \quad \lim_{n\to\infty} \mathsf{P}^n([M^n]_{T_n} > \sigma_n^2) = 0,$$

then it holds that

$$\sup_{t\in[0,T_n]} \max_{1\leq i\leq p_n} |M_t^{n,i}| \xrightarrow{\mathsf{P}^n} 0.$$

(ii) When M^n is a p_n-dimensional locally square-integrable martingale starting from zero, if

$$\lim_{n\to\infty} \mathsf{P}^n(\langle M^n\rangle_{T_n} > \sigma_n^2) = 0,$$

then it holds that

$$\sup_{t\in[0,T_n]} \max_{1\leq i\leq p_n} |M_t^{n,i}| \xrightarrow{\mathsf{P}^n} 0.$$

Proof of Theorem A1.1.3. Introduce a localizing sequence (T_n) of stopping times corresponding to the local martingale M' in Lemma A1.1.1 (i). Then, for any bounded stopping time T, the optional sampling theorem yields that

$$\mathsf{E}\left[\max_{1\leq i\leq p} |M_T^{i,T_n}| \right] \leq \mathsf{E}\left[\left(2\sigma + \frac{[M]_{T\wedge T_n}}{\sigma} \right) \sqrt{\log(1+p)} \right].$$

So the adapted process $\max_{1\le i\le p}|M^{i,T_n}|$ is L-dominated by the adapted process

$$t \rightsquigarrow \left(2\sigma + \frac{[M]_{t\wedge T_n}}{\sigma}\right)\sqrt{\log(1+p)}.$$

Now, given any stopping time T and any constants $\eta,\sigma > 0$, for any sufficiently large $m \in \mathbb{N}$ so that $\eta_m = \eta - m^{-1} > 0$, apply Theorem 6.6.2 with "$\eta = \eta_m$" and $\delta = (3+m^{-1})\sigma\sqrt{\log(1+p)}$ to deduce that

$$\begin{aligned}
&\mathsf{P}\left(\sup_{t\in[0,T\wedge T_n]}\max_{1\le i\le p}|M_t^i| > \eta_m\right)\\
&\le \frac{\{(3+m^{-1})\sigma + (\mathsf{E}[\sup_{t\in[0,T\wedge T_n]}\Delta[M]_t]/\sigma)\}\sqrt{\log(1+p)}}{\eta_m}\\
&\quad +\mathsf{P}\left(\left(2\sigma+\frac{[M]_{T\wedge T_n}}{\sigma}\right)\ge (3+m^{-1})\sigma \text{ and } [M]_{T\wedge T_n}\le \sigma^2\right)\\
&\quad +\mathsf{P}([M]_{T\wedge T_n} > \sigma^2)\\
&= \frac{\{(3+m^{-1})\sigma + (\mathsf{E}[\sup_{t\in[0,T\wedge T_n]}\Delta[M]_t]/\sigma)\}\sqrt{\log(1+p)}}{\eta_m}\\
&\quad +0+\mathsf{P}([M]_{T\wedge T_n} > \sigma^2).
\end{aligned}$$

Let $n\to\infty$ to have that

$$\begin{aligned}
&\mathsf{P}\left(\sup_{t\in[0,T]}\max_{1\le i\le p}|M_t^i| > \eta - m^{-1}\right)\\
&\le \frac{\{(3+m^{-1})\sigma + (\mathsf{E}[\sup_{t\in[0,T]}\Delta[M]_t]/\sigma)\}\sqrt{\log(1+p)}}{\eta - m^{-1}}\\
&\quad +\mathsf{P}([M]_T > \sigma^2).
\end{aligned}$$

Finally, let $m\to\infty$ to obtain the desired inequality.

The proof of (ii) is similar to (and easier than) that of (i). □

When we are interested in proving only that $\max_{1\le i\le p_n}|M_{T_n}^{n,i}| \xrightarrow{\mathsf{P}^n} 0$, rather than that $\sup_{t\le T_n}\max_{1\le i\le p_n}|M_t^{n,i}| \xrightarrow{\mathsf{P}^n} 0$, the following results give some other sets of sufficient conditions.

Exercise A1.1.1 For every $n\in\mathbb{N}$, let p_n be a positive integer, and let a p_n-dimensional local martingale $M^n = (M^{n,1},\ldots,M^{n,p})^{\mathrm{tr}}$ starting from zero and a finite stopping time T_n, both defined on a stochastic basis $\mathbf{B}^n$, be given.

Prove that either of the following condition (a) or (b) implies that

$$\lim_{n\to\infty}\mathsf{E}^n\left[\max_{1\le i\le p_n}|M_{T_n}^{n,i}|\right] = 0, \quad \text{and, in particular, that} \quad \max_{1\le i\le p_n}|M_{T_n}^{n,i}| \xrightarrow{\mathsf{P}^n} 0:$$

(a) $\lim_{n\to\infty} \mathsf{E}^n[[M^n]_{T_n}]\log(1+p_n)=0$;

(b) Each M^n is a p_n-dimensional locally square-integrable martingale, and

$$\lim_{n\to\infty} \mathsf{E}^n[\langle M^n\rangle_{T_n}]\log(1+p_n)=0.$$

Prove also that a sufficient condition for (a) and that for (b) are (a') and (b'), respectively, given as follows:

(a') The random sequence $(X_n)_{n=1,2,...}$ given by

$$X_n=[M^n]_{T_n}\log(1+p_n)$$

is asymptotically uniformly integrable and satisfies that $X_n \xrightarrow{\mathsf{P}^n} 0$;

(b') Each M^n is a p_n-dimensional locally square-integrable martingale, and the random sequence $(Y_n)_{n=1,2,...}$ given by

$$Y_n=\langle M^n\rangle_{T_n}\log(1+p_n)$$

is asymptotically uniformly integrable and satisfies that $Y_n \xrightarrow{\mathsf{P}^n} 0$.

Exercise A1.1.2 For every $n\in\mathbb{N}$, let N^n be a counting process with the intensity process λ^n, let W^n be a standard Wiener process, let $H^n=(H^{n,1},...,H^{n,p_n})^{\mathrm{tr}}$ and $K=(K^{n,1},...,K^{n,p_n})^{\mathrm{tr}}$ be p_n-dimensional predictable processes such that $|H^{n,i}|\le\overline{H}^n$ and $|K^{n,i}|\le\overline{K}^n$, for all i, for some predictable processes $\overline{H}^n$ and $\overline{K}^n$ such that the process $t\leadsto\int_0^t\{(\overline{H}_s^n)^2\lambda_s^n+(\overline{K}_s^n)^2\}ds$ is locally integrable, and let T_n be a finite stopping time, all defined on a stochastic basis $\mathbf{B}^n$.

Prove that, if

$$\lim_{n\to\infty}\mathsf{E}^n\left[\int_0^{T_n}\{(\overline{H}_s^n)^2\lambda_s^n+(\overline{K}_s^n)^2\}ds\right]\log(1+p_n)=0,$$

then it holds that

$$\lim_{n\to\infty}\mathsf{E}^n\left[\max_{1\le i\le p_n}\left|\int_0^{T_n}H_s^{n,i}(dN_s^n-\lambda_s^n ds)+\int_0^{T_n}K_s^{n,i}dW_s^n\right|\right]=0.$$

A1.1.2 Discrete-time case

In this subsection, we deduce some inequalities in discrete-time case from the corresponding results presented in the previous subsection. Let a discrete-time stochastic basis $\mathbf{B}=(\Omega,\mathcal{F};(\mathcal{F}_n)_{n\in\mathbb{N}_0},\mathsf{P})$ be given (except the last corollary).

Lemma A1.1.5 (Stochastic maximal inequality) Let p be any positive integer. Let a p-dimensional martingale difference sequence $\xi=(\xi^1,...,\xi^p)^{\mathrm{tr}}$ on $\mathbf{B}$ be given.

(i) For any constant $\sigma>0$ there exists a (1-dimensional) discrete-time martingale M' starting from zero such that

$$\max_{1\le i\le p}\left|\sum_{k=1}^n\xi_k^i\right|\le\left(2\sigma+\sigma^{-1}\sum_{k=1}^n\max_{1\le i\le p}(\xi_k^i)^2\right)\sqrt{\log(1+p)}+M'_n,\quad\forall n\in\mathbb{N},\quad\text{a.s.}$$

(ii) If $\mathsf{E}[(\xi_k^i)^2] < \infty$ for all i,k, then for any constant $\sigma > 0$ there exists a (1-dimensional) discrete-time martingale M'' starting from zero such that

$$\max_{1\leq i\leq p}\left|\sum_{k=1}^{n}\xi_k^i\right| \leq \left(2\sigma+\sigma^{-1}\sum_{k=1}^{n}\mathsf{E}\left[\max_{1\leq i\leq p}(\xi_k^i)^2\middle|\mathcal{F}_{k-1}\right]\right)\sqrt{\log(1+p)}+M_n'', \quad \forall n\in\mathbb{N}, \quad \text{a.s.}$$

Theorem A1.1.6 (Maximal inequality) Let p be any positive integer. Let a p-dimensional martingale difference sequence $\xi = (\xi^1,...,\xi^p)^{\mathrm{tr}}$ on $\mathbf{B}$ such that $\mathsf{E}[(\xi_k^i)^2] < \infty$ for all i,k be given. Then, it holds for any finite stopping time T that

$$\mathsf{E}\left[\max_{1\leq i\leq p}\left|\sum_{k=1}^{T}\xi_k^i\right|\right] \leq 2\sqrt{2}\sqrt{\mathsf{E}\left[\sum_{k=1}^{T}\max_{1\leq i\leq p}(\xi_k^i)^2\right]}\sqrt{\log(1+p)}$$

and that

$$\mathsf{E}\left[\max_{1\leq i\leq p}\left|\sum_{k=1}^{T}\xi_k^i\right|\right] \leq 2\sqrt{2}\sqrt{\mathsf{E}\left[\sum_{k=1}^{T}\mathsf{E}\left[\max_{1\leq i\leq p}(\xi_k^i)^2\middle|\mathcal{F}_{k-1}\right]\right]}\sqrt{\log(1+p)}.$$

Theorem A1.1.7 (Lenglart's inequality for multi-dimensional case) Let p be any positive integer. Let a p-dimensional martingale difference sequence $\xi = (\xi^1,...,\xi^p)^{\mathrm{tr}}$ on $\mathbf{B}$ such that $\mathsf{E}[(\xi_k^i)^2] < \infty$ for all i,k be given. Then, it holds for any stopping time T and any constants $\eta,\sigma > 0$ that

$$\mathsf{P}\left(\sup_{n\leq T}\max_{1\leq i\leq p}\left|\sum_{k=1}^{n}\xi_k^i\right|\geq\eta\right) \leq \frac{\left\{3\sigma+\sigma^{-1}\mathsf{E}\left[\sup_{k\leq T}\max_{1\leq i\leq p}(\xi_k^i)^2\right]\right\}\sqrt{\log(1+p)}}{\eta} +\mathsf{P}\left(\sum_{k=1}^{T}\max_{1\leq i\leq p}(\xi_k^i)^2>\sigma^2\right)$$

and that

$$\mathsf{P}\left(\sup_{n\leq T}\max_{1\leq i\leq p}\left|\sum_{k=1}^{n}\xi_k^i\right|\geq\eta\right) \leq \frac{3\sigma\sqrt{\log(1+p)}}{\eta}+\mathsf{P}\left(\sum_{k=1}^{T}\mathsf{E}\left[\max_{1\leq i\leq p}(\xi_k^i)^2\middle|\mathcal{F}_{k-1}\right]>\sigma^2\right).$$

Corollary A1.1.8 (Corollary to Lenglart's inequality for high-dimensional case) For every $n \in \mathbb{N}$, let p_n be a positive integer, and let a p_n-dimensional martingale difference sequence $\xi^n = (\xi^{n,1},...,\xi^{n,p})^{\mathrm{tr}}$ and a stopping time T_n, both defined on a discrete-time stochastic basis $\mathbf{B}^n = (\Omega^n, \mathcal{F}^n; (\mathcal{F}^n_m)_{m\in\mathbb{N}_0}, \mathsf{P}^n)$, be given. Suppose that a given sequence of positive constants σ_n satisfies that

$$\lim_{n\to\infty} \sigma_n \sqrt{\log(1+p_n)} = 0.$$

(i) If

$$\lim_{n\to\infty} \mathsf{E}^n \left[\sup_{k\le T_n} \max_{1\le i\le p_n} (\xi_k^{n,i})^2 \right] \frac{\sqrt{\log(1+p_n)}}{\sigma_n} = 0$$

and

$$\lim_{n\to\infty} \mathsf{P}^n \left(\sum_{k=1}^{T_n} \max_{1\le i\le p_n} (\xi_k^{n,i})^2 > \sigma_n^2 \right) = 0,$$

then it holds that

$$\sup_{m\le T_n} \max_{1\le i\le p_n} \left| \sum_{k=1}^{m} \xi_k^{n,i} \right| \xrightarrow{\mathsf{P}^n} 0.$$

(ii) If

$$\lim_{n\to\infty} \mathsf{P}^n \left(\sum_{k=1}^{T_n} \mathsf{E}^n \left[\max_{1\le i\le p_n} (\xi_k^{n,i})^2 \middle| \mathcal{F}^n_{k-1} \right] > \sigma_n^2 \right) = 0,$$

then it holds that

$$\sup_{m\le T_n} \max_{1\le i\le p_n} \left| \sum_{k=1}^{m} \xi_k^{n,i} \right| \xrightarrow{\mathsf{P}^n} 0.$$

When we are interested in proving only that $\sup_{1\le i\le p_n} |\sum_{k=1}^{T_n} \xi_k^{n,i}| \xrightarrow{\mathsf{P}^n} 0$ rather than that $\sup_{m\le T_n} \sup_{1\le i\le p_n} |\sum_{k=1}^{m} \xi_k^{n,i}| \xrightarrow{\mathsf{P}^n} 0$, the following results give some other sets of sufficient conditions.

Exercise A1.1.3 For every $n \in \mathbb{N}$, let p_n be a positive integer, and let a p_n-dimensional martingale difference sequence $\xi^n = (\xi^{n,1},...,\xi^{n,p})^{\mathrm{tr}}$ and a finite stopping time T_n, both defined on a discrete-time stochastic basis $\mathbf{B}^n = (\Omega^n, \mathcal{F}^n; (\mathcal{F}^n_m)_{m\in\mathbb{N}_0}, \mathsf{P}^n)$, be given.

Prove that either of the following condition (a) or (b) implies that

$$\lim_{n\to\infty} \mathsf{E}^n \left[\max_{1\le i\le p_n} \left| \sum_{k=1}^{T_n} \xi_k^{n,i} \right| \right] = 0, \quad \text{and, in particular, that} \quad \max_{1\le i\le p_n} \left| \sum_{k=1}^{T_n} \xi_k^{n,i} \right| \xrightarrow{\mathsf{P}^n} 0:$$

(a) It holds that

$$\lim_{n\to\infty} \mathsf{E}^n \left[\sum_{k=1}^{T_n} \max_{1\le i\le p_n} (\xi_k^{n,i})^2 \right] \log(1+p_n) = 0;$$

(b) It holds that

$$\lim_{n\to\infty} \mathsf{E}^n \left[\sum_{k=1}^{T_n} \mathsf{E}^n \left[\max_{1\leq i\leq p_n} (\xi_k^{n,i})^2 \middle| \mathcal{F}_{k-1}^n \right] \right] \log(1+p_n) = 0.$$

Prove also that a sufficient condition for (a) and that for (b) are (a') and (b'), respectively, given as follows:

(a') The random sequence $(X_n)_{n=1,2,...}$ given by

$$X_n = \sum_{k=1}^{T_n} \max_{1\leq i\leq p_n} (\xi_k^{n,i})^2 \log(1+p_n)$$

is asymptotically uniformly integrable and satisfies that $X_n \xrightarrow{\mathsf{P}^n} 0$;

(b') Each M^n is a p_n-dimensional locally square-integrable martingale, and the random sequence $(Y_n)_{n=1,2,...}$ given by

$$Y_n = \sum_{k=1}^{T_n} \mathsf{E}^n \left[\max_{1\leq i\leq p_n} (\xi_k^{n,i})^2 \middle| \mathcal{F}_{k-1}^n \right] \log(1+p_n)$$

is asymptotically uniformly integrable and satisfies that $Y_n \xrightarrow{\mathsf{P}^n} 0$.

Exercise A1.1.4 Let $(\mathcal{X},\mathcal{A})$ be a measurable space, and let $X_1, X_2, ...$ be an independent sequence of $\mathcal{X}$-valued random variables identically distributed to a probability law P on $(\mathcal{X},\mathcal{A})$. Let a sequence $h_1, h_2, ...$ of elements of $\mathcal{L}_2(P)$[3] with an envelop function $H \in \mathcal{L}_2(P)$ be given; that is, it is assumed that $|h_i| \leq H$ for all i.

(i) Prove that for any $n \in \mathbb{N}$ and any $p \in \mathbb{N}$,

$$\begin{aligned} &\mathsf{E}\left[\max_{1\leq i\leq p} \left| \frac{1}{\sqrt{n}} \sum_{k=1}^{n} \left\{ h_i(X_k) - \int_{\mathcal{X}} h_i(x)P(dx) \right\} \right| \right] \\ &\leq\ 8\sqrt{2}\sqrt{\int_{\mathcal{X}} H(x)^2 P(dx)}\sqrt{\log(1+p)}. \end{aligned}$$

(ii) Prove that, as $n\to\infty$, if $n^{-1}\log(1+p_n)\to 0$ then it holds that

$$\max_{1\leq m\leq n} \max_{1\leq i\leq p_n} \left| \frac{1}{n} \sum_{k=1}^{m} \left\{ h_i(X_k) - \int_{\mathcal{X}} h_i(x)P(dx) \right\} \right| \xrightarrow{\mathsf{P}} 0$$

and that

$$\lim_{n\to\infty} \mathsf{E}\left[\max_{1\leq i\leq p_n} \left| \frac{1}{n} \sum_{k=1}^{n} \left\{ h_i(X_k) - \int_{\mathcal{X}} h_i(x)P(dx) \right\} \right| \right] = 0.$$

[3] When a probability space $(\mathcal{X},\mathcal{A},P)$ is given, for any $q \geq 1$, we denote by $\mathcal{L}_q(P)$ the space of real-valued, $\mathcal{A}$-measurable functions h on $\mathcal{X}$ such that $\int_{\mathcal{X}} |h(x)|^q P(dx) < \infty$.

A1.2 Supplementary Tools for the Main Parts

The proof of Theorem 5.7.2 (ii) is based on the following lemma.

Lemma A1.2.1 If a real-valued $\mathcal{F}$-measurable random variable ξ defined on a probability space $(\Omega, \mathcal{F}, \mathsf{P})$ is integrable then the class $\{\mathsf{E}[\xi|\mathcal{G}];\ \mathcal{G} \subset \mathcal{F}\}$ is uniformly integrable, that is,

$$\lim_{K\to\infty} \sup_{\mathcal{G}\subset\mathcal{F}} \mathsf{E}[|\mathsf{E}[\xi|\mathcal{G}]|1\{|\mathsf{E}[\xi|\mathcal{G}]| > K\}] = 0.$$

See, e.g., Lemma 6.5 of Kallenberg (2002) for a proof.

The proof of Theorem 7.4.1 is based on the following inequality.

Lemma A1.2.2 (Gronwall's inequality) Let $t \mapsto m(t)$ be continuous on $[0,T]$, and let $t \mapsto \alpha(t)$ be integrable on $[0,T]$. For a constant $\beta > 0$, assume that

$$0 \le m(t) \le \alpha(t) + \beta \int_0^t m(s)ds, \quad \forall t \in [0,T]. \tag{A1.1}$$

Then it holds that

$$m(t) \le \alpha(t) + \beta \int_0^t \alpha(s)e^{\beta(t-s)}ds, \quad \forall t \in [0,T].$$

Proof. It follows from (A1.1) that

$$\frac{d}{dt}\left(e^{-\beta t}\int_0^t m(s)ds\right) = \left(m(t) - \beta\int_0^t m(s)ds\right)e^{-\beta t} \le \alpha(t)e^{-\beta t},$$

which implies that

$$\int_0^t m(s)ds \le e^{\beta t}\int_0^t \alpha(s)e^{-\beta s}ds.$$

Substitute this into (A1.1) to obtain the desired inequality. □

The following lemma, which provides a bound for a function that is sufficiently smooth at the origin, is useful for the proofs of CLTs.

Lemma A1.2.3 Let $n \ge 1$ be an integer. Suppose that $f : \mathbb{R} \to \mathbb{C}$ is an n-times continuously differentiable function, in the sense that both $\mathrm{Re}(f)$ and $\mathrm{Im}(f)$ are n-times continuously differentiable, such that

$$f^{(k)}(0) = \left.\frac{d^k}{dx^k}f(x)\right|_{x=0} = 0, \quad \text{for every } k = 0, \dots, n-1,$$

where $f^{(0)} = f$. Then it holds that for any $a \in (0,\infty]$

$$|f(x)| \le \frac{\sup_{|y|\le a}|f^{(n)}(y)|}{n!}|x|^n, \quad \forall x \in \mathbb{R} \text{ such that } |x| \le a.$$

The proof is a simple application of the Taylor expansion. As its consequences, we easily get:

$$\left| e^{ix} - 1 - ix + \frac{x^2}{2} \right| \le x^2 \wedge \frac{|x|^3}{6}, \quad \forall x \in \mathbb{R};$$

$$|e^{-x} - 1 + x| \le \frac{e^a}{2} x^2, \quad \text{if } |x| \le a;$$

$$|e^{ix + x^2/2} - 1 - ix| \le \frac{2(3 + a^2)e^{a^2/2}}{6} |x|^3, \quad \text{if } |x| \le a.$$

A2

Notes

The notes below are not intended to produce a complete list of acknowledgments of the original founder's achievements for the theories, methods, or ideas described in this monograph. The notes for some sections or chapters aim at giving readers useful information for further study, rather than exhaustive historical reviews.

Chapter 1. For further study on statistics of diffusion processes and related topics, readers should proceed to Kutoyants (2004), Kessler *et al.* (2012) and Aït-Sahalia and Jacod (2014). More details on counting process approach to survival analysis are well explained by Andersen *et al.* (1993), Kalbfleisch and Prentice (2002) and Aalen *et al.* (2008).

Section 2.1. Among a lot of good textbooks on the Lebesgue integral theory from different perspectives, I dare to quote only Pollard (2002) as a unique and excellent guide to measure theoretic probability.

Section 2.2. It might not be easy to understand the definition of conditional expectation. While the explanation given in this section is one of the possible approaches to interpret the definition, another approach via L^2-projections can be found in Jacod and Protter (2003).

Section 2.3. A good summary for stochastic convergence towards asymptotic statistics is found in van der Vaart (1998). The explanation of Lévy's continuity theorem given by Pollard (2002) is insightful.

Chapter 3. The somewhat hasty introduction to statistics of stochastic processes given in this chapter is a revised version of Chapter 4 in Nishiyama's (2011) monograph written in Japanese; it should be remarked that the phrase "core of statistics" is *not* a widely approved term academically but a direct translation from a Japanese term which I am personally using in my own lectures.

Chapters 4–6. Many of the proofs for the theorems presented in these chapters have been skipped in this small monograph; readers should consult with the authoritative books cited in the main texts to deepen their study.

Sections 7.1 and 7.2. The central limit theorems (CLTs) for discrete-time martingales are owing to the contributions by a lot of authors since the early 1960's, who considered stationary and ergodic sequences at first stages. Brown (1971) showed the crucial condition is not stationarity or ergodicity but the convergence of

DOI: 10.1201/9781315117768-A2

"conditional variance" (i.e., the condition (7.3)), while McLeish (1974) introduced elegant techniques to prove new CLTs and invariance principles. The CLTs for stochastic integrals related to counting processes are due to Aalen (1977) and Kutoyants (1979); the innovative proof via Itô's formula, presented as that for Theorem 7.1.5, has been taken from Kutoyants (1984). The functional CLT for (general) local martingales was established by Rebolledo (1980). On the other hand, the study of limit theorems for semimartingales in terms of the "characteristics" was started also around 1980 by a lot of authors, and it has been systematically explained by Jacod and Shiryaev (1987, 2003).

Section 7.3. The idea presented in this section has its roots in a proof of the Glivenko-Cantelli theorem; see Chapter 2.4 of van der Vaart and Wellner (1996) for the modern versions of the theorem.

Section 7.4. The collection of the tools presented in this section is merely a minimal package to deal with high-frequency data of diffusion processes. Consult with Jacod and Protter (2012) and Aït-Sahalia and Jacod (2014) for further study.

Chapter 8. It seems that van der Vaart and Wellner (1996) are the first to have used the terminology "Z-estimator". See Chapter 3.3 of their book, as well as Chapter 5 of van der Vaart (1998), for the theory of Z-estimators (also for infinite-dimensional parametric models, but mainly for i.i.d. random sequences), where the differentiability of the random fields $\theta \rightsquigarrow \mathbb{Z}_n(\theta)$ (in our notations) is not assumed. The approach has been developed up to semi-parametric models; see Chapter 25 of van der Vaart (1998) for a comprehensive study in i.i.d. models and the paper of Nishiyama (2009) for an attempt to treat stochastic process models, among many others. The discussion on the method of moment estimators in Subsection 8.5.1 has been taken from Chapter 5 of van der Vaart (1998). Results and techniques concerning the quasi-likelihoods for ergodic diffusion models in Subsections 8.5.2 and 8.5.3 are based on the contributions of some authors' works started from the late 1980's, including Prakasa Rao (1988), Florens-Zmirou (1989), Yoshida (1992) and Kessler (1997). The asymptotic theory for Cox's regression model presented in Subsection 8.5.4 is due to Andersen and Gill (1982). The discussion on the cases of different rates of convergence mentioned in the last section of this chapter has been taken from the paper of Negri and Nishiyama (2017b).

Chapter 9. The concept of the local asymptotic normality (LAN) was introduced by Le Cam (1960). The convolution theorem (Theorem 9.2.1) was established by Hájek (1970) in the LAN set-up; Inagaki's (1970) work in the i.i.d. set-up should also be acknowledged. The asymptotic minimax theorem is due to Hájek (1972); Le Cam (1972) showed similar theorems for non-Gaussian limit experiments. The current version of the theorem (Theorem 9.2.3) and its extension to infinite-dimensional cases are due to van der Vaart and Wellner (1996). On the other hand, Ibragimov and Has'minskii (1981) developed a deep theory of asymptotic efficiency, involving also the moment convergence of rescaled residuals for maximum likelihood and Bayes estimators, and Kutoyants (1984) successfully applied their theory to the study of parameter estimation for stochastic processes.

Chapter 10. There are huge literature for change point problems. Among several approaches, the idea for the Z-process method developed in this chapter comes from the study of "Fisher-score change process" originally due to Horváth and Parzen (1994). The general perspective explained in this chapter has been taken from the paper of Negri and Nishiyama (2017a).

Section A1.1. The inequalities in this section are new; they have been presented there with my highest responsibility for their correctness. They would hopefully act as some new devices in high-dimensional statistics, especially for stochastic process models (cf., e.g., the papers of Fujimori and Nishiyama (2017a,b) who took an approach to the "short, fat data" problems in such models via the Dantzig selectors based on the classical inequality of Bernstein). Use the new ones, if you like, at your own risk!

A3

Solutions/Hints to Exercises

Solution to Exercise 2.2.1. (ii) $\mathsf{E}[(X_1+\cdots+X_n)^2] = n\sigma^2$. $\mathsf{E}[(X_1+\cdots+X_n)^2|\mathcal{G}_k] = (X_1+\cdots+X_k)^2+(n-k)\sigma^2$ a.s. for $k=1,...,n$. (i) is a special case of (ii).

Solution to Exercise 2.2.2. $\mathsf{E}[L_n|\mathcal{G}_0] = 1$ and $\mathsf{E}[L_n|\mathcal{G}_k] = L_k$ a.s. for $k=1,...,n$.

Solution to Exercise 2.3.1.

$$\begin{aligned}\limsup_{n\to\infty} \mathsf{E}^n[|X_n|1\{|X_n|>K\}] &\leq \limsup_{n\to\infty} \mathsf{E}^n\left[|X_n|\left(\frac{|X_n|}{K}\right)^{\delta} 1\{|X_n|>K\}\right] \\ &\leq \frac{\limsup_{n\to\infty} \mathsf{E}^n[|X_n|^{1+\delta}]}{K^{\delta}} \\ &\to 0, \quad \text{as } K\to\infty.\end{aligned}$$

Solution to Exercise 2.3.2. Any uniformly bounded sequence of real-valued random variables is asymptotically uniformly integrable due to, e.g., Exercise 2.3.1. Thus the claim follows from Theorem 2.3.12.

Solution to Exercise 2.3.3. For any $\varepsilon>0$, it holds that

$$\begin{aligned}\mathsf{P}^n(|X_n|>\varepsilon) &\leq \mathsf{E}^n\left[\frac{|X_n|}{\varepsilon}1\{|X_n|>\varepsilon\}\right] \\ &\leq \frac{\mathsf{E}^n[|X_n|]}{\varepsilon}\to 0, \quad \text{as } n\to\infty.\end{aligned}$$

Solution to Exercise 2.3.4. Fix any $\varepsilon>0$. It holds for any $K>0$ that

$$\begin{aligned}\limsup_{n\to\infty} \mathsf{P}^n(|X_n|>K) &\leq \limsup_{n\to\infty} \mathsf{E}^n\left[\frac{|X_n|}{K}1\{|X_n>K\}\right] \\ &\leq \frac{\limsup_{n\to\infty} \mathsf{E}^n[|X_n|]}{K}.\end{aligned}$$

By choosing a large $K=K_\varepsilon>0$, it is possible to make the right-hand side smaller than ε.

Solution to Exercise 2.3.5. Fix any $\varepsilon>0$. Since X is a real-valued random variable, there exists a positive integer $m=m_\varepsilon$ such that $\mathsf{P}(|X|>m)<\varepsilon$; as a matter of fact, since $\{|X|>m\}\downarrow\emptyset$ as $m\to\infty$, it holds that $\lim_{m\to\infty}\mathsf{P}(|X|>m) = \mathsf{P}(\lim_{m\to\infty}\{|X|>m\}) = \mathsf{P}(\emptyset)=0$. Now, choose any bounded continuous function

DOI: 10.1201/9781315117768-A3

f such that $1\{|x|>m+1\}\le f(x)\le 1\{|x|>m\}$. (Such a function f indeed exists; it is even possible to construct such a function f which is piecewise linear.) Then, we have that

$$\begin{aligned}\limsup_{n\to\infty}\mathsf{P}^n(|X_n|>m+1) &\le \limsup_{n\to\infty}\mathsf{E}^n[f(X_n)]\\ &= \mathsf{E}[f(X)]\\ &\le \mathsf{P}(|X|>m)\\ &< \varepsilon.\end{aligned}$$

Solution to Exercise 2.3.6. Apply Slutskey's lemma repeatedly to deduce that $(X_n^1,\ldots,X_n^p)^{\mathrm{tr}}\xrightarrow{\mathsf{P}}(c^1,\ldots,c^p)^{\mathrm{tr}}$. Thus the first claim is immediate from the continuous mapping theorem. To show the second claim, it is sufficient to prove that the function $f(\mathbf{x})=f(x_1,\ldots,x_p)=\max_{1\le i\le p}x_i$ is continuous. In fact, it holds for any $\mathbf{x},\mathbf{y}\in\mathbb{R}^p$ that

$$\max_i x_i\le\max_i y_i+\max_i(x_i-y_i)\quad\text{and}\quad \max_i y_i\le\max_i x_i+\max_i(y_i-x_i),$$

hence

$$|f(\mathbf{x})-f(\mathbf{y})|=|\max_i x_i-\max_i y_i|\le\max_i|x_i-y_i|\le||\mathbf{x}-\mathbf{y}||.$$

Solution to Exercise 2.3.7. $AX_n\overset{\mathsf{P}}{\Longrightarrow}\mathcal{N}_q(A\mu,A\Sigma A^{\mathrm{tr}})$ in $\mathbb{R}^q$, if the matrix $A\Sigma A^{\mathrm{tr}}$ is positive definite.

Solution to Exercise 4.1.1. Each X_n is clearly $\mathcal{F}_n$-measurable and integrable. For every $n\in\mathbb{N}$, it holds that $\mathsf{E}[X_n|\mathcal{F}_{n-1}]=\mathsf{E}[\mathsf{E}[Y|\mathcal{F}_n]|\mathcal{F}_{n-1}]=\mathsf{E}[Y|\mathcal{F}_{n-1}]=X_{n-1}$ a.s.

Solution to Exercise 4.1.2. (i) Each M_n is clearly $\mathcal{F}_n$-measurable and integrable. For every $n\in\mathbb{N}$, it holds that

$$\mathsf{E}[M_n|\mathcal{F}_{n-1}]=\mathsf{E}\left[\prod_{k=1}^{n}(1+\xi_k)\,\middle|\,\mathcal{F}_{n-1}\right]=\prod_{k=1}^{n-1}(1+\xi_k)\mathsf{E}[1+\xi_n|\mathcal{F}_{n-1}]=M_{n-1},\quad\text{a.s.}$$

(ii) is a special case of (iii) with $H_k\equiv 1$.

(iii) It is easy to check the adaptedness and the integrability of the term "$(M_n'')_{n\in\mathbb{N}_0}$". Since $M_k-M_{k-1}=M_{k-1}\xi_k$ for every $k\in\mathbb{N}$, recalling Theorem 4.1.3 (ii), the first term on the right-hand side of the current decomposition should be

$$\sum_{k=1}^{n}H_{k-1}^2\mathsf{E}[(M_{k-1}\xi_k)^2|\mathcal{F}_{k-1}]=\sum_{k=1}^{n}H_{k-1}^2M_{k-1}^2\mathsf{E}[\xi_k^2|\mathcal{F}_{k-1}],\quad\text{a.s.}$$

Solution to Exercise 4.3.1. It holds for every $n\in\mathbb{N}_0$ that $\{R\le n\}=\bigcup_{k=1}^n\{X_k>\eta\}\in\mathcal{F}_n$, thus R is a stopping time. Recalling $S\ge 1$, $(S-1)$ takes values in $\{0,1,\ldots,\infty\}$, and it holds for every $n\in\mathbb{N}_0$ that

$$\{(S-1)\le n\}=\{S\le n+1\}=\{A_{n+1}\ge\delta\}\in\mathcal{F}_n,$$

because A is a predictable process, and thus $(S-1)$ is a stopping time.

Solution to Exercise 5.1.1. Choose a countable, dense subset S of $[0,\infty)$. Choose a P-null set N such that for every $\omega \in \Omega \setminus N$, paths $t \mapsto X_t(\omega)$ and $t \mapsto Y_t(\omega)$ are right-continuous, and that $X_s(\omega) = Y_s(\omega)$ for all $s \in S$. For any $t \in [0,\infty)$, it holds that $X_t(\omega) = \lim_{s_n \searrow t} X_s(\omega) = \lim_{s_n \searrow t} Y_s(\omega) = Y_t(\omega)$, where the limit operations "$\lim_{s_n \searrow t}$" are taken along a decreasing sequence $(s_n) \subset S$ converging to t. Thus we have proved that $X_t(\omega) = Y_t(\omega)$ for all $t \in [0,\infty)$, for every $\omega \in \Omega \setminus N$.

Solution to Exercise 5.1.2. An example is the following. Let ξ be a $(0,\infty)$-valued random variable with a Lebesgue density. and define $X_t(\omega) = 1\{\xi(\omega) \leq t\}$ and $Y_t(\omega) = 1\{\xi(\omega) < t\}$ for every $t \in [0,\infty)$. Then, $\mathsf{P}(X_t = Y_t) = \mathsf{P}(t \neq \xi) = 1$ for all $t \in [0,\infty)$, while $\mathsf{P}(X_t = Y_t, \forall t \in [0,\infty)) = \mathsf{P}(\emptyset) = 0$.

Solution to Exercise 5.1.3. Introduce a localizing sequence (T_n) of stopping times such that $X^{T_n} \in \mathcal{M}$. Then, it follows from Theorem 5.1.3 that there exists a càdlàg martingale $\widetilde{X}^n$ which is indistinguishable from X^{T_n}. Put $\widetilde{X}_t = X_0 + \sum_{(n)} (\widetilde{X}^n_{t\wedge T_n} - \widetilde{X}^n_{t\wedge T_{n-1}})$, where $T_0 := 0$. Then, the process $(\widetilde{X}_t)_{t\in[0,\infty)}$ is a càdlàg local martingale which is indistinguishable from X.

Solution to Exercise 5.1.4. (i) See Theorem 23.1 of Billingsley (1995).

(ii) First construct a Poisson process $N^* = (N^*_t)_{t\in[0,\infty)}$ with the intensity parameter 1, and then put $N_t = N^*_{\Lambda(t)}$, where $\Lambda(t) = \int_0^t \lambda(s)ds$, for every $t \in [0,\infty)$.

Solution to Exercise 5.5.1. (i) $\{T + c \leq t\} = \{T \leq t - c\} \in \mathcal{F}_{(t-c)\vee 0} \subset \mathcal{F}_t$.

(ii) If T is a stopping time, then $\{T < t\} = \bigcap_{(n)} \{T \leq t + (1/n)\} \in \bigcap_{(n)} \mathcal{F}_{t+(1/n)} = \mathcal{F}_t$ by the right-continuity of the filtration. The proof of the converse is the following: $\{T \leq t\} = \{T > t\}^c = (\bigcup_{(n)} \{T \geq t + (1/n)\})^c = \bigcap_{(n)} \{T < t + (1/n)\} \in \bigcap_{(n)} \mathcal{F}_{t+(1/n)} = \mathcal{F}_t$.

(iii) $\{\bigvee_{(n)} T_n \leq t\} = \bigcap_{(n)} \{T_n \leq t\} \in \mathcal{F}_t$ for every $t \in [0,\infty)$.

As for the other claim, it is *wrong* to argue that "$\{\bigwedge_{(n)} T_n \leq t\} = \bigcup_{(n)} \{T_n \leq t\} \in \mathcal{F}_t$"; a counter example is that $T_n = 1 + (1/n)$ and $t = 1$. A correct proof is the following. Recalling (ii), $\{\bigwedge_{(n)} T_n < t\} = \bigcup_{(n)} \{T_n < t\} \in \mathcal{F}_t$ for every $t \in [0,\infty)$.

(iv) $\{(\omega,t) : 0 \leq t < T(\omega)\}^c = \{(\omega,t) : T(\omega) \leq t\} \in \mathcal{P}$, so the càdlàg process $X_t = 1\{T \leq t\}$ is adapted, which means that T is a stopping time.

(v) The first claim is clear from that $\{(\omega,t) : 0 \leq t < \bigvee_{(n)} T_n(\omega)\} = \{(\omega,t) : \bigvee_{(n)} T_n(\omega) \leq t\}^c = (\bigcap_{(n)} \{(\omega,t) : T_n(\omega) \leq t\})^c = \bigcup_{(n)} \{(\omega,t) : 0 \leq t < T_n(\omega)\} \in \mathcal{P}$.

As for the second claim, the hypothesis $\bigcup_{(n)} \{S = T_n\} = \Omega$ implies that

$$
\begin{aligned}
&\{(\omega,t) : 0 \leq t < S(\omega)\} \\
&= \{(\omega,t) : 0 \leq t < S(\omega)\} \cap \bigcup_{(n)} \{(\omega,t) : S(\omega) = T_n(\omega)\} \\
&= \bigcup_{(n)} (\{(\omega,t) : 0 \leq t < S(\omega)\} \cap \{(\omega,t) : S(\omega) = T_n(\omega)\}) \\
&= \bigcup_{(n)} (\{(\omega,t) : 0 \leq t < T_n(\omega)\} \cap \{(\omega,t) : S(\omega) = T_n(\omega)\})
\end{aligned}
$$

$$= \bigcup_{(n)}\{(\omega,t): 0 \le t < T_n(\omega)\} \cap \bigcup_{(n)}\{(\omega,t): S(\omega) = T_n(\omega)\}$$
$$= \bigcup_{(n)}\{(\omega,t): 0 \le t < T_n(\omega)\} \in \mathcal{P}.$$

Solution to Exercise 5.7.1.

	(general) S.T.	finite S.T.	bounded S.T.	$T = t$
martingale	No	No	Yes	Yes
$\mathcal{M}$	Yes	Yes	Yes	Yes
$\mathcal{M}^2$	Yes	Yes	Yes	Yes
$\mathcal{M}_{\text{loc}}$	No	No	No	No
$\mathcal{M}^2_{\text{loc}}$	No	No	No	No

Solution to Exercise 5.8.1. It suffices to consider the case where X is a locally integrable, increasing process. For a localizing sequence (T_n), we have $(X - X^p)^{T_n} \in \mathcal{M}$. So we can apply the optional sampling theorem to get $\mathsf{E}[X_{T \wedge T_n}] = \mathsf{E}[X^p_{T \wedge T_n}]$. Use the monotone convergence theorem to obtain the claim by letting $n \to \infty$.

Solution to Exercise 5.9.1. It follows from Doob's inequality that $\sup_{t \in [0,\infty)} X_t^2$ is integrable. Thus we have that

$$\begin{aligned}
\sup_{T \in \mathcal{T}} \mathsf{E}[X_T^2 1\{X_T^2 > K\}] &\le \sup_{T \in \mathcal{T}} \mathsf{E}\left[\sup_{t \in [0,\infty)} X_t^2 1\left\{\sup_{t \in [0,\infty)} X_t^2 > K\right\}\right] \\
&= \mathsf{E}\left[\sup_{t \in [0,\infty)} X_t^2 1\left\{\sup_{t \in [0,\infty)} X_t^2 > K\right\}\right] \\
&\to 0, \quad \text{as } K \to \infty.
\end{aligned}$$

Solution to Exercise 5.9.2. We may assume that X is a càdlàg process. Introduce a localizing sequence (T_n) to make $X^{T_n} \in \mathcal{M}^2$. Since $X^{T \wedge T_n} \in \mathcal{M}^2$, it follows from Doob's inequality that

$$\mathsf{E}\left[\sup_{t \in [0, T \wedge T_n]} (X_t)^2\right] = \mathsf{E}\left[\sup_{t \in [0,\infty)} (X_t^{T \wedge T_n})^2\right] \le 4\mathsf{E}[(X_{T \wedge T_n})^2] = 4(\mathsf{E}[X_0^2] + \mathsf{E}[\langle X \rangle_{T \wedge T_n}]).$$

Letting $n \to \infty$, we obtain $\mathsf{E}[\sup_{t \in [0,T]} (X_t)^2] \le 4(\mathsf{E}[X_0^2] + \mathsf{E}[\langle X \rangle_T]) < \infty$ by the monotone convergence theorem.

Solution to Exercise 5.9.3. (i) We shall prove only that $W_t^2 - t$ is a martingale. For any $0 \le s < t$, it holds that

$$\begin{aligned}
\mathsf{E}[W_t^2 - t | \mathcal{F}_s] &= \mathsf{E}[(W_t - W_s)^2 + 2(W_t - W_s)W_s + W_s^2 - t | \mathcal{F}_s] \\
&= (t - s) + 2 \cdot 0 W_s + W_s^2 - t \\
&= W_s^2 - s.
\end{aligned}$$

(ii-a) Let us prove only that $(N_t - \lambda t)^2 - \lambda t$ is a martingale. For any $0 \le s < t$, it holds that

$$\begin{aligned} &\mathsf{E}[(N_t - \lambda t)^2 - \lambda t | \mathcal{F}_s] \\ &= \mathsf{E}[(N_t - N_s - \lambda(t-s))^2 + 2(N_t - \lambda t)(N_s - \lambda s) - (N_s - \lambda s)^2 | \mathcal{F}_s] - \lambda t \\ &= \lambda(t-s) + 2(N_s - \lambda s)^2 - (N_s - \lambda s)^2 - \lambda t \\ &= (N_s - \lambda s)^2 - \lambda s. \end{aligned}$$

(ii-b) Put $M^k = N^k - A^k$. It follows from the formula for integration by parts that

$$M_t^k M_t^{k'} = \int_0^t M_{s-}^{k'} dM_s^k + \int_0^t M_{s-}^k dM_s^{k'} + \sum_{s \le t} \Delta N_s^k \Delta N_s^{k'}.$$

The predictable compensators for the first and the second terms of the right-hand side are zero. When $k = k'$, the third term coincides with $\sum_{s \le t} \Delta N_s^k = N_t^k$, whose predictable compensator is A^k. On the other hand, when $k \ne k'$, the third term is zero due to the assumption that N^k's have no simultaneous jump.

Solution to Exercise 5.9.4. (i) Notice that $(X^T)^2 - \langle X^T \rangle \in \mathcal{M}_{\mathrm{loc}}$. On the other hand, $X^2 - \langle X \rangle \in \mathcal{M}_{\mathrm{loc}}$ implies that $t \rightsquigarrow X_{t \wedge T}^2 - \langle X \rangle_{t \wedge T}$ is a local martingale. By the uniqueness of the predictable quadratic variation, $\langle X^T \rangle_\cdot$ and $\langle X \rangle_{\cdot \wedge T}$ coincide up to indistinguishability.

(ii) By using the result of (i), we have that

$$\begin{aligned} \langle X^T, Y^T \rangle_t &= \frac{1}{4}\{\langle X^T + Y^T \rangle_t - \langle X^T - Y^T \rangle_t\} \\ &= \frac{1}{4}\{\langle (X+Y)^T \rangle_t - \langle (X-Y)^T \rangle_T\} \\ &= \frac{1}{4}\{\langle X+Y \rangle_{t \wedge T} - \langle X-Y \rangle_{t \wedge T}\} \\ &= \langle X, Y \rangle_{t \wedge T}. \end{aligned}$$

Solution to Exercise 6.4.1. Apply Itô's formula to the function $f(x) = x_i x_j$. Noting that $D_i = x_j$, $D_{i,i} = 0$ and $D_{i,j} = 1$, we have

$$\begin{aligned} X_t^i X_t^j - X_0^i X_0^j &= \int_0^t X_{s-}^j dX_s^i + \int_0^t X_{s-}^i dX_s^j + \int_0^t d\langle X^{i,c}, X^{j,c} \rangle_s \\ &\quad + \sum_{s \le t} (X_s^i X_s^j - X_{s-}^i X_{s-}^j - X_{s-}^j \Delta X_s^i - X_{s-}^i \Delta X_s^j) \\ &= \int_0^t X_{s-}^j dX_s^i + \int_0^t X_{s-}^i dX_s^j + \langle X^{i,c}, X^{j,c} \rangle_t + \sum_{s \le t} \Delta X_s^i \Delta X_s^j \\ &= \int_0^t X_{s-}^j dX_s^i + \int_0^t X_{s-}^i dX_s^j + [X^i, X^j]_t. \end{aligned}$$

Solution to Exercise 6.4.2. Note that $X_0 > 0$ by assumption. In order to obtain the representation of the stochastic process $Y_t = \frac{X_t}{X_0}$, apply Itô's formula to the semimartingale $\beta t + \sigma W_t$ and the function $f(x) = e^x$.

Solution to Exercise 6.4.3. As it was mentioned in Hint, the necessity has already been proved. To show the sufficiency, choose any $0 = t_0 < t_1 < \cdots < t_n$ and $z = (z_1,...,z_n)^{\mathrm{tr}} \in \mathbb{R}^n$, and put $H(s) = \sum_{j=1}^n z_j 1\{s \in (t_{j-1},t_j]\}$. Let us apply Itô's formula to the 2-dimensional semimartingale $(X^{(1)},X^{(2)}) = (\frac{1}{2}\int_0^\cdot H(s)^2 ds, \int_0^\cdot H(s)dM_s)$ and the function $f(x_1,x_2) = \exp(x_1+ix_2)$. Since $\frac{\partial}{\partial x_1}f(x_1,x_2) = e^{x_1+ix_2}$, $\frac{\partial}{\partial x_2}f(x_1,x_2) = ie^{x_1+ix_2}$ and $\frac{\partial^2}{\partial x_2^2}f(x_1,x_2) = -e^{x_1+ix_2}$, we have for any $t \in [0,\infty)$

$$
\begin{aligned}
&\exp\left(\frac{1}{2}\int_0^t H(s)^2 ds + i\int_0^{t_n} H(s)dM_s\right) - 1 \\
&= \int_0^t \exp(X_{s-}^{(1)} + iX_{s-}^{(2)})\frac{1}{2}H(s)^2 ds + i\int_0^t \exp(X_{s-}^{(1)} + iX_{s-}^{(2)})H_s dM_s \\
&\quad + \frac{1}{2}\int_0^t -\exp(X_{s-}^{(1)} + iX_{s-}^{(2)})H(s)^2 d\langle M\rangle_s \\
&= i\int_0^t \exp(X_{s-}^{(1)} + iX_{s-}^{(2)})H_s dM_s.
\end{aligned}
$$

Introduce a localizing sequence (T_m) for the continuous local martingale on the right-hand side, and consider the above equation with t being replaced by $t_n \wedge T_m$. Then, since the expectation of the right-hand side is zero, it holds that

$$
\mathsf{E}\left[\exp\left(\frac{1}{2}\int_0^{t_n\wedge T_m} H(s)^2 ds + i\int_0^{t_n\wedge T_m} H(s)dM_s\right)\right] = 1.
$$

By Lebesgue's convergence theorem, as $m \to \infty$ we have

$$
\mathsf{E}\left[\exp\left(\frac{1}{2}\int_0^{t_n} H(s)^2 ds + i\int_0^{t_n} H(s)dM_s\right)\right] = 1,
$$

which implies that

$$
\begin{aligned}
\mathsf{E}\left[\exp\left(i\sum_{j=1}^n z_j(M_{t_j} - M_{t_{j-1}})\right)\right] &= \mathsf{E}\left[\exp\left(i\int_0^{t_n} H(s)dM_s\right)\right] \\
&= \exp\left(-\frac{1}{2}\int_0^{t_n} H(s)^2 ds\right) \\
&= \exp\left(-\frac{1}{2}\sum_{j=1}^n z_j^2 |t_j - t_{j-1}|\right) \\
&= \exp\left(-\frac{1}{2}z^{\mathrm{tr}}\Sigma z\right),
\end{aligned}
$$

where $\Sigma = \mathrm{diag}(|t_1 - t_0|,...,|t_n - t_{n-1}|)$. Hence $(M_{t_1} - M_{t_0},...,M_{t_n} - M_{t_{n-1}})^{\mathrm{tr}}$ is distributed to $\mathcal{N}_n(0,\Sigma)$.

Solution to Exercise 6.4.4. As it was mentioned in Hint, the necessity has already been proved. To show the sufficiency, choose any $z \in \mathbb{R}$, and for any $s \leq t$ put

$$G(s,t) = \exp\left(-(e^{iz} - 1) \int_s^t \lambda(u)du + iz(N_t - N_s) \right).$$

Then it follows from Itô's formula that

$$G(s,t) = G(s,s) + \int_s^t (e^{iz} - 1)G(s,u-)(dN_u - \lambda(u)du),$$

and thus $\mathsf{E}[G(s,t)|\mathcal{F}_s] = G(s,s) = 1$, a.s. Hence, we have that

$$\mathsf{E}\left[\exp(iz(N_t - N_s))|\mathcal{F}_s\right] = \exp\left((e^{iz} - 1) \int_s^t \lambda(u)du \right), \quad \text{a.s.}$$

By taking the expectation, we conclude that $N_t - N_s$ is distributed to the Poisson distribution with mean $\int_s^t \lambda(u)du$.

On the other hand, we can deduce that, for any $0 = t_0 < t_1 < t_2 < \cdots < t_n$, the random variables $\{N_{t_j} - N_{t_{j-1}}\}_{j=1,2,\ldots,n}$ are independent from the computation of the characteristic function given as follows: for any $z = (z_1, \ldots, z_n)^{\mathrm{tr}} \in \mathbb{R}^n$, it holds that

$$\begin{aligned}
&\mathsf{E}\left[\exp\left(i\sum_{j=1}^n z_j(N_{t_j} - N_{t_{j-1}})\right)\right] \\
&= \mathsf{E}\left[E\left[\exp(iz_n(N_{t_n} - N_{t_{n-1}}))\middle|\mathcal{F}_{t_{n-1}}\right]\exp\left(i\sum_{j=1}^{n-1} z_j(N_{t_j} - N_{t_{j-1}})\right)\right] \\
&= \exp\left((e^{iz_n} - 1)\int_{t_{n-1}}^{t_n} \lambda(u)du\right)\mathsf{E}\left[\exp\left(i\sum_{j=1}^{n-1} z_j(N_{t_j} - N_{t_{j-1}})\right)\right] \\
&= \cdots \\
&= \prod_{j=1}^n \exp\left((e^{iz_j} - 1)\int_{t_{j-1}}^{t_j} \lambda(u)du\right) \\
&= \prod_{j=1}^n \mathsf{E}\left[\exp(iz_j(N_{t_j} - N_{t_{j-1}}))\right].
\end{aligned}$$

Solution to Exercise 6.6.1. The claim is immediate from Lenglart's inequality (Theorem 6.6.2 (ii)).

Hint to Exercise 6.6.2. For both (i) and (ii), Lenglart's inequality is helpful to prove the "right" claims. A counter example to both of the converses in (i) and in (ii) is the following. Let N be the homogeneous Poisson process with intensity 1, and put $X_t^n = nN_t$, $A_t^n = nt$ and $T_n = n^{-1/2}$; then, $X_{T_n}^n \xrightarrow{a.s.} 0$ and $A_{T_n}^n \to \infty$.

Solution to Exercise 7.1.1. It holds for any $\varepsilon > 0$ that

$$\sum_{k=1}^{T_n} \mathsf{E}^n[||\xi_k^n||^\alpha 1\{||\xi_k^n|| > \varepsilon\}|\mathcal{F}_{k-1}^n] \leq \sum_{k=1}^{T_n} \mathsf{E}^n\left[\frac{||\xi_k^n||^{\alpha+\delta}}{\varepsilon^\delta} 1\{||\xi_k^n|| > \varepsilon\}\middle|\mathcal{F}_{k-1}^n\right]$$

$$\leq \quad \frac{\sum_{k=1}^{T_n} \mathsf{E}^n[||\xi_k^n||^{\alpha+\delta}|\mathcal{F}_{k-1}^n]}{\varepsilon^\delta} \xrightarrow{\mathsf{P}^n} 0.$$

Solution to Exercise 7.1.2. To apply Cramér-Wold's device, choose any $c \in \mathbb{R}^p \setminus \{0\}$, and consider the "1-dimensional case" given by $c^{\mathrm{tr}}\sum_{k=1}^{T_n}\xi_k^n = \sum_{k=1}^{T_n} c^{\mathrm{tr}}\xi_k^n$. The condition (a) in Theorem 7.1.2 is reduced to

$$\begin{aligned}
\sum_{k=1}^{T_n} \mathsf{E}^n[(c^{\mathrm{tr}}\xi_k^n)(c^{\mathrm{tr}}\xi_k^n)^{\mathrm{tr}}|\mathcal{F}_{k-1}^n] &= \sum_{k=1}^{T_n} \mathsf{E}^n[c^{\mathrm{tr}}\xi_k^n(\xi_k^n)^{\mathrm{tr}}c|\mathcal{F}_{k-1}^n] \\
&= c^{\mathrm{tr}}\left(\sum_{k=1}^{T_n} \mathsf{E}^n[\xi_k^n(\xi_k^n)^{\mathrm{tr}}|\mathcal{F}_{k-1}^n]\right)c \\
&\xrightarrow{\mathsf{P}^n} c^{\mathrm{tr}}Cc.
\end{aligned}$$

Regarding the condition (b), since $|c^{\mathrm{tr}}\xi_k^n| \leq ||c|| \cdot ||\xi_k^n||$ we have that

$$\begin{aligned}
&\sum_{k=1}^{T_n} \mathsf{E}^n[(c^{\mathrm{tr}}\xi_k^n)^2 1\{|c\mathrm{tr}\xi_k^n| > \varepsilon\}|\mathcal{F}_{k-1}^n] \\
&\leq \quad ||c||^2 \sum_{k=1}^{T_n} \mathsf{E}^n\left[||\xi_k^n||^2 1\left\{||\xi_k^n|| > \frac{\varepsilon}{||c||}\right\}\middle|\mathcal{F}_{k-1}^n\right] \\
&\xrightarrow{\mathsf{P}^n} \quad 0, \quad \forall \varepsilon > 0.
\end{aligned}$$

Hence, Theorem 7.1.2 yields that

$$c^{\mathrm{tr}} \sum_{k=1}^{T_n} \xi_k^n \xRightarrow{\mathsf{P}^n} \mathcal{N}(0, c^{\mathrm{tr}}Cc) \quad \text{in } \mathbb{R}.$$

The limit is the distribution of $c^{\mathrm{tr}}G$, where G is a random variable distributed to $\mathcal{N}_p(0,C)$. Since the choice of c is arbitrary, we obtain that $\sum_{k=1}^{T_n}\xi_k^n \xRightarrow{\mathsf{P}^n} G$ in $\mathbb{R}^p$ by Cramér-Wold's device.

Hint to Exercise 7.1.3. Repeat the same argument as that for Exercise 7.1.1.

Solution to Exercise 7.2.1. Choose any finite points in $[0,T] \times \{1,...,d\}$, namely, $(t_1,i_1),...,(t_r,i_r)$. In order to apply Theorem 7.1.8 to the r-dimensional local martingale $\widetilde{M}^n_\cdot = (\widetilde{M}^{n,i_1}_\cdot, ..., \widetilde{M}^{n,i_r}_\cdot)^{\mathrm{tr}} = (M^{n,i_1}_{\cdot\wedge t_1}, ..., M^{n,i_r}_{\cdot\wedge t_r})^{\mathrm{tr}}$ and the stopping time $T(\geq \max\{t_1,...,t_r\})$, we can verify that, for any $\varepsilon > 0$,

$$\begin{aligned}
&\sum_{t\leq T} \sqrt{\sum_{j=1}^r (\Delta M^{n,i_j}_{t\wedge t_j})^2} 1\left\{\sqrt{\sum_{j=1}^r (\Delta M^{n,i_j}_{t\wedge t_j})^2} > \varepsilon\right\} \\
&\leq \sum_{t\leq T} \sqrt{\sum_{j=1}^r (\Delta M^{n,i_j}_t)^2} 1\left\{\sqrt{\sum_{j=1}^r (\Delta M^{n,i_j}_t)^2} > \varepsilon\right\} \\
&\leq \sum_{t\leq T} ||\Delta M^n_t|| 1\{||\Delta M^n_t|| > \varepsilon\} \xrightarrow{\mathsf{P}^n} 0,
\end{aligned}$$

and that

$$[\widetilde{M}^{n,i_p},\widetilde{M}^{n,i_q}]_T = [M^{n,i_p},M^{n,i_q}]_{t_p\wedge t_q} \xrightarrow{\mathsf{P}^n} C^{i_p,i_q}(t_p\wedge t_q), \quad \forall (p,q)\in\{1,...,r\}^2.$$

Hint to Exercise 7.2.2. For given $\varepsilon,\eta>0$, choose a large $m\in\mathbb{N}$, and then for every $i=1,...,d$ choose some time grids $0=t_1^i<t_2^i<\cdots<t_{N_m}^i=T$ such that $C^{i,i}(t_j^i)-C^{i,i}(t_{j-1}^{i,i})\le m^{-1}$ for every $j=1,...,N_m$ and that $N_m\le K_i m$ for a constant K_i depending only on $C^{i,i}(T)$, as in the proof of Theorem 7.1.8. Introduce the finite partition $[0,T]\times\{1,...,d\}=(\bigcup_{i,j}A_{i,j})\cup(\{0\}\times\{1,...,d\})$ by

$$A_{i,j}=(t_{j-1}^i,t_j^i]\times\{i\}, \quad j=1,...,N_m, \ i=1,...,d.$$

Hint to Exercise 8.7.1. Lemma 8.5.1 is helpful.

Hint to Exercise 10.3.1. $Q_n=\sqrt{t_n^n}I_p$. The matrix $J(\theta_*)=(J^{i,j}(\theta_*))_{(i,j)\in\{1,...,p\}^2}$ is given by

$$J^{i,j}(\theta_*)=\int_{\mathbb{R}}\frac{\partial_i\beta(x;\theta_*)\partial_j\beta(x;\theta_*)}{\sigma(x)^2}P_{\theta_*}^{\circ}(dx).$$

A natural estimator is given by $\widehat{J}_n=(\widehat{J}_n^{i,j})_{(i,j)\in\{1,...,p\}^2}$, where

$$\widehat{J}_n^{i,j}=\frac{1}{n}\sum_{k=1}^{n}\frac{\partial_i\beta(X_{t_{k-1}^n};\widehat{\theta}_n)\partial_j\beta(X_{t_{k-1}^n};\widehat{\theta}_n)}{\sigma(X_{t_{k-1}^n})^2}.$$

Hint to Exercise 10.3.2. $Q_n=\sqrt{n}I_p$. The matrix $J(\theta_*)=(J^{i,j}(\theta_*))_{(i,j)\in\{1,...,p\}^2}$ is given by

$$J^{i,j}(\theta_*)=2\int_{\mathbb{R}}\frac{\partial_i\sigma(x;\theta_*)\partial_j\sigma(x;\theta_*)}{\sigma(x;\theta_*)^2}P_{\theta_*}^{\circ}(dx).$$

A natural estimator is given by $\widehat{J}_n=(\widehat{J}_n^{i,j})_{(i,j)\in\{1,...,p\}^2}$, where

$$\widehat{J}_n^{i,j}=\frac{2}{n}\sum_{k=1}^{n}\frac{\partial_i\sigma(X_{t_{k-1}^n};\widehat{\theta}_n)\partial_j\sigma(X_{t_{k-1}^n};\widehat{\theta}_n)}{\sigma(X_{t_{k-1}^n};\widehat{\theta}_n)^2}.$$

Hint to Exercise 10.3.3. $Q_n=\mathrm{diag}(\sqrt{t_n^n},\sqrt{n})$. The matrix $J(\theta_*)=(J^{i,j}(\theta_*))_{(i,j)\in\{1,2\}^2}$ is given by

$$J^{1,1}(\theta_*)=\int_{\mathbb{R}}\frac{(\partial_1\beta(x;\theta_{*1}))^2}{\sigma(x;\theta_{*2})^2}P_{\theta_*}^{\circ}, \quad J^{2,2}(\theta_*)=2\int_{\mathbb{R}}\frac{(\partial_1\sigma(x;\theta_{*2}))^2}{\sigma(x;\theta_{*2})^2}P_{\theta_*}^{\circ},$$

and $J^{1,2}(\theta_*)=J^{2,1}(\theta_*)=0$. A natural estimator is given by $\widehat{J}_n=(\widehat{J}_n^{i,j})_{(i,j)\in\{1,2\}^2}$, where

$$\widehat{J}_n^{1,1}=\frac{1}{n}\sum_{k=1}^{n}\frac{(\partial_1\beta(X_{t_{k-1}^n};\widehat{\theta}_{n1}))^2}{\sigma(X_{t_{k-1}^n};\widehat{\theta}_{n2})^2}, \quad \widehat{J}_n^{2,2}=\frac{2}{n}\sum_{k=1}^{n}\frac{(\partial_2\sigma(X_{t_{k-1}^n};\widehat{\theta}_{n2}))^2}{\sigma(X_{t_{k-1}^n};\widehat{\theta}_{n2})^2},$$

with $\widehat{\theta}_n = (\widehat{\theta}_{n1}, \widehat{\theta}_{n2})^{\mathrm{tr}}$, and $\widehat{J}_n^{1,2} = \widehat{J}_n^{2,1} = 0$.

Solutions to Exercises A1.1.1 and A1.1.3. The results are easy consequences from Theorems A1.1.2 and A1.1.6, respectively.

Hint to Exercise A1.1.2. Apply Exercise A1.1.1.

Hint to Exercise A1.1.4. Apply Theorem A1.1.6, Corollary A1.1.8 and Exercise A1.1.3.

Bibliography

[1] Aalen, O.O. (1975). *Statistical Inference for a Family of Counting Processes.* Ph.D. thesis, University of California, Berkeley.

[2] Aalen, O.O. (1977). Weak convergence of stochastic integrals related to counting processes. *Z. Wahrsch. verw. Geb.* **38**, 261–277. Correction: **48**, 347 (1979).

[3] Aalen, O.O. (1978). Nonparametric inference for a family of counting processes. *Ann. Statist.* **6**, 701–726.

[4] Aalen, O.O., Borgan, Ø. and Gjessing, H.K. (2008). *Survival and Event History Analysis: A Process Point of View.* Springer, New York.

[5] Aït-Sahalia, Y. and Jacod, J. (2014). *High-Frequency Financial Econometrics.* Princeton University Press, Princeton.

[6] Andersen, P.K., Borgan, Ø., Gill, R.D. and Keiding, N. (1993). *Statistical Models Based on Counting Processes.* Springer, New York.

[7] Andersen, P.K. and Gill, R.D. (1982). Cox's regression models for counting processes: A large sample study. *Ann. Statist.* **10**, 1100–1120.

[8] Billingsley, P. (1968, 1999). *Convergence of Probability Measures. (1st and 2nd editions.)* Wiley, New York.

[9] Billingsley, P. (1995). *Probability and Measure. (3rd edition.)* Wiley, New York.

[10] Black, F. and Scholes, M. (1973). The pricing of options and corporate liabilities. *J. Political Economy* **81**, 637–659.

[11] Brown, B. (1971). Martingale central limit theorems. *Ann. Math. Statist.* **42**, 59–66.

[12] Cox, D.R. (1972). Regression models and life-tables (with discussion). *J. Roy. Statist. Soc. B* **34**, 187–220.

[13] Dellacherie, C. and Meyer, P.-A. (1978). *Probabilities and Potential.* North-Holland, Amsterdam New York Oxford.

[14] Donsker, M. (1951). An invariance principle for certain probability limit theorems. *Mem. Amer. Math. Soc.* **6**.

[15] Doob, J.L. (1953). *Stochastic Processes.* Wiley, New York.

[16] Dzhaparidze, K. and van Zanten, J.H. (2001). On Bernstein-type inequalities for martingales. *Stochastic Process. Appl.* **93**, 109–117.

[17] Florens-Zmirou, D. (1989). Approximate discrete-time schemes for statistics of diffusion processes. *Statistics*, **20**, 547–557.

[18] Fujimori, K. and Nishiyama, Y. (2017a). The l_q consistency of the Dantzig selector for Cox's proportional hazards model. *J. Statist. Plann. Inference* **181**, 62–70.

[19] Fujimori, K. and Nishiyama, Y. (2017b). The Dantzig selector for diffusion processes with covariates. *J. Japan Statist. Soc.* **47**, 59–73.

[20] Hájek, J. (1970). A characterization of limiting distributions of regular estimators. *Z. Wahrsch. verw. Geb.*, **14**, 323–330.

[21] Hájek, J. (1972). Local asymptotic minimax and admissibility in estimation. *Proc. Sixth Berkley Symp. Math. Statist. Probab.* **1**, 175–194.

[22] Hall, P. and Heyde, C.C. (1980). *Martingale Limit Theory and Its Application.* Academic Press, San Diego.

[23] Hawkes, A.G. and Oakes, D. (1974). A cluster representation of a self-exciting process. *J. Appl. Probab.* **11**, 493–503.

[24] Horváth, L. and Parzen, E. (1994). Limit theorems for Fisher-score change processes. In: Change-point problems (edited by Carlstein, E., Müller, H.-G. and Siegmund, D.). IMS Lecture Notes — Monograph Series **23**, 157–169.

[25] Ibragimov, I.A. and Has'minskii, R.Z. (1981). *Statistical Estimation: Asymptotic Theory.* Springer, New York Heidelberg Berlin.

[26] Ikeda, N. and Watanabe, S. (1989). *Stochastic Differential Equations and Diffusion Processes. (2nd edition.)* North-Holland/Kodansha, Amsterdam Oxford New York Tokyo.

[27] Inagaki, N. (1970). On the limiting distribution of a sequence of estimators with uniformity property. *Ann. Inst. Statist. Math.* **22**, 1–13.

[28] Itô, K. (1944). Stochastic integral. *Proc. Imp. Acad. Tokyo* **20**, 519–524.

[29] Jacod, J. and Protter, P.E. (2003). *Probability Essentials. (2nd edition.)* Springer, Berlin Heidelberg New York.

[30] Jacod, J. and Protter, P.E. (2012). *Discretization of Processes.* Springer, Berlin Heidelberg.

[31] Jacod, J. and Shiryaev, A.N. (1987, 2003). *Limit Theorems for Stochastic Processes. (1st and 2nd editions.)* Springer, Berlin Heidelberg.

[32] Kalbfleish, J.D. and Prentice, R.L. (2002). *The Statistical Analysis of Failure Time Data. (2nd edition.)* Wiley, Hoboken.

[33] Kallenberg, O. (2002). *Foundations of Modern Probability. (2nd edition.)* Springer, New York Berlin Heidelberg.

[34] Kessler, M. (1997). Estimation of an ergodic diffusion from discrete observations. *Scand. J. Statist.*, **24**, 211–229.

[35] Kessler, M., Lindner, A. and Sørensen, M. (Eds.) (2012). *Statistical Methods for Stochastic Differential Equations.* CRC Press, Boca Raton London New York.

[36] Kunita, H. and Watanabe, S. (1967). On square-integrable martingales. *Nagoya J. Math.* **30**, 209–245.

[37] Kutoyants, Yu.A. (1979). Local asymptotic normality for processes of Poisson type. (In Russian.) *Izv. Akad. Nauk Arm. SSR Mathematika*, **14**, 3–20.

[38] Kutoyants, Yu.A. (1984). *Parameter Estimation for Stochastic Processes.* Heldermann, Berlin.

[39] Kutoyants, Yu.A. (2004). *Statistical Inference for Ergodic Diffusion Processes.* Springer, London.

[40] Le Cam, L. (1960). Locally asymptotically normal families of distributions. *University of California Publications in Statistics* **2**, 207–236.

[41] Le Cam, L. (1972). Limits of experiments. *Proc. Sixth Berkley Symp. Math. Statist. Probab.* **1**, 245–261.

[42] Lenglart, E. (1977). Relation de domination entre deux processus. *Ann. Inst. Henri Poincaré (B)*, **13**, 171–179.

[43] McLeish, D.L. (1974). Dependent central limit theorems and invariance principles. *Ann. Probab.* **2**, 620–628.

[44] Negri, I. and Nishiyama, Y. (2017a). *Z*-process method for change point problems with applications to discretely observed diffusion processes. *Stat. Methods Appl.* **26**, 231–250.

[45] Negri, I. and Nishiyama, Y. (2017b). Moment convergence of *Z*-estimators. *Stat. Inference Stoch. Process* **20**, 387–397.

[46] Nishiyama, Y. (2009). Asymptotic theory of semiparametric *Z*-estimators for stochastic processes with applications to ergodic diffusions and time series. *Ann. Statist.* **37**, 3555–3579.

[47] Nishiyama, Y. (2011). *Statistical Analysis by the Theory of Martingales. (In Japanese.)* Kindaikagakusha, Tokyo.

[48] Ogata, Y. (1988). Statistical models for earthquake occurrences and residual analysis for point processes. *J. Amer. Statist. Assoc.*, **83**, 9–27.

[49] Pollard, D. (1984). *Convergence of Stochastic Processes.* Springer, Berlin Heidelberg New York.

[50] Pollard, D. (2002). *A User's Guide to Measure Theoretic Probability.* Cambridge University Press, Cambridge.

[51] Prakasa Rao, B.L.S. (1988). Statistical inference from sampled data for stochastic processes. *Contemporary Mathematics*, **80**, 249–284.

[52] Protter, P.E. (2005). *Stochastic Integration and Differential Equations. (2nd edition, Version 2.1.)* Springer, Berlin Heidelberg.

[53] Rebolledo, R. (1980). Central limit theorems for local martingales. *Z. Wahrsch. verw. Geb.*, **51**, 269–286.

[54] Revuz, D. and Yor, M. (1999). *Continuous Martingales and Brownian Motion. (3rd edition.)* Springer, Berlin Heidelberg.

[55] Shiryaev, A.N. (1996). *Probability. (2nd edition.)* Springer, New York.

[56] Shorack, G.R. and Wellner, J.A. (1986). *Empirical Processes with Applications to Statistics.* Wiley, New York.

[57] van de Geer, S.A. (1995). Exponential inequalities for martingales, with application to maximum likelihood estimation for counting processes. *Ann. Statist.* **23**, 1779–1801.

[58] van de Geer, S.A. (2000). *Empirical Processes in M-Estimation.* Cambridge University Press, Cambridge.

[59] van der Vaart, A.W. (1998). *Asymptotic Statistics.* Cambridge University Press, Cambridge.

[60] van der Vaart, A.W. and Wellner, J.A. (1996). *Weak Convergence and Empirical Processes: With Applications to Statistics.* Springer, New York.

[61] Yoshida, N. (1992). Estimation for diffusion processes from discrete observations. *J. Multivariate Anal.*, **41**, 220–242.

Index